*Zamia standleyi.*

# CYCADS OF THE WORLD

*Dioon edule*

**David L. Jones**

Smithsonian Institution Press
Washington, D.C.

Foreword by Dennis Wm. Stephenson
The New York Botanical Gardens

# ACKNOWLEDGEMENTS

Many people made valuable contributions to this book and I am grateful to them all.

I particularly thank my wife Barbara for her solid support and valuable assistance with numerous tasks over the protracted period of compilation of this work. I am also greatly indebted to Dennis Stevenson for encouragement, checking and commenting on the whole text, providing photographs and writing the foreword. Others who commented on parts of the text include Piet Vorster (*Encephalartos, Stangeria*), Paul Kennedy and Craig Thompson (*Macrozamia*), Ken Hill (*Cycas*) and Andrew Vovides (*Ceratozamia, Dioon, Zamia*). Peter Jones supplied climatic details of the habitats of Australian cycads.

Special thanks go to Catherine Jordan for unstinted and cheerful help in locating elusive texts. I am also grateful to John Bolger for preparing some of the drawings used in the book.

I thank the editors of the following scientific journals for permission to reproduce certain cycad drawings from their publications; Memoirs of the *New York Botanical Garden, Flora de Veracruz* and *Lyonia.* These drawings are credited where used.

Many people provided photographs for the project and in this aspect I am especially grateful to the contribution of Piet Vorster. Ken Hill made his excellent collection of Australian *Cycas* slides freely available for the project. Others who supplied slides include Bob Dressler, Roy Osborne, Mrs Brien Bosworth, Dennis Stevenson, Alec Pridgeon, Mark Clements and Aldo Moretti.

I thank Don Evans, Superintendent at Fairchild Tropical Garden, Florida, for cheerful guidance while there. Stan and Jane Walkley gave me free access to their special palm and cycad collection and provided material for drawings. Gary Wilson made details of his studies into cycad-insect relationships available. Many others supplied plant material or information including Julie and John Roach, Clyde Dunlop, Bart Schutzman, Bob Ornduff and Mark Clements.



First published in Australia
by Reed Books 1993

Published in the United States of America
by Smithsonian Institution Press

ISBN 1-56098-220-9

Library of Congress Catalog Number 93-83068

Edited by James Young
Designed by Robert Taylor
Typeset in Sydney by Character
Printed in Singapore
for Imago Productions (F.E.) Pte Ltd

# Contents

## PART ONE
## *Structure, Biology and Cultivation*



## PART TWO
## *The Living Cycads*



Male cone of *Stangeria eriopus.*

# Preface

Although initially somewhat hesitant to write a book on cycads, I was greatly motivated by the palm and cycad enthusiasts, both local and international, that I have met since writing the well accepted *Palms in Australia*.
Upon attending the Cycad 90 conference held in Queensland, Australia during 1990, I quickly realised there was a cult following of this unique group of plants. The enthusiasm of these disciples, both scientists and amateurs alike, is out of all proportion to the paucity of living cycads.

Cycads have interested people for centuries and it is apparent, from the extensive bank of literature compiled about them, that they have long attracted the attention of scientists, ethnologists, physicians and horticulturists. The vast accumulation of knowledge contained in hundreds of written accounts, both scientific and general, contributed much to this book. For me this abundance was a relatively novel situation, quite unlike that for other books I have written, where a lack of literature has been the norm. The quantity of literature created its own problems, chiefly because of the sheer volume of material to be sifted through, but also not infrequently because of conflicting views contained therein.

Given the special regard with which these plants are held, it is perhaps surprising that no comprehensive text has previously been written about them. This book deals with all living species currently recognised and is designed to be of interest both to the scientist and the general reader. I have attempted — successfully I hope — to achieve that difficult compromise of producing a scientifically accurate text which is at the same time easy to read and understand. An alphabetical layout has been used where dealing with the individual cycads and while relationships between some species may be detailed, there has been no attempt at taxonomic grouping. Botanical terms, often explained when first used, have been kept to a minimum and a detailed glossary is provided. Botanical keys have been given for some genera but these have not been attempted for the large ones or those, such as *Cycas*, which are badly in need of botanical study.

Apart from over 250 colour photos, the book is enriched by a number of illustrations. These include some fine, old, colour engravings previously hidden away in old texts, detailed botanical drawings from scientific publications and a moderate number of line drawings by John Bolger and myself.

It is sad to realise what heavy pressure the living cycads are under; in many cases their very survival in the wild is in doubt, with extinction close for many species. A recent upsurge of scientific interest has resulted in the recognition or discovery of many new species of cycads on at least three different continents. Some of these new species are threatened before they are even named. It would be a sorry testament indeed if these ancient plants, which have managed to survive the rigours of great geological and climatic changes over the eons, were to be made extinct by us.

David L. Jones
Canberra, Australia 1993

# Foreword

Cycads have been around for at least 250 million years. They have seen the dinosaurs come and go. During the Jurassic Period, they were a prominent component of the earth's vegetation and played an important role as a food source for herbivores of that time.

Archaeological evidence indicates that human interest in cycads has a history of at least 4,000 years. They have been used for ceremonial purposes and they have been planted at burial sites. When cycadologists are unsuccessfully looking for these plants in habitat, the native people always know exactly where to find them. And without taking one to palms!

Cycads have also been used as a food source in many regions, including Africa, Asia, Australia and the Americas. More recently they have been used as an alternative food source in times of deprivation, such as during World War II when shipping routes where disrupted in the South Pacific and the West Indies. This led to the discovery of Guam dementia, a neurological disorder caused by cycad toxicity. They have, of course, long been known to cause neurotoxicity in domesticated grazing animals.

In the early 17th century, the early explorers brought the first cycad plants to Europe. In recent years renewed effort has gone into understanding and cultivating these plants because of their relative scarcity in the wild and because of their archaic features.

There are few books on cycads and none on cycad horticulture. Not since 1919, when Charles Joseph Chamberlain published a general popular account of the cycads, has such a book as David Jones' *Cycads of the World* appeared. Given that all cycads are listed as endangered species and protected by CITES (International Convention on Trade in Endangered Species) and many private and public gardens are trying to propagate these plants, there is an obvious need for such a book. Fortunately, David Jones, who has written well illustrated books on ferns, palms, and orchids, has recognised this need. David brings many years of expertise to this book, based upon his horticultural and scientific interest in these plants.

In this book cycads are placed in their evolutionary and historical perceptive by a concise treatment of their fossil history. This in turn is augmented by a chapter on cycad biology with particular reference to reproductive biology, distribution, and ecology. All these aspects are necessary to achieve a level of understanding about these plants in their habitat if one is going to successfully propagate them. Another important feature of this book is its treatment of all cycad species: a formidable task given the lack of modern systematic treatments of some genera. Each genus is covered with respect to its distribution and general information and keys to the species are given for the smaller, well-known genera. The species descriptions are concise and informative. They include the etymology of the name, an accurate description of the plant, habitat and distribution, and horticultural information on seed germination, growing conditions and propagation methods where known. Most species entries are accompanied by excellent photographs.

This book is of interest and value to a wide audience. Anyone interested in cycads, whether from a conservation, scientific, or horticultural perspective, will find that this books suits their needs. Enjoy!

Dennis Wm. Stevenson
Director, Harding Laboratory
New York Botanical Garden

**Cycas siamensis**

From *L'Illustration Horticole*,
vol. 28, plate CCCCXXXIII, p.157 (1863)

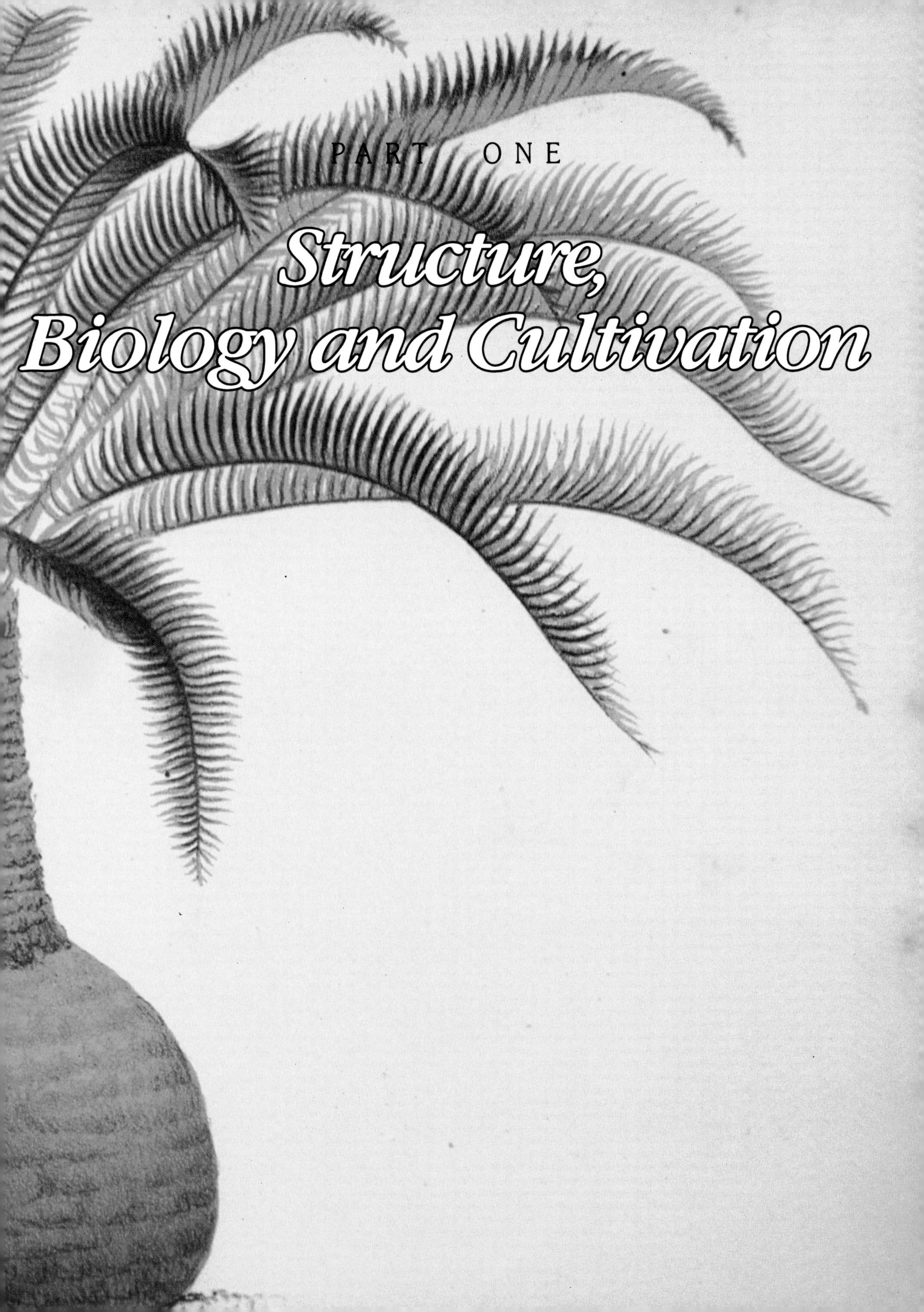

PART ONE

# Structure, Biology and Cultivation

# WORLD DISTRIBUTION OF CYCADS

CHAPTER 1

# Introduction to Cycads

Cycads are woody plants which produce seeds. Although they may have a general appearance which is readily identifiable by most people, they are usually wrongly linked to palms and ferns, when in fact they are not related to either. Cycads are actually a unique assemblage of plants and although they are grouped with the gymnosperms they are in point of fact unrelated to any other group of living plants.

Cycads certainly have a distinctive appearance which is related to their primitiveness or antiquity. Within the living seed plants they are nearly unique in that they produce motile sperm cells, and thus are an important link to the earliest of the ancient seed plants. Cycads flourished in eons past and reached their peak in the Mesozoic Era some 150 million years ago. While the fossil story may still be unclear, it is certain that cycads were more varied and profuse in earlier times and more widely distributed. Today cycads may be regarded as relicts consisting of small populations distributed disjunctly on many continents.

## NUMBER OF SPECIES

The living cycads are a relatively small group of plants presently consisting of 185 species in 11 genera. The number of species has been climbing steadily over recent years, parallelling detailed taxonomic studies of various genera. Undescribed species still await naming and undoubtedly there are new species which still await discovery. These are not numerous, however, and the final number of cycads will probably not exceed two hundred and ten. In total, as a proportion of the world flora, modern cycads are not a significant group.

## WORLD DISTRIBUTION

Living cycads are found in the tropical, subtropical and warm temperate regions of both the north and south hemispheres. While substantial numbers exist on the continents of South Africa, Australia and South America they are also prominent in Central America (which has the greatest diversity) and the Caribbean Islands. *Cycas* is the most widespread genus, with representatives occurring as far north as Japan and others being scattered through various Pacific islands, China and India to Madagascar and the east coast of Africa. The accompanying world map shows the present-day distribution of cycads in more detail.

# CYCAD CLASSIFICATION

Cycads produce seeds and are thus included in that phylum of the plant kingdom known as the Spermatophyta or seed plants. As their seeds are produced on exposed sporophylls arranged in a cone (not enclosed within an ovary), cycads are also included with other cone-bearing plants in the class Gymnospermae. The gymnosperms are an ancient group of plants with strong fossil links. Of the eight orders of gymnosperms, three are now extinct (Cordaitales, Bennettitales, Glossopteridales) and five still contain extant genera (Gingkoales, Cycadales, Pinales, Taxales, Gnetales). Cycads, which are placed in the order Cycadales, are the most primitive of living gymnosperms and thus cycads are among the most ancient of all plants surviving today.

Formerly all the living genera of cycads were classified in the one family Cycadaceae. However in 1959, L. A. S. Johnson in the 'The Families of Cycads and the Zamiaceae of Australia', in the *Proceedings of the Linnaean Society of New South Wales* 84: 64–117, showed that three distinct groups of cycads were better accommodated in separate families. This approach was universally accepted until 1981 when Dennis W. Stevenson in 'A Proposed Classification of the Cycadales', in the *American Journal of Botany* 72: 1971–1985, presented evidence for the creation of a fourth family, Boweniaceae, to accommodate the genus *Bowenia*. Further studies into relationships between cycad genera are still underway and some researchers have suggested that the recognition of further families may be warranted.

More recently, however, Dennis Stevenson has carried out a series of detailed cladistic analyses on the living cycads. The first of these, based on the analysis of thirty characters, resulted in three disparate cladograms, one of which was used by Stevenson as the basis for a provisional classification of the living cycads (for more details see 'Morphology and Systematics of the Cycadales', *Memoirs of the New York Botanical Garden* 57 (1990): 8–55). A later more detailed analysis using fifty-two characters resulted in a single cladogram and basically confirmed the earlier provisional classification with some slight modifications. (See 'A Formal Classification of the Extant Cycads', *Brittonia* 44, no.2: (1992): 220–223. That classification is as follows:

**Order Cycadales**
- **Suborder Cycadineae**
  - **Family Cycadaceae**
    - **Genus *Cycas***
- **Suborder Zamiineae**
  - **Family Stangeriaceae**
    - **Subfamily *Stangerioideae***
      - **Genus *Stangeria***
    - **Subfamily Bowenioideae**
      - **Genus *Bowenia***
  - **Family Zamiaceae**
    - **Subfamily Encephalartoideae**
      - **Tribe Diooeae**
        - **Genus *Dioon***
      - **Tribe Encephalarteae**
        - **Subtribe Encephalartinae**
          - **Genus *Encephalartos***
        - **Subtribe Macrozamiinae**
          - **Genus *Macrozamia, Lepidozamia***

**Subfamily Zamioideae**
**Tribe Ceratozamieae**
**Genus *Ceratozamia***

**Tribe Zamieae**
**Subtribe Microcycadinae**
**Genus *Microcycas***

**Subtribe Zamiinae**
**Genus *Zamia, Chigua***

## CYCAD FAMILIES

Using the above classification the living cycads are placed in three families, details of which follow.

### Family Cycadaceae

This family, described by Carolus Linnaeus in 1753, contains the solitary genus *Cycas* which is distributed in South-East Asia, southern China, Malaysia, tropical Australia and various islands of the western Pacific, with a disjunct species from Africa and Madagascar. The type species of the family is *Cycas circinalis* L. Family characters include:

- female cones consisting of loosely organised sporophylls and lacking a central axis;
- sporophylls of female cones with a leaf-like apex and bearing two or more ovules on their margins;
- the growing point of the female plant grows through the developing crown of megasporophylls;
- young leaves with straight rhachis and leaflets coiled like watch springs;
- leaves pinnate;
- leaflets with a midrib of a single vein and no side veins;
- lower leaflets reduced and spine-like; and
- seeds platyspermic.

### Family Stangeriaceae

This family, described by Lawrence Johnson in 1959, contains two genera; *Stangeria* in the subfamily Stangerioideae; and *Bowenia* in the subfamily Bowenioideae. The type species of the family is *Stangeria eriopus* (Kunze) Baillon. Family characters include:

- rootstock much branched;
- root buds present;
- cataphylls absent or produced erratically;
- stipules containing a vascular bundle present on leaf bases;
- cones terminal, solitary either on the main stem or branches arising from it;
- seeds attached below the sporophyll stalk;
- seeds radiospermic; and
- sarcotesta plum or purplish in colour.

Of the two genera occurring in this family, *Bowenia* can be readily distinguished by its bipinnate leaves which are coiled when young, whereas the leaves of *Stangeria* are pinnate and folded.

### Subfamily Stangerioideae

This subfamily includes the type which is the monotypic African genus *Stangeria*. Subfamily characters include:

- leaves folded when young;
- leaves pinnate, fern-like; and
- cones terminal on the main stem.

### Subfamily Bowenioideae

This subfamily, described by Robert Pilger in 1926, contains only the Australian genus *Bowenia*. Subfamily characters include:

- leaves coiled when young;
- leaves bipinnate; and
- cones terminal on slender branches of the main stem.

## Family Zamiaceae

This family, described by Heinrich Reichenbach in 1837, is the largest family of cycads and is divided into two subfamilies (Encephalartoideae and Zamioideae) each of which is further subdivided into tribes (Diooeae, Encephalarteae, Ceratozamieae, Zamieae) and subtribes (Encephalartinae, Macrozamiinae, Microcycadinae, Zamiinae). Some researchers suggest that elevation of one or more of the tribes to family rank may be warranted. The type species of the family is *Zamia pumila* L. Family characters include:

- cataphylls present on vegetative shoots;
- stipules present or absent but always lacking a vascular bundle;
- young leaves straight or inflexed;
- leaflets flat and overlapping during development;
- cones borne terminally or laterally on the main stem;
- seeds attached above the sporophyll stalk;
- seeds radiospermic; and
- sarcotesta red, orange or yellow.

### Subfamily Encephalartoideae

This subfamily, described by Dennis Stevenson in 1992, contains two tribes, Diooeae and Encephalarteae. Subfamily characters include:

- stipules absent;
- leaf bases persistent;
- leaflets not jointed to the rhachis;
- lower leaflets reduced and spine-like;
- spinose prickles absent from petiole;
- megasporophylls not peltate;
- leaflet veins anastomosing at the tips; and
- cone peduncles bearing decurrent catatphylls.

**Tribe Diooeae:** This tribe, described by Julius Schuster in 1932, contains the single genus *Dioon* which is found in Mexico, Nicaragua and Honduras. Tribal characters include:

- female sporophylls with an expanded, domed apex but lacking thickening or an apical lobe; and
- seeds attached to the sporophyll by a basal stalk.

**Tribe Encephalarteae:** This tribe, described by Friedrich Miquel in 1861, has been further subdivided into two subtribes, Encephalartinae and Macrozamiinae. Tribal characters include:

- female sporophylls thickened on the upper surface or with apical lobes; and
- seeds attached directly to the sporophyll.

SUBTRIBE ENCEPHALARTINAE: This subtribe, described by George Bentham and Joseph Hooker in 1880, contains the single genus *Encephalartos* which is widely distributed in Africa. Subtribal characters include:

- leaflets lacking a coloured, basal callous area; and
- megasporophylls with outer facets but lacking a spine-like apical lobe.

SUBTRIBE MACROZAMIINAE: This subtribe, described by Dennis Stevenson in 1992, contains two Australian genera, *Macrozamia* and *Lepidozamia*. Subtribal characters include:

- leaflets with a coloured, callous area in the upper basal margin; and
- megasporophylls with a spine-like apical lobe.

## Subfamily Zamioideae

This subfamily, containing the type, includes two tribes, Ceratozamieae and Zamieae. Subfamily characters include:

- stipules present, at least in seedlings;
- leaf bases deciduous;
- leaflets jointed to the rhachis;
- lower leaflets not reduced and not spine-like;
- prickles present or absent on petiole;
- megasporophylls peltate; and
- cone peduncles lacking cataphylls.

**Tribe Ceratozamieae:** This tribe, described by Dennis Stevenson in 1992, contains the single genus *Ceratozamia* which is distributed in Mexico, Guatemala and Belize. Tribal characters include:

- female sporophylls with two prominent horns;
- spinose prickles present on the petiole and rhachis; and
- trichomes with an oblong basal cell.

**Tribe Zamieae:** This tribe, containing the type, has been further subdivided into two subtribes Microcycadinae and Zamiinae. Tribal characters include:

- female sporophylls lacking horn-like projections;
- equally branched trichomes present; and
- presence or absence of spinose prickles on the petiole and rhachis.

SUBTRIBE MICROCYCADINAE: This subtribe, described by Dennis Stevenson in 1992, contains the monotypic Cuban genus *Microcycas*. Subtribal characters include:

- truncated leaves;
- reflexed leaflets; and
- absence of spinose prickles on the petiole and rhachis.

SUBTRIBE ZAMIINAE: This subtribe, containing the type, includes the genera *Zamia* and *Chigua*. Subtribal characters include:

- spinose prickles on the petiole and rhachis;
- leaves not truncated; and
- leaflets not reflexed.

## CYCAD GENERA

The living cycads comprise eleven genera, details of which follow.

***Bowenia:*** Consists of three living species all endemic in Australia. Cycads of this genus are readily recognised by their bipinnate leaves which when young are coiled like watch springs.

***Ceratozamia:*** Consists of eleven living species distributed in Mexico, Guatemala and Belize. Although similar to some species of *Zamia*, these cycads can be recognised by the paired, horn-like structures on the sporophylls.

***Chigua:*** Consists of two living species both endemic to Colombia in South America. Although superficially similar to species of *Zamia*, plants of *Chigua* are readily recognised by the prominent midrib in the leaflets.

***Cycas:*** About thirty living species distributed in Asia, South-East Asia, Malaysia, Philippines, Indonesia, New Guinea, tropical Australia, various islands of the western Pacific, Africa and Madagascar. Species of *Cycas* can be readily recognised by the loosely aggregated sporophylls of the female cone, the prominent midrib in each leaflet and the lower leaflets often reduced to paired, spine-like processes.

***Dioon:*** Consists of ten living species found in Mexico and Honduras. Members of this genus have non-jointed leaflets, which are broadly decurrent on the rhachis at the base, lack prickles on the petiole, have leaf-like female sporophylls and the ovules are borne on short stalks.

***Encephalartos:*** The second largest genus of cycads consisting of about fifty living species all endemic to the African continent. These are small to large cycads with non-jointed leaflets and the lower leaflets often reduced and spine-like.

***Lepidozamia:*** Consists of two living species both endemic to eastern Australia. These species are readily recognised by their large female cones and the way the leaflets are attached to the upper surface of the rhachis.

***Macrozamia:*** About twenty-five living species all endemic to Australia. They can be recognised by their leaflets, which have a prominent, often colourful, callous area at the base, being attached to the lateral margins of the rhachis and the cones being basically green.

***Microcycas:*** A monotypic genus which is endemic to Cuba. It is close to *Zamia* but can be distinguished by the reflexed leaflets and the truncated leaves.

***Stangeria:*** A monotypic genus endemic to Africa. Plants of this genus have an unusual fern-like appearance and the leaflets have a prominent midrib and side veins.

***Zamia:*** The largest genus with about sixty living species distributed in the southern States of North America, Central America, South America and the Caribbean Islands. Distinguished by the jointed leaflets which lack a midrib, the absence of persistent leaf bases on the trunk and the more or less hexagonal sporophylls of the cones.

## GENERALISED GROWTH FEATURES

Cycads are basically woody plants which have roots, a stem, leaves and reproductive structures known as cones.

The main roots of cycads are thickened and fleshy and as they may have storage capacities they are often termed tuberous. Along with the stem they may have contractile properties which serve to regulate the level of the stem in the ground. Specialised, upright-growing, branched roots, known as coralloid roots, are also produced by all species. These roots contain symbiotic blue-green algae which can fix nitrogen from the atmosphere.

The stems of cycads may be completely subterranean or emerge from the ground and be trunk-like. Soil depth may influence this development and in shallow, stony soils, species which normally have a subterranean stem may develop an above-ground trunk.

The leaves of most cycads are once-divided (pinnate) and often develop in an attractive, palm-like crown. Those of *Stangeria* bear a strong resemblance to the fronds of a fern, whereas in the Australian genus *Bowenia* the leaves are twice-divided (bipinnate).

Cycads reproduce when mature by the production of cones. A plant is either male or female and the cones of each sex are usually quite different in size and shape and to a much lesser extent colour. Specialised woody growths on the cones, called sporophylls, bear the sexual parts with those on a male cone producing pollen and on a female cone they bear large ovules which if fertilised develop into seeds. The seeds of cycads are relatively large and have an outer layer (sarcotesta) which is often colourful.

Further details on the structure of cycads can be found in Chapter 4.

## GENERALISED BIOLOGY

Cycads are long-lived, perennial, unisexual plants which develop cones and reproduce by seeds. Some species of cycad produce cones regularly (often annually), whereas others cone very sporadically sometimes with ten to fifteen years or more between coning events. In some cycads coning may be regulated by summer fires.

For reproduction to occur, male and female plants must produce cones at the same time. In most species the transference of pollen from a male cone to a female cone seems to be via insect vectors (beetles or small bees), but in certain species wind may also be involved or perhaps even both mechanisms operate. At maturity male and female cones may both produce heat and also often develop odours. Cycads are one of the very few seed plants to produce motile sperm cells called spermatozoids. These are generated in the pollen tube after the pollen grains germinate following pollination and these male gametes are released into the proximity of the ovules to effect fertilisation.

The seeds of cycads have a fleshy outer layer known as the sarcotesta. This is eaten by a diverse range of birds and animals which aid in the dispersal of the species. The seeds of most cycad species cannot germinate immediately on maturity for the embryo has an after-ripening period, the length of which varies with the species.

For more details of cycad biology see Chapter 6.

## CYCAD HABITATS

Cycads are found naturally in a variety of habitats ranging from mesic to semi-arid. It is known that cycads thrived in the warm moist conditions of the Mesozoic Era and researchers postulate that primitive members of the living cycads still grow in similar habitats today. The commonest warm, moist habitat of this type is rainforest, particularly those where the rainfall is evenly distributed over most of the year and warm humid conditions prevail. In habitats of this type species of *Chigua, Zamia, Bowenia, Cycas* and *Lepidozamia hopei* thrive. Slightly less congenial are the forests of the littoral zones where *Cycas* species and *Stangeria paradoxa* occur along with deciduous mesophyll forests and evergreen sclerophyll forests. Rainfall here is strictly seasonal and cycads have to cope with dry periods of limited duration. Many cycads are found in such habitats including representatives of most genera.

Perhaps the most adapted of the cycads are to be found in the xeric habitats where rainfall is low and extremely seasonal, long hot periods are common and burning may take place annually. Such habitats include grassland, sparse forests and woodlands, and rocky escarpments and gorges. In these habitats a few species of *Cycas, Dioon, Encephalartos* and *Macrozamia* cling tenaciously to life. In Africa a few species of *Encephalartos* occur at moderately high altitudes where very heavy frosts and snowfalls are common occurrences in winter.

## TOXICITY

Although dealt with more fully in Chapter 5, it should be mentioned at this early stage that all cycads contain chemicals which are highly toxic and if ingested in sufficient quantities can cause disability or death. These materials are variously distributed in the plant but are not always found in the same tissues; for example toxins are present in the outer seed coat of species of *Macrozamia* but appear to be absent from this region in many *Encephalartos*.

## NATURAL HYBRIDS

Natural hybrids between different species of cycad occur sporadically in nature where two species grow intermingled or in close proximity. Such hybrids however are generally uncommon to rare and as with other groups of plants there are indications that natural hybrids may be more common on disturbed sites, particularly those influenced by humans. Natural hybrids can only occur if the two species share the same pollinator and if their cones reach maturity at the same time.

In South Africa a number of suspected natural hybrids in *Encephalartos* have been recorded. These include *E. altensteinii* x *E. trispinosus* (hybridises fairly frequently and may develop hybrid swarms with bewildering variations), *E. altensteinii* x *E. villosus*, *E. altensteinii* x *E. latifrons*, *E. altensteinii* x *E. arenarius*, *E. horridus* x *E. longifolius* and *E. lebomboensis* x *E. villosus*.

In Australia the following sporadic hybrids within the genus *Macrozamia* have been noted; *M. communis* x *M. flexuosa*, *M. diplomera* x *M. heteromera*, *M. communis* x *M. secunda*, *M. lucida* x *M. miquelii* and *M. moorei* x *M. fearnsidei*.

Hybrids appear to be rare or unreported in other genera. Some researchers believe that hybrids in *Dioon* may be relatively frequent but are largely overlooked. Of interest is a suspicion that *Zamia loddigesii* may have arisen as an ancient stable hybrid between *Z. furfuracea* and *Z. spartea*. In this genus the following natural hybrids have been reported; *Z. loddigesii* x *Z. spartea*, *Z. loddigesii* x *Z. furfuracea* (in disturbed sites only).

It should be noted that natural hybrids between species in different genera are unknown.

## ARTIFICIAL HYBRIDS

Artificial hybridisation, which has played a major role in the development of commercial crop plants, has been carried out on a very limited scale with cycads. With the tremendous current interest in this group of plants and the number of comprehensive collections of cycads in existence, however, it is very likely that many more artificial hybrids will be produced in the near future. Such a trend, which has occurred in many other plant groups, is acceptable providing that the genetic diversity of the original species is never lost or compromised. With so much pressure on the wild populations of cycads the first priority must always be to produce seeds of a natural species rather than an artificial hybrid.

Hybridisation by enthusiasts is often sporadic and is more likely to involve haphazard crossing between species which happen to cone at the same time rather than by following a planned program. Researchers may carry out deliberate crossings in order to elucidate relationships between species. Horticulturists may have the goal of producing more adaptable cycad plants which fill a specific landscaping niche.

Artificial hybridisation may also be involved in the conservation of rare genetic resources. Thus in South Africa a far-sighted program has commenced crossing *Encephalartos natalensis* and *E. woodii*. The latter species, now extinct in the wild, is known only from male plants and it is hoped that after a number of generations of back crossing the hybrids to *E. woodii*, plants will be produced which should be identical with *E. woodii*. Female plants are of course needed to ensure the future of this species.

## HORTICULTURAL APPEAL

Cycads along with palms, tree ferns and many other foliage plants are very popular subjects for cultivation. Although other groups of plants attract an enthusiastic following, the attention paid to cycads is well out of proportion to the numbers of species found in the world. For example the South African Cycad Society has a current membership of more than 500 and yet the whole continent of Africa contains only about sixty species of cycad.

Cycads are very popular plants with the collector and cycad enthusiasts will go to extraordinary lengths to obtain new plants (see also Chapter 3). Rare species and specimen plants are expensive and hence large collections represent a considerable outlay in time and money. A wide range of species are propagated by specialist nurserymen and new species and variants are constantly being introduced into cultivation.

Cycads are excellent horticultural subjects and are of value to groups other than collectors. Landscapers find their primitive shapes and silhouettes appealing and their predictable dimensions extremely useful. Many species make very decorative container plants and are excellent for indoor use where there is sufficient light. As a bonsai subject, cycads are only rivalled by their gymnosperm relatives.

## CYCAD CONSERVATION

Coincident with the tremendous upsurge in interest in all aspects of cycads, including their cultivation, has been a plundering of the wild populations of many species with unfortunate consequences. Some species are now extinct in the wild and many others reduced to the point where natural reproduction is no longer possible. In some cases there are more plants of a particular species in cultivation than are left in the wild.

It behoves all cycad enthusiasts to take an active role in the conservation of these plants. The regular propagation of seedlings from cultivated plants is one easy way which growers can contribute. All enthusiasts should hope that in the next decade positive progress can be made towards arresting human's greed and conserving the remaining wild populations. See also Chapter 3.

## LITERATURE

Cycads are extremely popular plants which engender an amazing degree of interest in people, whether they be amateurs or professionals and in a wide range of fields. Some idea of this interest can be gauged if one surveys the vast amount of literature that has been written about these plants. Articles range from popular accounts to botanical revisions and learned tomes dealing with economic aspects, in particular the matter of toxicity. For the latter subjects see J. W. Thieret, 'Economic Botany of the Cycads', *Economic Botany* 12, no.1 (1958): 3–41, M. G. Whiting, 'Toxicity of Cycads', *Economic Botany* 17, no.4 (1963): 271–302 and M. G. Whiting, 'Neurotoxicity of Cycads, an Annotated Bibliography for the Years 1829–1989', *Lyonia* 2, no.5 (1989): 201–270. The fact that together the first two of these works are supported by some 550 references gives an indication of the volume of material available on the subjects.

Easily the most comprehensive modern bibliography compiled on cycads is the 'Bibliography of the Living Cycads' (annotated), compiled by R. W. Read and M. L. Solt (see *Lyonia* 2, no.4: (1986): 33–199). Not only is this work all-encompassing but it is user-friendly with the references listed alphabetically by author and cross-referenced under selected subjects as well as cycad genera and species. A useful recent publication for serious students of the subject is 'The Biology, Structure, and Systematics of the Cycadales', edited by D. W. Stevenson, *Memoirs of the New York Botanical Garden* 57: (1990): 1–210. Finally mention must be made of *Encephalartos*, the excellent journal of the South African Cycad Society. This publication, which appears quarterly, keeps growers and researchers up to date on all aspects of these plants.

Surprisingly, given the popularity of the subject, few books have been produced on cycads. The classic introduction to the subject is undoubtedly provided by C. J. Chamberlain in *The Living Cycads*, Hafner Publishing Coy, Inc., New York. This easily read, often entertaining account, which contains details of early explorations and botanical studies, was published in 1919 and is unfortunately now out of print. The book which brought South African cycads to the attention of the world, and which is still so deservedly popular, is *Cycads of South Africa* by Cynthia Giddy (Purnell & Sons, Cape Town). This book was revised in 1989 after being first published in 1974. Another publication, although much more technical in its content and arrangement, is *Cycas and the Cycadales*, second edition 1973, Central Book Depot, Allahabad, by D. D. Pant, who is Professor of Botany at the University of Allahabad, India. The most recent book on the cycad scene originates from South Africa and it deals with all African cycads. This superb publication, by Douglas Goode and titled *Cycads of Africa* (C. Struik Publishers, Cape Town, 1989), contains not only magnificent paintings of each species but is supplemented by an excellent, comprehensive text which shows clearly the author's familiarity with his subject.

## CURRENT RESEARCH

It is comforting to realise that cycads are being actively studied by research groups in various parts of the world. Taxonomic aspects in particular are receiving detailed consideration and revisions are being undertaken into most of the major genera. Other research includes detailed anatomical studies, biochemical and toxicity studies, research into aspects of cycad biology, ecological studies and propagation by tissue culture. Two excellent papers summarising current cycad research are: R. Osborne, 'Cycad Research in the 80s', *Encephalartos* 6 (1986): 26–34; and R. Osborne, N. Grobelaar and P. Vorster, 'South African Cycad Research: Progress and Prospects', *South African Journal of Science* 84 (1988): 891–6.

## CYCADS IN RELIGION

In India, species of *Cycas* (particularly *C. circinalis*) are often called 'Church Palms' because their fronds are used to decorate places of worship. In some Christian countries the fronds of cycads may be used in Palm Sunday processions. In Mexico and Honduras the fronds of species of *Dioon* are used for decorating altars, and are regarded as being particularly important for some ceremonies and this appears to be a pre-Columbian tradition. They are also used in ceremonies in the Philippines, New Hebrides and the Solomon Islands. In South Africa, fronds of *Encephalartos lanatus* have been used for Church decoration at Christmas.

Cycad fronds are also often displayed at funerals and used in ceremonies to venerate the dead. The fronds of *Dioon mejiae* are used in this manner in Honduras and in Mexico other species of *Dioon*, including *D. merolae*, are used for the same purpose. In South Africa fronds of *Encephalartos lanatus* have been used to decorate graves. The fronds of *Cycas revoluta* are commonly used in the preparation of wreaths. In parts of Indonesia *Cycas* species may be planted in cemeteries.

## NAME CHANGES

Name changes, or more accurately corrections to names, are the inevitable result of detailed taxonomic studies carried out by botanists. Such name changes are particularly frustrating to nurserymen and amateur growers and may even cause confusion to other botanists who are unfamiliar with the group. With the plethora of names which have been applied to cycads, particularly by botanists in the late eighteenth and early nineteenth centuries, it is inevitable that detailed studies of the various genera will result in changes to the application of names.

Like all organisms, cycads are classified by a binomial system which was introduced by the Swedish scientist Carolus Linnaeus in 1753. Thus the generic name and the specific epithet make up the binomial and one pair of names is correctly applied to each species: for example, *Cycas media, Zamia pygmaea*.

The basic unit in any biological classification system is the species. A species may be defined as a group of plants with a common set of characters which sets them apart from another group of closely related organisms. While such a definition is useful it fails to account for the fact that species are constantly evolving to meet changes in their habitats. This factor is of particular significance in cycads because so many species consist of disjunct relict populations which have been isolated from each other for long periods of time. Changes in morphology, both major and minor, which result from adaptations to each environment can cause problems in identification.

A genus is usually defined as being made up of a group of closely related species which are too distinctive to be grouped with any other. Within cycads, the majority of genera are readily recognisable as distinctive natural groups. Some researchers who have a conservative taxonomic view however, may question the segregation of both *Microcycas* and the recently described *Chigua* from the large genus *Zamia*.

Botanical nomenclature or the naming of plants is governed by a set of rules laid down by botanical authorities and revised every five or six years. Most changes to plant names occur as the result of detailed morphological studies, from examination of type specimens or from someone applying the botanical rules correctly.

Two of these rules are of major significance and concern priority of publication. A plant can have only one correct binomial and that is usually the oldest name available unless this name contravenes the rules. Thus if two or more correctly published names are available for one species, then the first one published (as a species under whatever generic name) is the correct name and all others are synonyms. Similarly if the same name is used for two different plants then the second use is incorrect, for example as in *Cycas pectinata*. As well as understanding and correctly applying the botanical rules, detailed studies of the original type specimens are of paramount importance in sorting out the nomenclature of any genus. Often, particularly with early collections, these type specimens are poor, were never retained or have been subsequently lost or destroyed.

CHAPTER 2

# History and Prehistory of the Cycads

Cycads, because of their great antiquity, have variously been described as living fossils and as the coelacanths of the plant world. They have also been linked to dinosaurs, for their long history can be traced back to the time when those stupendous creatures roamed the earth. Unlike the dinosaurs however, the cycads have direct living remnants which still cling tenaciously to life today.

The modern cycads are but a fraction of their ancestors (both in total numbers and diversity of species) which reached their peak in the Jurassic and Early Cretaceous Periods, dominating the vegetation of the day. That the living species share characters with these early plants is obvious from an examination of fossils. It is tempting to speculate that these plants have survived more than 200 million years of evolution, with very little change. After all, parallels exist in the animal kingdom with sharks, turtles and crocodiles changing little in form over the eons. The coelacanth, that primitive vertebrate fish which still swims today in the warm waters of the Comoros, off the coast of south-east Africa, has a fossil record which goes back 360 million years. This is nearly twice that of the cycads so it is perhaps not beyond the realms of possibility that these plants could have survived their period of history in a relatively unchanged state.

The long history of these plants tends to place in people's minds the idea that cycads are now static and are incapable of evolving to meet environmental change. This is quite erroneous, as the modern cycads are diverse and have successfully occupied a wide range of habitats, including littoral zones, grasslands, rainforests, deciduous mesophyll forests, evergreen sclerophyll forests, gorges, relict situations on ancient escarpments and mangrove swamps. There is evidence that cycads have actively occupied recent available niches in the environment and their ability to adapt is supported by modern taxonomic studies into a wide range of genera. Thus these remarkable plants are obviously still evolving and are capable of responding to new environmental conditions.

Cycads are paradoxical in that they have obviously retained some primitive features (e.g. the presence of motile sperm cells), and yet have other characteristics which could be considered as advanced. In fact their motile sperm cells are the most complex known in plants. The apparent reliance of a large number of species on fairly close relationships with insects to achieve pollination is extremely interesting and perhaps challenges the views of the primitiveness of these plants (for more details see pages 56–59). Certainly these relationships are with ancient groups of insects but they represent a major evolutionary advance over using fickle physical systems such as the wind to achieve the same purpose. The success of such pollination systems must have contributed greatly to the survival of these plants over the ages and has more than likely aided their adaptation in modern habitats.

Cycads then are survivors from the past, but they now face the very real threat imposed by modern human greed (see Chapter 3). An examination of cycad history and possible relationships to other plant groups (mostly extinct), is pertinent at this stage. It should be pointed out however, that these topics are still the subject of considerable debate among palaeobotanists and disagreements occur. Modern views may differ considerably from those expounded in the textbooks which were published earlier this century.

## CYCAD PREHISTORY

To comprehend the evolution of plants over the history of the earth, it is essential to understand that there has been a continually changing succession of new plant groups evolving. A group may disappear if unsuccessful, or else they diversify into new habitats and climates, explode in numbers and then decline as the environment changes, sometimes becoming extinct in the process. Often those with successful adaptations survived the changing conditions and were then able to exploit the new environment.

Cycad-like plants and perhaps even true cycads seem to have arisen some time in the early Permian Period, possibly about 230 million years ago. Some fossils which have been classified as being related to cycads have been collected even earlier than this, from deposits laid down in the Carboniferous Period, but recent studies have suggested that these may in fact not be true cycads or even cycad relatives. The Permian Period lasted for about 65 million years and was the last epoch in the Palaeozoic or Ancient Era during which plants first became established on the earth. Whether cycads arose from the plant group which flourished in the Permian Period, the Glossopteridales, or developed from some other line, is unclear, but current opinion links them with the Pteridosperms.

Cycads reached their peak of development and diversity in the Mesozoic or Median Era. They apparently survived the transition from the Permian Period to the Triassic Period without apparent change and became a well-established component of the world's flora. Examples of cycad fossils in Mesozoic rocks have been found in Australia, South Africa, North America, India, England, Europe, Alaska, Greenland and Antarctica. During the Jurassic Period and until the middle of the Cretaceous Period, the world's climate was warm to hot with an abundance of moisture, and as a result plant growth thrived in these conditions. Plants, including cycads, grew luxuriantly at both poles of the earth, for the prevailing climate in these regions was warm. The Jurassic Period was the age of the dinosaurs and it was also the period when cycads achieved their pinnacle of abundance and diversity. It spanned 57 million years and occurred between 193 million and 136 million years ago.

The period which followed the Jurassic is known as the Cretaceous and this was the time of change for cycads and many other plants. During this period a newcomer arrived on the scene, the group known as the angiosperms, or flowering plants. These plants were faster growing and quicker maturing than the prevailing gymnosperms (which had possibly reached their zenith and were on the decline anyway) and they quickly diversified and expanded into all of the significant plant habitats. No major plant groups in existence at the time became extinct but all declined in significance. The angiosperms thrived and because of their major adaptive feature, the enclosure of their seeds in an ovary, they were able to exploit all available environmental niches and are still the dominant plant group today.

### The Influence of Continental Drift

Geological and fossil evidence suggests that for much of the earth's history there existed a single large continent which was named Pangaea. About 180 million years ago this supercontinent divided into two sections which drifted northwards and southwards respectively. The northern land mass was known as Laurasia and the southern one Gondwana. Cycads had evolved prior to this split and were present in both landmasses.

Over millions of years the landmasses of Laurasia and Gondwana have changed shape and have divided again, forming the continents as we know them today. These continents continue to drift apart at a rate of a few centimetres each year.

The theory of continental drift is supported not only by the similarities of fossils on different continents but also by relationships between groups within the present floras and faunas. Whereas the continent of Laurasia changed shape over time, no major landmasses, except North America, were split off. By contrast Gondwana broke up into a series of landmasses which are recognisable today as the continents of Australia, Africa, South America, Antarctica, India, New Guinea and New Zealand.

Most modern cycads are confined to the tropics and subtropics with a few species extending to temperate zones. Carbon dating of fossils, biochemical studies and research into plant evolutionary systems suggests that cycads may have originated in eastern Gondwana and then dispersed to North America, England and Sweden via Antarctica. Migration patterns apparently taken by certain groups of cycads are often circuitous and puzzling. For example there is evidence that the family Stangeriaceae may have originated in Argentina (where it is now extinct) and then migrated to Africa where it is still represented today. Perhaps instead of migration such phenomena are better explained by adaptation and survival. The ancestor of *Stangeria* may have been widespread in Gondwana and the modern African population is the sole survivor. *Androstrobus zamioides*, a fossil cycad known only by a fossilised male cone from Middle Jurassic sediments in North Yorkshire, is almost certainly a *Stangeria* relative and emphasises the relictual nature of *Stangeria*. The same is true for *Dioon*, which has Jurassic and earlier fossils in Asia and Europe and Tertiary fossils in Alaska and California. After glaciation in the Pleistocene Period it remains as a Mexican–Nicaraguan relict. Much of the present day relationships can be discerned from previous continent proximity with migrations being influenced by continental drift and climate changes, particularly ice ages, glaciation, the uplift of mountain ranges and periods of aridity.

The present-day situation with cycads shows a great reduction in abundance and diversity from their peak in the Jurassic Period. That they still survive indicates they are a stable life form capable of adapting to change.

Of the cycad genera in existence today, *Cycas* is probably the oldest with fossil links in China extending back to the Lower Permian Period and is almost certainly of Gondwanan origin. Members of this genus retain the most primitive morphological and biochemical characters of all cycads, indicating that the group separated very early on from other cycad genera as a distinct line of evolution.

Within the living cycads, the more advanced, or modern, genera have adapted to drier marginal habitats whereas the more primitive genera, such as *Cycas*, have many species which are found in warm humid habitats similar to those which prevailed in the Jurassic Period. Even within this genus it is apparent that evolutionary adaptation has occurred. Thus a number of species from Australia and India are found in harsh, seasonally dry climates, whereas other species from Australia, Asia and Africa still occupy their traditional warm humid environments. Studies of chromosome patterns indicate that *Cycas thouarsii* from Africa, Madagascar and some adjacent islands is possibly the most primitive species in the genus, although this view may be contradicted by morphological evidence. For example the megasporophylls of *Cycas revoluta* and *C. siamensis* are very similar morphologically to those from Lower Permian fossils in China, in particular *Crossozamia chinensis*. Continental drift patterns suggest that *Cycas thouarsii* may have existed more or less in its present form for about 140 million years.

The genus *Dioon* is also regarded by many researchers as having numerous primitive features. Although members of this genus usually grow in humid climates, the plants are rarely found within a forest canopy, mostly growing in the open on rocky scarps.

## Evolution of Sporophylls and the Cycad Cone

Because of general morphological similarities between cycad leaves and the megasporophylls of the genus *Cycas*, many researchers have proposed that cycad megasporophylls could have arisen from the fertile leaves of an ancestral group of plants such as the pteridosperms. Some fossil pteridosperms from the Upper Carboniferous Period, such as *Spermatopteris*, *Phasmatocycas* and *Taeniopteris*, have two rows of small ovules on the upper side of a leaf, usually towards its base. A reduction in the size of the leaf lamina (mostly simple and entire in these plants), and in the number of ovules could have given rise to a structure similar to the megasporophyll of modern *Cycas*. Some researchers postulate that only the basal area of fertile leaves may have become reduced, others that the whole leaf lamina was reduced in size with a third group suggesting a reduction by the formation of divisions and lobes.

Most serious cycad students regard both the leaf-like structure of *Cycas* megasporophylls and their loose arrangement in a crown as being primitive within the Cycadales. This has led some workers to suggest that the well-structured female cones of other extant genera such as *Encephalartos*, *Ceratozamia* and *Zamia* may have evolved from the loose *Cycas* crown, with those of *Dioon* having somewhat intermediate features and perhaps forming a link.

Fossils of the genus *Crossozamia* are significant because of the great similarity between their female sporophylls and those of modern Cycas species, particularly *C. siamensis*. The fact that the female sporophylls of *Crossozamia minor* are clearly borne in a cone-like structure is highly significant from an evolutionary viewpoint. The presence of such a structure in Permian fossils contrasts with the above views which hold that the loose, poorly organised female fruiting body of *Cycas* is primitive. In fact the cone-like structure of *Crossozamia minor* may represent the ancestral form of the female reproductive organs of all cycads, since these structures are the oldest known cycad megasporophylls and are of a comparable age to some pteridosperms such as *Phasmatocycas*.

Thus it could be postulated that a reduction in the size of the cone scales of a *Crossozamia* structure and either an increase in their number, or reduction in the length of the rhachis, could give rise to the compact female cones found in ten of the eleven genera of living cycads. Similarly a reduction in the central axis of such a cone could give rise to the loose open crown of *Cycas*. While most speculation of this type is based on the female cone, it is worth noting that the male cones found in all genera of living cycads, including *Cycas*, are basically of a similar morphology.

# A SELECTION OF FOSSIL CYCADS

The literature is well-endowed with descriptions of fossil cycads, not only of species in modern genera but also in genera which are now long extinct. Most of these fossils have been discribed from fragments, usually of leaves or leaflets and rarely have the remains of trunks or cones been found. While a large number of fossils have been assigned to the cycads, especially by researchers in the last century and early this century, modern studies have often shown that many of these properly belong to other plant groups, such as the

Bennettiales. In studies of fossil cycads, details of the cuticle have often proved to be of great value in determining placement of taxa and relationships.

The inclusion of fossil cycads in this chapter, of great interest in relation to modern cycads, is mainly for convenience and is by no means a comprehensive coverage of the large number of species described. It is obvious from the literature, however, that a comprehensive systematic study of fossil cycads is badly needed, as much of the information available is scanty or fragmentary and there are often contradictory views as to the systematics and relationships of the various taxa. Indeed it has become apparent from recent studies that some fossil cycads may have been described in more than one genus after the discovery of different organs at different times or in different places. Thus the female cones described as *Beania gracilis*, male cones as *Androstrobus manis*, leaves as *Nilssonia compta* and scale leaves as *Deltolepis crepidota*, in fact are all parts of the same cycad species. For a recent synopsis on the subject of fossil cycads see 'The Fossil History and Phylogeny of the Cycadales' by D. D. Pant, *Geophytology* 1987:17(2):125–162.

## GENUS *Androstrobus*

This genus consists of about eight species of fossil cycad from England.
**Derivation:** This genus was described in 1872 by W. P. Schimper and is derived from the Greek *andros*, 'male', and *strobus*, *strobilis*, 'a cone'; it apparently refers to the predominance of male cones in the fossils.
**Generic Features:** See species description.

### *Androstrobus balmei*
C. R. Hill
(after Dr B. E. Balme, Australian palynologist)
**Description:** Known from cone fragments, sporophylls and pollen grains; male cones to 7 cm long, 1.9–2.9 cm across; sporophylls 0.7–1.4 cm x 0.1–1 cm, ovate to wedge-shaped.
**Distribution:** Known from Jurassic deposits in the Wrack Hills, Yorkshire, England.
**Notes:** This species was described in 1990.

### *Androstrobus manis*
T. M. Harris
(from an apparent resemblance of the cone-scales to the appearance of a pangolin)
**Description:** Known from fragments of male cones which are at least 5 cm x 2 cm; sporophylls with a rhomboidal apex about 1 cm wide; pollen grains ovoid.
**Distribution:** Known from Jurassic deposits in Yorkshire, England.
**Notes:** About seven specimens of this fossil cycad were discovered prior to its description in 1946.

### *Androstrobus wonnacottii*
T. M. Harris
(after F. M. Wonnacott, discoverer of this fossil)
**Description:** Known from a single specimen which is a male cone about 5 cm x 1 cm; sporophylls with a more or less square apex; pollen grains ovoid.
**Distribution:** Known from Jurassic deposits in Yorkshire, England.
**Notes:** This species was described in 1946.

### *Androstrobus zamioides*
Schimper
(resembling a Zamia)
**Description:** Known from a male cone which is cylindrical, has overlapping scales and sessile sporangia on the underside of the scales.
**Distribution:** Known from Jurassic deposits in Yorkshire, England.
**Notes:** This species, the type of the genus, was described in 1872.

## GENUS *Antarcticycas*

This genus consists of a single species from Antarctica.
**Derivation:** Literally a *Cycas* from Antarctica, although the relationship between this genus and *Cycas* is very tenuous. The genus was described by E. L. Smoot, T. N. Taylor and T. Delevoryas in 1985.
**Generic Features:** See species description.

## Antarcticycas schopfii

E. L. Smoot, T. N. Taylor & T. Delevoryas.
(after Dr James M. Schopf, researcher on fossil plants from Antarctica).
**Description:** Known from silicified fragments of stems up to 4 cm diameter, the largest piece being 13.4 cm long; mucilage canals present; cortex composed of thin-walled parenchyma cells; two stems are branched.
**Distribution:** Known from Triassic sediments, collected on Fremouw Peak near the Beardmore Glacier in the Transantarctic Mountains of Antarctica.
**Notes:** The available material suggest that plants of this species were of a relatively small stature.

# GENUS *Aricycas*

This genus consists of a single species from Arizona, USA.
**Derivation:** This genus, described in 1991 by Sidney R. Ash, literally means a *Cycas* from Arizona.
**Generic Features:** See species description.

## Aricycas paulae

S. Ash
(after Paula Andress, who discovered the fossil locality)
**Description:** Known from leaf and leaflet fragments; leaves pinnate, to 40 cm x 14 cm; petiole to 7 cm x 0.8 cm; leaflets 2–7 cm x 0.3–0.5 cm, linear, with a mid-vein, contracted abruptly at the base into a narrow stalk 1–3 mm long, apex obtuse to acute; veins anastomosing; stomata mainly on the lower surface.
**Distribution:** Known from Late Triassic sediments in the Petrified Forest, Arizona, USA.
**Notes:** This species is distinctive for its narrow leaflets which have a basal stalk and distinct midrib. Altogether some sixty specimens of this cycad have been discovered.

# GENUS *Beania*

This genus consists of about six species of fossil cycads from England.
**Derivation:** The genus was described in 1869 by William Carruthers and honors a 19th century English fossil collector by the name of Bean.
**Generic Features:** Species of *Beania* are characterised by a cone-like structure which has widely spaced, stalked, peltate sporophylls.
**Notes:** The cone-like structures of this genus are not only highly interesting but have proved to be of major importance in elucidating the evolution of cycads.

## Beania carruthersii

Nathorst
(after William Carruthers, who described the genus *Beania*)
**Description:** Known from a cone-like structure which has widely spaced, stalked, peltate sporophylls bearing seeds.
**Distribution:** Known from Oolite deposits (Jurassic) at Helmsdale, Scotland.
**Notes:** This species was described in 1902.

## Beania gracilis

W. Carruthers
(slender, graceful)
**Description:** Known from fossilised fragments of cone-like structures (to 10 cm long), which have long-stalked (1.5–2.5 cm long), peltate sporophylls (about 18 mm x 6 mm), each bearing two ovoid, sessile seeds (to 16 mm x 13 mm), one on each side of the stalk; central axis ribbed.
**Distribution:** Known from Oolitic Shale deposits (Middle Jurassic) at Gristhorpe, near Scarborough, Yorkshire, England.
**Notes:** This fossil is remarkable for its cone-like structure which agrees in most respects with the structure of extant cycad cones, except for the widely spaced sporophylls. Some researchers believe that these cone-like structures may gave been pendulous from within the leafy crown of the plant.

## Beania kochi

T. M. Harris
(after M. Koch, the original collector)
**Description:** Known from fragments of sporophylls and seeds.
**Distribution:** Known from Jurassic deposits in Greenland.
**Notes:** This species was first described in the genus Nilssonia (in 1932) and later (1937) was described as belonging to the genus *Beania*.

# GENUS *Bowenia*

For the generic description and details of living species of this genus see page 105 . Two fossil species have been recorded.

## *Bowenia eocenica*

R. S. Hill
(from the Eocene Period)
**Description:** Known from a fragment of a pinnule which has forked veins and toothed margins; veins arise alternately from a large central vein; stomata occur on both surfaces but are rare on the upper surface.
**Distribution:** Known only from Eocene deposits in an open cut brown coalmine at Anglesea, Victoria, Australia.
**Notes:** This species has some similarities to *B. spectabilis* but is quite distinct in features of the cuticle and the veins appearing to arise alternately from a large central vein which resembles a midrib.

## *Bowenia papillosa*

R. S. Hill
(with small outgrowths or papillae)
**Description:** Known from the basal half of a pinnule which has forked veins and toothed margins; veins arise from base of pinnule; small papillae occur on the surface of the veins, particularly on the upper side of the leaflet; stomata occur on both surfaces but are rare on the upper surface.
**Distribution:** Known from Eocene deposits at Nerriga, New South Wales, Australia.
**Notes:** This species is readily distinguished by the numerous papillae occurring over the veins of both the upper and lower epidermis.

# GENUS *Ceratozamia*

For the generic description and details of living species see page 109. Two fossil species have been described, but one *C. hofmannii* Ettinghausen is probably a monocotyledon.

## *Ceratozamia wrightii*

A. Hollick
(after C. W. Wright, original collector)
**Description:** Known from leaf fragments; leaflets about 7.5 cm x 1.2 cm, linear-lanceolate to linear-elliptical, broadly attached at the base, opposite to sub-opposite, apex acute; venation parallel.
**Distribution:** Known from Eocene deposits on Kupreanof Island, Alaska.
**Notes:** The existence of this fossil supports the notion of a subtropical climate prevailing in these northern latitudes during the Tertiary Period.

# GENUS *Crossozamia*

This genus consists of eight species, four from Jurassic deposits in France and four from Permian deposits in China.
**Derivation:** The genus *Crossozamia* was described in 1849 by A. Pomel from material collected at St Mihiel in France. The derivation is uncertain.
**Generic Features:** Leaves pinnate; female sporophylls with a basal stalk and expanded apical lobe which is divided into marginal teeth and with an elongated, pinnately lobed apical extension; ovules attached to the sporophyll stalk; in one species (*C. minor*), the female sporophylls are attached to the central rhachis of a cone-like structure.
**Notes:** Although eight species have been recognised in this genus it should be noted that four species (all from France) have been found in Jurassic deposits, whereas the others are known from Permian deposits in China. Two of the latter were originally described in the genus *Primocycas*.

## *Crossozamia chinensis*

(Zhu & Du) Zhi-Feng & B. A. Thomas
(from China)
**Description:** Known from female sporophylls; stalk more than 5 cm x 1 cm; apical lobe about 5 cm x 6 cm, palmate, with uneven scales and transverse wrinkles, the margins with numerous long spreading teeth, the apex elongated and pinnately divided; ovules ten, about 1.4 cm x 1.2 cm, ovoid.
**Distribution:** Known from Lower Permian deposits of East Hill, Taiyuan, China.
**Notes:** This species was originally described as *Primocycas chinensis* in 1981 and transferred to the genus *Crossozamia* in 1989. It is known from about fourteen specimens, eight of which are more or less complete megasporophylls. The species appears to have some similarities to those of the extant species *Cycas siamensis*.

## *Crossozamia cucullata*

(Halle)
Zhi-Feng & B. A. Thomas
(hooded)
**Description:** Known from female sporophylls; stalk about 1.5 cm x 0.2 cm, widened at each end; apical lobe about 1.5 cm x 1.6 cm, fan-shaped, deeply divided into linear, pointed teeth each about 1 cm x 0.1 cm, the lower margin at right angles to the stalk and curved downwards; ovules two, about 0.4 cm x 0.3 cm, elliptical.
**Distribution:** Known from Lower Permian deposits of East Hill, Taiyuan, China.

**Notes:** This species was first described in the genus *Norinia* in 1927 and transferred to *Crossozamia* in 1989. It has also been described as *Primocycas muscariformis* Zhu & Du, which is a synonym.

### *Crossozamia minor*

Zhi-Feng & B. A. Thomas
(lesser)
**Description:** Known from female sporophylls; stalk 1 cm x 0.1 cm; apical lobe about 1 cm x 1 cm, palmate, deeply divided into about twelve linear, pointed teeth each about 0.6 cm x 0.1 cm, the apical area elongated and pinnately divided; ovule about 0.3 cm across, ovoid.
**Distribution:** Known from Lower Permian deposits of East Hill, Taiyuan, China.
**Notes:** This species, described in 1989, is known from several collections most of which are fragments of sporophylls. In the most significant specimen however a sporophyll can be seen clearly attached to a central axis in an arrangement which seems to closely parallel the female cones of all extant cycad genera with the exception of *Cycas*. The individual sporophylls however clearly have a very similar morphology to those of species of *Cycas* surviving today.

### *Crossozamia spadicea*

Zhi-Feng & B. A. Thomas
(with a spadix or sheath)
**Description:** Known from female sporophylls; stalk incomplete, about 0.1 cm wide; apical lobe about 0.8 cm x 0.7 cm, triangular, divided into about twenty-four linear, pointed teeth each about 0.3 cm long; veins thick, each ending in a tooth.
**Distribution:** Known from Lower Permian deposits of East Hill, Taiyuan, China.
**Notes:** This species, only a single specimen of which has been found, was described in 1989. The specimen has a distinct central vein running straight to the apex of the sporophyll.

## GENUS *Cycas*

For the generic description and details of living species see page 124 One fossil species has been described.

### *Cycas fujiiana*

Yokoyama
(Mt Fuji)
**Description:** Known from a fragment of a leaf; leaflets narrow, linear, spreading from the rhachis at about 45°, with a prominent midrib.
**Distribution and Habitat:** Known from Eocene deposits in Kyushu, Japan.
**Notes:** This species is very similar to *C. revoluta*.

## GENUS *Dioon*

For the generic description and details of living species see page 163. See also *Dioonopsis* following. Two fossil species have been described.

### *Dioon inopinus*

A. Hollick
(unexpected)
**Description:** Known from leaf fragments; leaves pinnate; leaflets to about 12.5 cm x 1.5 cm, linear-lanceolate, slightly falcate, expanded at the base, distal margins possibly with short fine teeth; veins simple, parallel.
**Distribution:** Known from Eocene deposits on Kupreanof Island, Alaska.

### *Dioon praespinulosum*

A. Hollick
(literally an early *Dioon spinulosum*, to which it apparently bears a close resemblance)
**Description:** Known from leaf fragments; leaves pinnate or nearly pinnatifid; leaflets about 2–6 cm x 0.5–0.8 cm, linear, tapered, margins with short spiny teeth; venation simple, parallel.
**Distribution:** Known from Eocene deposits on Kupreanof Island, Alaska.
**Notes:** This species bears a close superficial similarity to *Dioon spinulosum*.

## GENUS *Dioonopsis*

This genus consists of a single species from Japan.
**Derivation:** The genus *Dioonopsis* was described by J. Horiuchi and T. Kimura in 1987 and refers to its close resemblance to *Dioon* (*opsis*, 'like' or 'resembling').
**Generic Features:** See species description.

### *Dioonopsis nipponica*

J. Horiuchi and T. Kimura
(from Japan)
**Description:** Known from leaves and leaflet fragments; leaves pinnate; leaflets 1.5–12 cm x 0.7–2.8 cm, linear-elliptical, decurrent on the rhachis, tapered to an

acuminate apex, margins entire, wavy or with spines towards the apex; veins free, forked, rarely anastomosing; stomata mainly on the lower surface.
**Distribution:** Known from Palaeocene deposits at four localities in the vicinity of Kuji City, north-eastern Japan.
**Notes:** This species, described in 1987, appears to have been locally common as some seventy-nine specimens have been collected from the four sites. The leaves of this species have a similar general appearance to those of *Dioon spinulosum*.

## GENUS *Glandulataenia*

This genus consists of two species of fossil cycad from India and a third from Queensland, Australia.
**Derivation:** This genus, described in 1989 by D. D. Pant, literally means a *Taeniopteris* with glands. The new genus is actually segregated from *Taeniopteris* (which is non-glandular) and is distinguished by its glandular leaves.
**Generic Features:** Leaflets elongated, entire, with a prominent midrib; veins forked, rarely anastomosing, with characteristic round glands between them; stomata restricted to the lower surface.

### *Glandulataenia glandulatus*

(Srivastava) Pant
(bearing glands)
**Description:** Known from fragments of leaflets with the estimated dimensions to 12 cm x 1.2 cm, oblong, ovate or spathulate; midrib to 1.5 mm across; lateral veins once forked.
**Distribution:** Known from Triassic sediments at Nidhpuri, India.
**Notes:** This species was previously included in the genus *Taeniopteris*.

### *Glandulataenia triassicus*

Pant (from Triassic deposits)
**Description:** Known from fragments of leaflets with the estimated dimensions to 15 cm x 2.5 cm, elliptical to oblong-lanceolate, entire, apex blunt; midrib to 1.5 mm across; lateral veins once forked.
**Distribution:** Known from Triassic sediments at Nidhpuri, India.
**Notes:** Fossils of this species are very similar to those of *G. glandulatus*.

## GENUS *Lepidozamia*

For the generic description and details of living species see page 225. Two fossil species have been described. Fossils of this genus can usually be identified by features of the cuticle, particularly the arrangement of the epidermal cells in which their long axis is arranged obliquely or transversely to the long axes of the pinnae.

### *Lepidozamia hopeites*

(Cookson) L. Johnson
(resembling *Lepidozamia hopei*)
**Description:** Known from leaf fragments which had large leaflets to about 2 cm across; epidermal cells with their long axis oblique to the long axes of the pinnae; veins about twenty, each raised like a small rib; stomata on the lower surface only.
**Distribution:** Known from Tertiary sediments in a brown coal mine at Bacchus Marsh, Victoria, Australia.
**Notes:** This species was originally described in the genus *Macrozamia* in 1953 and transferred to *Lepidozamia* in 1959.

### *Lepidozamia foveolata*

R. S. Hill
(with foveoles or small pits, in reference to numerous pits in the walls of upper epidermal cells)
**Description:** Known from a fragment of a leaflet; epidermal cells with the long axis oblique to the veins; stomata restricted to lower surface.
**Distribution:** Known from Eocene deposits at Nerriga, New South Wales, Australia.
**Notes:** This species was described in 1980 from the same site where two other fossil cycads have been collected.

## GENUS *Leptocycas*

This genus consists of a single species from North America.
**Derivation:** The genus *Leptocycas* was described by T. Delevoryas and R. C. Hope in 1971 and is derived from the Greek, *leptos*, 'slender', and *Cycas*, another cycad genus; it particularly refers to the very slender stems of the species.
**Generic Features:** See species description.

### *Leptocycas gracilis*

Delevoryas & R. Hope
(slender, graceful)
**Description:** Stem slender, 3–5 cm in diameter; leaves longer than 30 cm, pinnate with persistent leaf bases; cataphylls mixed with leaves; leaflets about 4.5 cm x 4.5 mm, linear, attached broadly at the base; veins

parallel; stomata only on the lower surface.
**Distribution:** Known from Upper Triassic beds in North Carolina, North America.
**Notes:** This species was described in 1971. It is one of the few fossil cycads to have good stem material as well as leaf fragments, cataphylls and even part of a cone. An excellent reconstruction of the plant is included in the paper; see T. Delevoryas, R. C. Hope's 'A New Triassic Cycad and its Phyletic Implications', *Postilla* 150 (1971): 1–21. The authors postulate that cycads of the Mesozoic Period, unlike the modern ones, had very slender stems.

## GENUS *Lyssoxylon*

A genus of fossil cycads known by a single species from North America.
**Derivation:** The genus *Lyssoxylon*, described by L. H. Daugherty in 1941, is derived from the Greek *lusis*, 'loosening', 'freeing', *xylon*, 'wood'.
**Generic Features:** See species description.

### *Lyssoxylon grigsbyi*

Daugherty
(after M. Grigsby)
**Description:** Known from a petrified stem which has petiole bases attached, epidermal hairs and girdling leaf traces.
**Distribution:** Known from Upper Triassic deposits in Arizona and New Mexico, North America.
**Notes:** No material of leaves or cones of this species has been found.

## GENUS *Macrozamia*

For the generic description and details of living species of this genus see page 230. One fossil species has been described. Fossils of this genus can usually be identified by features of the cuticle, particularly the arrangement of the epidermal cells in which their long axis is arranged parallel to the long axes of the pinnae. See also text on *Pterosoma*.

### *Macrozamia australis*

R. J. Carpenter
(southern)
**Description:** Known from a leaflet fragment 1.9 cm long and 0.6 cm wide which contains twelve or thirteen veins; stomata restricted to lower surface.
**Distribution:** Known only from Oligocene deposits at Cethana, Tasmania, Australia.
**Notes:** This species was described in 1991. Although the material is fragmentary, the author (R. J. Carpenter) was able to conclude that the fossil species did not belong to either section in which living members of this genus are placed (section *Macrozamia* or section *Parazamia*).

### *Macrozamia hopeites* Cookson = *Lepidozamia hopeites.*

## GENUS *Michelilloa*

This genus consists of a single species from northern Argentina.
**Derivation:** The genus *Michelilloa* was described by S. Archangelsky and D. W. Brett in 1963 and honours Miquel Lillo, founder of the Lillo Institute at Tucumun, Argentina.
**Generic Features:** See species description.

### *Michelilloa waltonii*

S. Archangelsky & D. W. Brett
(after Professor John Walton)
**Description:** Known from silicified fragments of a stem about 10 cm across, the longest piece being about 8 cm long; medulla about 4 cm across consisting mainly of parenchyma cells; mucilage canals present; vascular cylinder with features matching those of a cycad; leaf bases persistent; filamentous hairs on the stem surface.
**Distribution:** Known from Triassic sediments in the San Juan Province in northern Argentina.
**Notes:** The relationships of this species are uncertain as the fossil material is scanty.

## GENUS *Pterosoma*

This genus consists of two species which are known only from fossil records in Australia.
**Derivation:** The genus *Pterosoma* was described in 1980 by Robert S. Hill and is derived from the Greek, *pteros*, 'a wing', '*soma*', a mouth, referring to cuticular wing-like structures which occur between stomatal guard and subsidiary cells.
**Generic Features:** Leaves large, pinnate; rhachis swollen

at the base into an abscission layer; leaflets numerous, elongate, entire, apex acute, attached on upper surface of the rhachis, hypostomatic; veins regularly forked, often also anastomosing.
**Notes:** Species of this genus have also erroneously been placed in *Macrozamia*. They differ from this genus by the regularly forked and anastomosing veins, the presence of an abscission layer in the leaf base, random stomatal arrangement and ornamentation on the surface of the cuticle.

## *Pterosoma anastomosans*

R. S. Hill
(anastomosing, referring to the net-like venation)
**Description:** Known from a basal fragment of a leaflet which has a conspicuously anastomosing venation; about fourteen veins per leaflet; stomata restricted to lower surface.
**Distribution:** Known only from Eocene deposits at Nerriga, New South Wales, Australia.
**Notes:** This species was described in 1980 from the same site where two other fossil cycads have been collected. *Pterosoma* specimens have also been recorded from two other Eocene deposits in Australia, Buckland and Cethana in Tasmania, but the species is not reported.

## *Pterosoma zamioides*

R. S. Hill
(like the genus *Zamia*)
**Description:** Fronds to more than 40 cm long with at least twenty-six leaflets evenly spaced along rhachis; basal leaflets 20 cm x 2 cm; apical leaflets 15 cm x 1.5 cm, veins rarely anastomosing; stomata restricted to lower surface.
**Distribution:** Known only from Eocene deposits in an open cut brown coalmine at Anglesea, Victoria, Australia.
**Notes:** This species, the type of the genus, was described in 1980.

# GENUS *Tianbaolinia*

This genus consists of a single species from China.
**Derivation:** The genus, described in 1989, by Gao Zhi-Feng and Barry A. Thomas is named after Professor Tian Baolin, Chinese palaeobotanist.
**Generic Features:** See species description.

## *Tianbaolinia circinalis*

Zhi-Feng & B. A. Thomas
(coiled like a watch spring)
**Description:** Known from leaf and leaflet fragments; leaf bipinnate at the apex; rhachis thick, tapered to the apex; leaflets circinnate when young, 0.3–0.8 cm x 0.1 cm, narrow-linear; vein solitary in each leaflet.
**Distribution:** Known from Lower Permian deposits of East Hill, Taiyuan, China.
**Notes:** This species, known from seven specimens, was described in 1989. Its authors postulate that the specimens may be representative of a juvenile stage of growth, possibly of some other fossil cycad species such as *Yuania chinensis*.

# GENUS *Yuania*

This genus consists of two species from China. The type of the genus, *Y. striata*, was collected from north-west Shaanxi and consists of fragmentary specimens of leaves.
**Derivation:** Described in 1953 by H. C. Sze, but the derivation in unknown.
**Generic Features:** See species description.

## *Yuania chinensis*

Du & Zhu
(from China)
**Description:** Known from leaf fragments; leaf about 1 m x 15 cm, pinnate; petiole expanded at the base, with paired, thorn-like reduced leaflets.
**Distribution:** Known from Lower Permian deposits of East Hill, Taiyuan, China.
**Notes:** This species, described in 1982, apparently has the lower leaflets reduced to paired thorn-like structures which are similar to those of living species of *Cycas*.

# GENUS *Zamia*

For the generic description and details of living species see page 269. At least seven fossil species have been described.

## *Zamia australis*

Berry
(southern)
**Description:** Known from portion of a leaf; leaflets linear, crowded, arising at about 40° to the rhachis, blunt.
**Distribution:** Known from Oligocene or Lower Miocene deposits in the Rio Negro region of Argentina, South America.
**Notes:** A distinctive fossil,which apparently resembles *Zamites arctitus*,known from deposits in Greenland.

## *Zamia collazoensis*
Hollick
(from the Collazo River)
**Description:** Known from leaflet fragments which show parallel venation.
Distribution; Known from Upper Eocene or Oligocene deposits on the Collazo River in Puerto Rico, Central America.
**Notes:** This species is similar to *Z. angustifolia.*

## *Zamia mississippiensis*
Berry
(from Mississippi)
**Description:** Known from fragments of leaves and leaflets; leaflets about 15 cm x 2 cm, linear-elliptical to linear-oblong, apex obtuse, contracted to base; venation free with limited forking.
**Distribution:** Known from Lower Eocene deposits in Lauderdale County, Mississippi, North America.
**Notes:** This species is similar to *Z. pumila.*

## *Zamia noblei*
Hollick
(after the original collector)
**Description:** Known from leaflets and leaflet fragments; leaflets obovate to spathulate, margins entire, apex broadly obtuse; venation free.
**Distribution:** Known from Upper Eocene or Oligocene deposits on the Collazo river in Puerto Rico, Central America.
**Notes:** This species, distinctive for its leaflet shape, is regarded as being similar to *Z. pumila.*

## *Zamia tennesseana*
Berry
(from Tennessee)
**Description:** Known from a single leaflet; about 15 cm x 0.5 cm, linear-lanceolate, lacking marginal teeth, apex more or less obtuse; venation free, some evidence of forking.
**Distribution:** Known from Lower Eocene deposits in Hardeman County, Tennessee, North America.
**Notes:** The single leaflet of this species resembles that of *Z. angustifolia.*

## *Zamia tertiaria*
Engelhardt
(from Tertiary deposits)
**Description:** Known from variable leaflet fragments; leaflets to about 10 cm x 2 cm, lanceolate, falcate, tapered to base, with numerous parallel veins.
**Distribution:** Known from Lower Miocene deposits in Chile, South America.
**Notes:** The collections of this species may represent more than one taxon.

## *Zamia wilcoxensis*
Berry
(from the Wilcox group of shales)
**Description:** Known from the basal part of a leaflet which is about 1.5 cm wide and tapers to the base; venation simple and parallel.
Distribution; Known from Lower Eocene deposits in Louisiana, North America.
**Notes:** This species is similar to *Z. pumila.*

CHAPTER 3

# Cycad Conservation

Cycads, with their great antiquity, obviously have all of the traits necessary to adapt to new environments and survive drastic environmental changes. Whether these remarkable plants will survive man's greed, however, which has reached such a peak of intensity over the last few decades, is debatable. For some reason, probably related to their long history, primitive appearance and compounded by their general slow growth, cycads have become extremely desirable plants to collect. Because of this popularity their natural populations have suffered a decline as a direct result of considerable poaching. Some have even become extinct in the wild. The passion for possessing cycad plants is largely to be found in the developed countries but its effects are felt wherever there are natural occurrences of these intriguing plants. Poaching has reached great significance in Africa, Mexico and Central America. The result has been a thriving national and international trade in plants, legal and otherwise. Because of the very high prices which some species fetch, there is also a trend for collectors to gather these plants as an investment for the future. They have also come to the attention of businessmen for this very reason.

Not all threats to the continued existence of natural populations of cycads come from poaching. Habitat destruction poses a major threat in all countries where they occur naturally. Conversion of land to farms, commercial developments, general urbanisation, drainage schemes and even roads and highways can pose significant threats to cycad populations and may even be a real danger to the survival of localised populations or rare species. Unfortunately many natural populations are destroyed before they have been adequately studied and the consequences of this may be far reaching. Biodiversity is a real feature of wild populations, especially those which have adapted to different soil types and habitats. Disjunct populations are often of great significance to taxonomists and geneticists. Perhaps they differ from other populations in some significant feature, and if they are lost to science then some valuable genetic diversity may be gone forever.

Some species of cycad may have never been particularly common in nature. Certainly this applies to a few species of *Encephalartos*, *Dioon* and *Zamia*, which consisted of small populations either concentrated in discrete areas or occurring as widely separated individuals. The classic example here is *Encephalartos woodii* which was known from a single clump (of the male sex) in Natal. This species, discovered in 1895, was extinct in the wild by 1907, when all of the clump had been transplanted, fortunately to a botanic garden. Rather than raising a general concern in the community for their welfare, the very rarity of these species seems to engender an even greater degree of greed in certain individuals, and the populations of such uncommon to rare species have suffered drastically as a result.

A particularly insidious factor is the concentration by poachers on female plants, thus altering the natural sex ratio of the remaining wild populations and reducing, even further their potential for regeneration. In the Middleburg district of the Transvaal, South Africa, *Encephalartos middelburgensis* is represented by a small population which has been reduced to scattered individuals by poaching. Seedlings are absent and the ability to regenerate is nearly zero because the male plants outnumber the females by ten to one. The natural population of *Encephalartos dolomiticus* has been reduced to less than twenty individuals and seed production in this species is now absent in the wild. Plants of

*Encephalartos latifrons* and *E. inopinus* have been so depleted by poaching that they are now too widely dispersed in the wild for reproduction to occur.

The survival of *Cycas beddomei*, from the Eastern Ghats in India, is threatened in a slightly different manner. In this case the male cones, which are used for medicinal purposes, are collected extensively by local tribes, thus reducing or eliminating natural pollination and possibly also posing a threat to the pollinating agent. This annual collection of the male cones is one significant factor which has contributed to the dramatic decline of natural populations of this species.

A comparatively common practice among modern botanists is to be deliberately vague when referring to the collecting localities and type localities of new species of cycad, for depletion of populations is an unfortunate consequence following the descriptions (and consequent pinpointing) of many new species. At least one botanist has been reluctant to name a new species because its population is so small and vulnerable that he fears its immediate extinction following the obvious interest that would be created by its formal description.

When describing *Encephalartos cupidus* in 1971, R. A. Dyer commented on the specific epithet which means 'desirous' and mentioned that because of the great interest in ornamental specimens of *Encephalartos*, the epithet could also have a tarnished meaning: 'implying a passionate desire, to the extent of greed or lust'. How prophetic he was, for today the populations of this species have suffered greatly, with all being eradicated except for one stand which is protected in a nature reserve. Nor is this unfortunate situation restricted to species in easily accessible areas. Wealthy collectors are prepared to finance expeditions into remote areas and may even strip a species completely so that its monetary value increases accordingly. *Zamia purpurea*, a native of Southern Mexico, may be one such unfortunate example for since its naming in 1983 it has virtually become almost extinct in the wild. A similar fate has befallen *Encephalartos cerinus* in Natal since its description in 1989. Of two colonies originally known, one has been completely eradicated and 75 per cent of the mature plants and all of the seedlings in the other colony have been poached. In Mozambique two species of *Encephalartos*, *E. pterogonus* and *E. munchii*, have suffered terribly at the hands of poachers, with *E. pterogonus* virtually disappearing from the wild since its discovery in 1949. *E. munchii* is known to resent disturbance, with a high proportion of transplants failing to survive. Despite this, or perhaps because of it, plants of this species are still eagerly sought after by collectors and its survival in the wild is at best precarious. In some cases even common cycads have been decimated. Thus *Zamia furfuracea*, despite originally being widespread and locally common, has been exterminated in most of its localities. It has been reported that in 1980 one single American nursery imported thirty thousand plants per month of this species to satisfy the demands of the ornamental plant industry. In 1984 some twenty thousand seedlings of this cycad, labelled as *Chamaedorea* palms, were seized by US officials at Miami airport and eventually returned to Mexico where they were distributed to local botanic gardens.

Cycads are ancient plants protected by human laws, but their conservation is haphazard and the need for strict adherence to all conservation measures must be realised by all cycad enthusiasts if these plants are to survive. In some areas even strict protective laws and penalties are insufficient to safeguard cycads from the predatory activities of poachers. Perhaps instead of linking these remarkable plants with the coelacanth or the dinosaur it would be better from a conservation viewpoint to associate them with the various species of rhinoceros, which are surely heading down the same path to total obliteration; and all because of human stupidity and greed.

# THE ROLE OF LEGISLATION

Control in the trade in cycads receives very little attention in all countries where they occur naturally, with the notable exception of South Africa. In this country the movement and trade in any cycad, either cultivated or wild, is strictly controlled. Even the movement of seed between provinces requires permits and an import permit will only be supplied after an export permit has been issued by the province from which the seed originates. The result of this restrictive legislation has been beneficial for African cycads. Seizures, substantial fines and well-publicised court cases have resulted in a greater awareness of the importance of these plants with the general public and must have also dissuaded the activities of some would-be poachers.

Apart from legislation which may be locally enacted for the protection of indigenous species, cycads are also protected by international law. The major international conservation organisation, the Threatened Plant Unit of the International Union for Conservation of Nature and Natural Resources (IUCN), has acted by listing those species of cycads which are regarded as being rare, endangered or vulnerable in nature. One species, *Encephalartos woodii*, is extinct and slightly less than half of the world's cycad species are regarded as being threatened in one way or another.

Listing the status of a rare species is one major step in its protection, the second is restricting the international trade in that species. This role is played by the Convention on International Trade in Endangered Species of Wild Fauna and Flora (CITES). Basically this is an international agreement between signatory countries that places restrictions on the import, export and re-export of listed species. Four genera of cycads (*Ceratozamia, Encephalartos, Microcycas, Stangeria*) are listed on Appendix 1 of CITES (the most restrictive list) and the remaining genera are included on Appendix 2.

Appendix 1 plants cannot be traded between signatory countries without permits from the management authorities of both the importing and exporting countries. The import permit must be obtained first, before the export permit is requested from the country of origin. All parties must agree that the trade in the specimens under question will not be detrimental to the survival of the species in the wild. Normally all parts and products of any species listed on Appendix 1 are afforded full protection, but in the case of cycads it was decided recently that trade in wild-collected seeds would be allowed as a further conservation measure to reduce pressure on natural populations. A secondary effect of this decision is the creation of local employment and foreign currency revenue from the sustainable use of natural resources. For similar reasons seeds and plants propagated from cultivated specimens may be judged under different criteria to those from wild collected plants. A proposal has been prepared to introduce a licensing system to allow registered nurseries to propagate and trade in Appendix 1 plants. Only those nurseries that comply with the CITES criteria would be allowed to trade internationally in these plants. Licensing and monitoring of these nurseries would be controlled by the various CITES scientific authorities in each signatory country.

Commercial trade is permitted in wild-collected plants listed in Appendix 2 of CITES, if the country of origin considers that the quantities traded do not represent a threat to the natural populations. More than one million Appendix 2 cycad plants are traded annually, about half of which are artificially propagated. This leaves a massive five hundred thousand cycad plants apparently being legally collected from the wild and traded internationally each year. In addition there is a massive trade in cycad seeds amounting to about six million seeds per annum. With such pressures on the natural populations of Appendix 2 cycads it is only a matter of time before all cycads become transferred to Appendix 1. It is apparent that the international trade in cycads (which is considerable) is controlled to a significant degree.

Loopholes still exist, however, and are exploited by determined, unscrupulous collectors and dealers. Document falsification is still possible and rare plants can be deliberately misidentified to mislead officials. For example two thousand leafless stems of what was purported to be *Ceratozamia mexicana* were seized at a border crossing in Texas in 1984 and returned to Mexico where they were distributed to official gardens. When the plants grew they were found to be *C. hildae*, a rare and endangered species of much greater commercial value than *C. mexicana*. It is also possible to camouflage trade in wild-collected plants by simply entering such exports under a broad category such as the family Zamiaceae. Although most of the major countries trading in cycads are signatories to the CITES agreement, there are still a number of important countries, a notable one being Mexico, who have not yet agreed to sign.

Finally the mere listing of a species on Appendix 1 can result in higher demand for plants which in turn creates higher prices and thus leads the way to more illegal collecting.

## THE ROLE OF CULTIVATION

One significant consequence of the demand for cycads, is that substantial collections exist in various parts of the world. These are a genetic resource which should not be ignored. At the very least every female cone borne on these plants should be pollinated so that seedlings can be raised to satisfy some of the considerable demand that exists for cycads in general. (For notes on hand pollination see Chapter 9.)

Certain difficulties arise with cultivated plants however, which should be borne in mind. Often details of the original collection or locality have been lost so that the exact genetic resource cannot be pinpointed. This is particularly true of cycads in private collections which may have been traded several times or passed on to later generations, with the vital details being lost in the process. Generally plants in botanic gardens have a more accurate recording of their pedigree but unfortunately this is not always the case.

Another problem which exists in both private and public collections is the possibility of hybridisation occurring between adjacent plants. Although horticultural experience suggests that this is a fairly unlikely process, especially within insect pollinated species, it must be taken into consideration when the conservation of pure genes is the goal. Species in collections should be arranged so as to minimise the risk of hybridisation. The prediliction for some cycad enthusiasts to deliberately cross different species is another factor which must also be taken into account.

After considering the above factors, cultivated plants would appear to play two important roles for cycad conservation. By helping to satisfy demand, the production of seedlings from cultivated plants also helps to lessen pressure on plants in the wild. If we are serious about this aspect then no female cone in cultivation should be left unpollinated. A small amount of pollen is sufficient to pollinate a cone and each is capable of producing between two hundred and one thousand seeds depending on the species. Staff at the Fairchild Tropical Garden in Florida, USA, have successfully propagated *Microcycas calocoma*, the endemic cycad from Cuba which is regarded as being endangered in the wild. The hand pollination of two cones in 1984 resulted in the production of about three hundred viable seeds, many of which were distributed to botanic gardens and kindred institutions in sixteen countries.

With the optimising of growth rates using modern horticultural techniques (see Chapter 7), it is possible for some cycads to produce female cones regularly each year and in greater numbers than in the wild. Young plants also mature faster so that the time between generations is reduced and a significant multiplying effect can be achieved.

One problem often encountered in collections is that pollen is not always available when the female cones are mature. In a unique approach to this problem, the South African Cycad Society recently established a Pollen Bank in Natal. Pollen of known species is collected and stored in a deep frozen state with a corresponding entry being made on the database of a computer. Upon emergence of a female cone of any species, a suitable source of pollen can be located via the computer, prior to the female cones' maturity. Studies have shown that deep frozen pollen deteriorates within about seventy-two hours of being removed from the freezer hence the pollen bank is mainly of local significance. The compilation of various collections onto a computer database can also assist in the location of suitable pollen when coning of a female plant becomes obvious. In such instances pollen can be readily transported by private individuals within a country. One successful way is to cut a male cone just before it commences pollen shedding and despatch it by mail. As the pollen is released successively over several days, fresh pollen can usually be obtained for pollination purposes.

Another positive step towards reducing the pressure on plants in the wild has also been taken in South Africa. Here confiscated plants are grown in a nursery and used as parents to produce seed. This is then germinated and the resultant plants are grown on for sale.

Cultivated cycads may also play an important conservation role by producing seedlings which can be replanted into the wild. Such replanting programs may offer the only chance for the wild survival of some species. In these cases, extreme care must be taken to ensure the genetic purity of the parent plants so used, and all chances of accidental cross-pollination must be eliminated. Another significant factor, the survival of the pollinating insects, must be considered for species whose populations have been greatly depleted by poaching. If the pollinator is host-specific, and has become extinct, then no amount of replanting will ensure the future survival of the cycad in the wild.

## RESCUES AND RESETTLEMENT SCHEMES

Seedlings and mature plants of the majority of cycads are generally easy to transplant. This response, which is unfortunately also an aid to poachers, has been used by conservation authorities in South Africa, when populations of cycads have been threatened by development.

The most notable rescue operation of this type seems to be that of 'Operation Wild Flower'. During this operation over six thousand specimens of *Encephalartos lebomboensis* were rescued from an area to be flooded by the completion of the Jozini Dam across the Pongola Poort between Transvaal and Natal. These plants, often quite large, were floated from the north to the south bank of the waterway or moved by rafts. In addition to the mature plants, large numbers of seedlings were also removed. The rescued plants were not replanted in the wild but were distributed to public gardens and interested growers.

Two other resettlement schemes, both involving *Encephalartos humilis*, have been undertaken following threats to wild populations by the expansion of adjacent plantations. The first occurred in 1986 when one thousand plants of the cycad were successfully resettled in a nature forestry reserve. The second resettlement was in 1989 when nine hundred plants of this species were transferred to a natural heritage area on a private farm.

By contrast to these successful rescues, one finds numerous instances in the literature of missed opportunities. Whether this be from the construction of freeways in Florida, the clearing of rainforest in Mexico or the deliberate poisoning of species in Australia, the result is always the same.

CHAPTER 4

# The Structure of Cycads

Cycads range in size from small species with subterranean stems as in *Zamia pygmaea* to lofty giants with woody columnar trunks such as *Lepidozamia hopei, Encephalartos transvenosus* and *Dioon spinulosum*. Although basically having a similar structure to other plants, cycads also have some unique features and an examination of their structure is presented in this chapter. Technical terms are explained where used and additional terms which may be encountered in papers on the subject have been included in the glossary.

## ROOTS

Cycads basically produce three types of roots, although one of these is restricted to species of *Cycas*.

The primary root system of a cycad is equivalent to that of the tap root system which is found in most types of plants. The primary tap root is the main elongated structure so prominent in cycad seedlings, and is present throughout the life of the plant. The primary tap roots of all cycads are contractile (see page 65).

Lateral roots develop from the tap root. These are mostly diminutive but in some species the early lateral roots develop into vigorous structures similar to the primary roots. In many species of cycad which develop large trunks, a number of vigorous lateral roots may develop as accessory tap roots, probably as an aid in supporting the large trunk. In species of *Zamia* some of these early lateral roots become swollen and probably also act as storage organs. The essential functions of this primary root system (tap roots and lateral roots) are plant anchorage and the uptake of nutrients and water.

A second highly specialised type of lateral root grows erect, forks repeatedly and develops into coral-like structures which are termed coralloid roots. Such roots are unique to cycads and they contain blue-green algae living in a symbiosis that results in the fixation of nitrogen from the atmosphere (see also page 63).

A third type of root is found on the trunks of species of *Cycas*. These roots, which are adventitious, arise from the lower side of trunk offsets and swollen bulbils and grow downwards through the air in close proximity to the trunk and often also down through the cortex.

## STEMS AND TRUNKS

Cycads have a woody stem or trunk which enlarges as the plant gets older. This trunk varies from being subterranean and almost bulbous, to a large columnar structure which emerges from the ground and in some species, may grow as tall as sixteen metres. Cycad trunks are described as being pachycaulous; this means they are more or less succulent with a well-developed pith and cortex and very little secondary wood. Secondary growth and thickening in cycad stems occurs by successive development of cambia.

Cycad stems are unbranched or branch sparingly. Forking or dichotomous branching of the trunk apex is sometimes seen in large plants but is generally rare (except in *Stangeria* and some species of *Zamia* which have subterranean trunks). Adventitious branching is more common, resulting from the development of bulbils and offsets on the above-ground portion of the trunk and suckers on that part of the trunk which is below ground. Since cycads lack axillary buds, these adventitious growths arise as adventitious buds produced from the callus. Similar buds may also be produced on damaged tissue, particularly on the apex of decapitated stems, and also on major wounds.

The cycad trunk is terminated by the apical meristem or growing apex. Cycads have the largest growing apex of all vascular plants with that of species of *Cycas* being more than three millimetres across. Most cycads produce regular flushes of growth with short dormant periods in between. Some species such as *Dioon edule* may have prolonged rest periods and these periods of inactivity are revealed as narrow regions in the trunk.

The surface of a cycad trunk is generally roughened with some species gaining protection from the persistent bases of fallen leaves (*Cycas, Dioon*). This is not so in all genera as leaf bases are shed completely in species of *Zamia, Chigua* and *Microcycas calocoma*.

The stems of those cycads which are largely subterranean are contractile and this feature is unique among vascular plants. This character, combined with contractile roots, results in the shoot apex of these largely subterranean species always remaining buried, despite the regular apical extensions of the trunk which occur along with each new flush of growth. Thus despite regular size increments, cycad plants are able to maintain their largely subterranean status by virtue of these remarkable contractile systems (see also page 65).

## Trunk Branching

Branching mechanisms are not well developed in those cycads which have aerial trunks. Branching does occur but it tends to be haphazard rather than organised. Axillary buds do not exist in cycads but branching may occur following damage to a trunk or from specialised structures.

Specialised growths called bulbils can develop on the trunks of all large species of *Cycas*, particularly *Cycas circinalis, C. revoluta* and *C. thouarsii*. These bulbils, which may develop from parenchyma tissue of the leaf bases, can be seen as small bumps on the trunk if the leaf bases are removed. Not all of these bulbils develop further and in fact very few actually sprout leaves and eventually grow into a branch. Undeveloped bulbils may act as a fallback mechanism if the major growing point(s) on the trunk are damaged or destroyed. It is noticeable that bulbils may frequently develop on the upper side of a leaning trunk indicating a possible role of hormones in their sprouting.

Branching as the result of damage to a trunk is common in some cycads, for example *Encephalartos transvenosus* and *Lepidozamia hopei*. By contrast to the bulbils discussed above, branches resulting from injury develop from meristems within the trunk which give rise to the wood and bark. Occasionally damage to an apical meristem may result in forking or paired new growths.

Bulbils and branches on trunks are of interest to cycad growers for these structures can be used for propagation purposes (see page 91).

## Multiple Trunk Development

Some species of cycad produce basal suckers freely, but the end result of the development of these suckers on the plant's appearance is somewhat variable depending on its growth habit.

Many species of *Encephalartos* produce basal suckers. In some species a certain number of these suckers develop into trunks resulting in a multi-trunked clump all of which arose from a single seed. Those growths which develop as trunks, dominate the basal suckers and these persist as dwarfed growths around their base. Other species of *Encephalartos* may typically develop a single trunk with dwarfed suckers present at the base (*E. ituriensis*, *E. manikensis* and *E. tegulaneus*). It would seem that the basal suckers remain dwarfed and in a state of low activity as long as the large trunks are healthy. If these are badly damaged or destroyed then one or more of the basal suckers may develop into a new trunk.

Species which have subterranean trunks may also form suckers freely, the result being a dense clump of many crowns usually separated spatially at ground level. Each crown is genetically identical to its parent and produces leaves and is also capable of coning. It is possible that the plant tissue uniting the subterranean trunks may decay with time leaving them as genetically identical, but separate individuals. Species which have the propensity to form large clumps by this method include *Zamia furfuracea*, *Z. pumila*, *Encephalartos cupidus*, *E. horridus* and *E. schmitzii* (one clump of this latter species has been recorded with about fifty separate crowns).

Species of *Macrozamia* are usually considered to be single-stemmed, but some species which have subterranean stems may branch and form multiple clumps. These branches are not suckers as they appear to arise from an endogenous meristem on the stems. Two species which frequently have this habit are *M. pauli-guilielmi* and the recently described *M. fearnsidei*. Often specimens of *Macrozamia* growing along road verges may develop multiple clumps artificially as the result of damage from machinery such as graders.

The underground stems of *Stangeria eriopus* are much-branched but a close examination will show a distinct pattern of branches not present in other cycads. Each growing point or meristem forks, resulting in two equal branches. This branching occurs regularly at right angles to the preceding branch, resulting in a more ordered appearance to subterranean stems of *Stangeria* than is seen in other cycads.

# LEAVES

Cycad leaves (sometimes called fronds) are carried in a crown which terminates the trunk. The number of leaves in a crown varies greatly with the species and the age of the plant concerned. Whereas many of the larger growing species can bear luxuriant crowns of up to one hundred leaves (or more), some of the smaller types with subterranean trunks may have less than five leaves in the crown of a mature plant. Large-growing cycads usually produce a number of leaves together in a flush whereas, in sparse-leaved species such as *Stangeria eriopus* and *Zamia wallisii* a single leaf is produced at a time. If the individual leaves of a flush are examined carefully it will be seen that they are arranged in a spiral. It is interesting to note that all the leaves of a flush mature almost simultaneously.

Cycads which have subterranean trunks usually have erect or obliquely spreading leaves. In cycads with an arborescent trunk, it is noticeable that as a flush of new, erect leaves is produced, so the older leaves are pushed outwards and downwards. This is particularly noticeable in large-growing species of *Encephalartos*. Cone production, particularly on a female plant, can result in a similar change of leaf orientation.

The leaves of all cycads, except species of *Bowenia* are once-divided or pinnate and lack a terminal segment. The divisions of leaves of this type are termed leaflets or pinnae. The leaves of *Cycas micholitzii* and some *Macrozamia* species such as *M. stenomera*, although still pinnate, differ from other cycads in that the leaflets themselves are dichotomously forked. By contrast to the pinnate-leaved cycads, the leaves of *Bowenia* species are twice-divided or bipinnate, and the segments are strictly called pinnules, although most growers still refer to them commonly as leaflets.

The basal woody structure of all cycad leaves that extends from the trunk to the first leaflet or segment is known as the petiole. The extension of the petiole on which the leaflets are borne is the rhachis (often spelt rachis). These organs may be smooth but in some species they bear spines or surface prickles. Spines are reduced leaflets and occur in parallel rows towards the base of the leaf. These spines may be much reduced as in species of *Cycas*, or exhibit an increasing transition between much reduced spines and true leaflets (*Dioon, Encephalartos, Lepidozamia* and *Macrozamia*). Prickles, which are a hard, sharp extension of the epidermal cells, usually occur at random and show no transitional stages with leaflets. They are common on the petioles and rhachises of species of *Ceratozamia, Chigua* and *Zamia*, although they may be absent from some species in the latter genus (e.g. *Z. inermis, Z. pumila*).

The basal part of the petiole of a cycad is swollen into a semi-sheathing structure. This leaf base persists as a protective layer around the trunks of species of *Cycas* and *Dioon*, but is shed with the dehiscing leaf in species of *Zamia* and *Microcycas calocoma*. Leaf bases may be swollen and bulbous (*Cycas, Dioon*) or nearly flat (*Zamia*). In some species stipules are present on the upper surface of the petiole just above the swollen base. In *Stangeria eriopus* and *Bowenia* species, these stipules are large and fleshy with a vascular strand. They persist as a protective mechanism on the stem of these cycads after the leaf has died. Small, non-vascular stipules are found on the leaf bases of species of *Chigua, Ceratozamia, Zamia* and juvenile plants of *Microcycas calocoma*.

## Diagnostic Leaflet Features

The shape and arrangement of the leaflets are important features useful for the identification of cycads. Of primary importance is whether the leaflet is articulated (jointed) to the rhachis (*Ceratozamia, Chigua, Zamia* and *Microcycas calocoma*) or attached with a decurrent base (*Cycas, Bowenia, Dioon, Encephalartos, Lepidozamia, Macrozamia* and *Stangeria*). In the former group the leaflets fall before the leaf, leaving the rhachis and petiole attached to the trunk. *Zamia manicata* is unusual because it has stalked leaflets with a glandular collar where the stalk joins the leaflet. The leaflets of *Zamia wallisii* are also stalked but lack the glandular collar. A coloured callous area at the base of the upper margin of species of *Macrozamia* is a useful feature for generic identification.

The presence of a midrib in the leaflet is a very significant diagnostic feature. This feature, so prominent in species of *Cycas*, also serves to separate species of *Chigua* from *Zamia*, and is present in *Stangeria eriopus*. Whereas the midrib of the latter species is formed by the unification of many parallel veins, that of *Cycas* is composed of a single vein.

Leaflet shape can be a useful diagnostic feature although most cycads tend to have narrow leaflets which are commonly much longer than wide. In some species of *Zamia*, notably *Z. skinneri* and *Z. wallisii*, the leaflets are proportionately broad, sometimes being nearly as broad as long. It is interesting to note that both of these species also have a distinctly corrugated upper surface to their leaflets.

While most cycads have entire leaflets, those in some species are divided, always by dichotomous forking. This is notable in *Cycas micholitzii* where it is the only species in the genus to have divided leaflets. Similarly many species of *Macrozamia* belonging to section

*Parazamia* have divided leaflets. The species with the most deeply dissected leaflets is *M. stenomera*. These leaflets fork many times and the narrow segments overlap with those of adjacent leaves to impart a very complex appearance to the leaf. It is noteworthy that leaflet division is not uniform in some species of *Macrozamia*. In *M. diplomera* for example it is not unusual to find whole leaves with entire leaflets and sometimes divided leaflets may be absent from whole plants.

The arrangement of the leaflets along the rhachis is distinctive in some species and is a useful character in *Encephalartos* for separating similar taxa. Leaflets may be regularly or irregularly spaced on the rhachis. In many species those towards the apex become crowded whereas those towards the base are widely spaced.

The angle of the leaflet to the rhachis can be significant. Commonly this is nearly a right angle but some species are arranged more acutely. Often this angle changes along a frond with the distal leaflets frequently being at a more acute angle than those near the middle and sometimes the lower (proximal) leaflets at an angle exceeding 90°.

In many species the leaflets are held in a flat plane along the frond whereas in others they are held obliquely erect to form a vee. The angle of the vee may be shallow or steep. The leaflets of *Microcycas calocoma* are deflexed at an oblique angle (like an inverted vee) and impart a distinctive drooping appearance to the leaves.

The outline of a flattened leaf can be characteristic. In most species the longest leaflets are found in the central portion of a leaf, with those towards either end decreasing in length. This gives a lanceolate to elliptical outline, whereas an oblong outline is obvious where the median leaflets are of similar length to the others. In *Microcycas calocoma* the terminal leaflets are of similar length to the median leaflets giving the leaf an unusual truncated or cut-off appearance.

Another useful diagnostic feature found in leaflets is whether they are attached in a plane parallel to that of the rhachis or whether they have a twist at the base which turns them at right angles to the rhachis and imparts an appearance similar to the slats of a venetian blind. In the latter arrangement the upper and lower margins of adjacent leaflets may overlap.

In most cycads the rhachis remains straight or curves gently, however in most species of section *Parazamia* in the genus *Macrozamia* a series of strong spiral twists in the rhachis results in a spiralling of the leaflets and imparts a whorled or plumose appearance to the leaves. In many cycads the rhachis is recurved near the apex to impart a hooked appearance to that part of the leaf.

## Vernation or Ptyxis

The shape of an emerging leaf before expansion is known as ptyxis or vernation. Fern leaves provide a classic, easily identified example of this aspect of development in that the young fronds of many species are coiled like watch springs. This arrangement is known as circinnate vernation and in members of the cycad genus *Bowenia*, the young leaves are coiled in the same way prior to expansion. Members of the genus *Cycas* are often also reported as having circinnate fronds but this is in fact erroneous as the developing leaves are erect and only the individual leaflets are coiled.

A second type of vernation, also shared by some species of ferns and cycads, occurs when the apex of the young leaf is inflexed like a shepherd's crook. This arrangement is termed inflexed vernation and is found in species of *Encephalartos*, *Zamia* and *Stangeria eriopus*. Some cycads may have the apex recurved rather than inflexed.

The coiling or recurving of the leaf apex is functional, particularly in species which have a subterranean stem, because the tender leaf tip and developing leaflets are protected by the coiled or humped rhachis during emergence through the soil in the early stages of their expansion.

## Juvenile Leaves

The first seedling leaves of all species of cycad show differences from those of the adult plant. In most species the leaflets are toothed on the margins, often coarsely so, and they are often of fairly uniform size and shape. They may also differ in other features such as colouration, texture, degree of pubescence, lack of reduction of the lower leaflets to spine-like processes and often a lack of differentiation of the terminal spine or leaflet. Juvenile leaves can provide useful characters for separating species of similar mature appearance, for example *Macrozamia moorei* and *M. johnsonii*.

Juvenile leaves persist on a plant for many years, in theory until it matures and begins to reproduce. Some odd leaves which have juvenile characteristics may appear sporadically on a mature plant, especially following severe structural damage or defoliation as may occur after insect predation or bushfires.

## Leaf Hairs

The juvenile leaves of all cycads bear small hairs (also called trichomes), which are outgrowths of epidermal cells. These structures are often not obvious on mature leaves because they are shed as the leaf matures. A comprehensive study on the leaf hairs of all ten genera of cycads was carried out by Dennis Stevenson of the New York Botanical Garden (see 'Observations on Ptyxis, Phenology and Trichomes in the Cycadales and their systematic implications' *American Journal of Botany* 68, no.8 (1981):1104–14). A summary of these results is presented here. Basically the hairs of cycads can be divided into six types: transparent branched; transparent unbranched; coloured branched; coloured unbranched; short curved coloured; short branched idioblastic coloured.

The distribution of hair types in the cycad genera is as follows:

***Bowenia:*** Transparent unbranched, coloured unbranched and short, curved coloured.

***Ceratozamia:*** Transparent unbranched and coloured unbranched (dark red-brown).

***Chigua:*** Unknown.

***Cycas:*** Transparent branched and transparent unbranched.

***Dioon:*** Transparent unbranched and coloured unbranched (dark red-brown).

***Encephalartos:*** Transparent unbranched and coloured unbranched (dark red-brown).

***Lepidozamia:*** Transparent unbranched, coloured unbranched and short, curved coloured.

***Macrozamia:*** Transparent unbranched, coloured unbranched short curved coloured and short branched coloured.

***Microcycas:*** Transparent branched, transparent unbranched, coloured branched and coloured unbranched.

***Stangeria:*** Transparent unbranched and coloured unbranched (light tan to yellow).

***Zamia:*** Transparent branched, transparent unbranched, coloured branched and coloured unbranched.

## CATAPHYLLS

Cataphylls are scale-like structures which are actually reduced leaves and may be referred to as scale leaves. Basically a protective device, they are produced on the vegetative shoots of most cycads, and there is an annual alternation of flushes of cataphylls and leaves. This alternation may cause a ribbing to develop in the trunks of some cycads. *Stangeria eriopus* and *Bowenia* species do not produce cataphylls but instead the prominent persistent stipules function in the same manner as cataphylls. Often cataphylls are found that have a small, undeveloped leaf at their tip. Sometimes a leaf may abort in *Bowenia* and *Stangeria* and leave what appears to be a cataphyll. In all other genera there are alternative flushes of cataphylls and leaves but (not in *Bowenia* and *Stangeria*, however).

Decurrent cataphylls also occur on the peduncles which support the cones in species of *Dioon, Encephalartos, Lepidozamia* and *Macrozamia*. These cataphylls are readily identifiable as scale-like structures and may terminate shallow ridges.

## CONES

Cycads do not form flowers but bear reproductive structures known as cones or strobili. While these structures bear a resemblance to the cones found in other gymnosperms, particularly conifers, they are not as fully developed as in these plants. One obvious difference is that the cone scales of cycads do not separate widely at maturity as they do in the conifers and the female cones of conifers are very complex whereas the female cycad cone is a relatively simple structure.

Cycad plants are strictly unisexual, either bearing male or female cones, never both. Similarly a cone is either male or female and never bears the structures of both sexes. Male plants bear male cones termed staminate strobili or microstrobili and cones on female plants are termed ovulate strobili or megastrobili.

Male and female cones of a cycad are generally dissimilar in shape and size (rarely are they of different colouration), although in a few species of *Encephalartos*, namely *E. aemulans*, *E. heenanii* and *E. manikensis*, the cones of each sex are nearly indistinguishable. Generally male cones are produced in greater abundance than are females. The extreme in this case is probably *Macrozamia moorei*, where up to one hundred cones have been counted on large male plants, whereas the females carry about five with a maximum of eight. However, female cones are much larger and longer lived and their formation causes a considerable drain in storage reserves. Consequently, intervals between coning may be longer in female plants than in males.

The basic structure of male and female cones is similar with each having a central axis and numerous spreading structures termed scales or sporophylls. These sporophylls are actually modified leaves and are arranged in columns that are distributed on the axis in a low spiral. Sporophylls on the male cones produce pollen and are termed microsporophylls and those of the female cones carry naked ovules and are termed megasporophylls.

At maturity the central axis of a cone elongates and this process causes the sporophylls to separate, thus enhancing pollen release in the case of the male and receptivity in the female. Whereas the lengthening of the central axis is obvious in male cones, it is generally much less obvious in the females.

The female cones of all species of *Cycas* differ markedly from the cones of all other cycads, as well as from the cones of male *Cycas* plants. All of these other types of cones exist as discrete units (having a central axis and spirally arranged sporophylls), and arise either from the apex of stems or stem branches or laterally from among the leaf bases. By contrast the female *Cycas* cone lacks a central cone axis and consists of a circle or crown of loosely arranged sporophylls. These appear to terminate the stem but after pollination has occurred the stem actually continues growth through the centre of the sporophylls. These are forced outwards as the stem expands and eventually form a ring around the crown.

## Female Cones

The female cones of most cycads are larger and bulkier than the male. In large cycads the female cones commonly weigh in excess of twenty kilograms. The heaviest seems to be those produced by *Lepidozamia peroffskyana* and *Encephalartos transvenosus* both of which can mature cones in excess of forty-five kilograms. By contrast the female cones of small-growing *Zamia* species may weigh less than thirty milligrams.

Each female sporophyll (megasporophyll) bears ovules near its base. These are attached directly to the stalk of the sporophyll in all genera except *Dioon* where each ovule has a small basal stalk. In *Bowenia* and *Stangeria* the two ovules are attached under the sporophyll stalk whereas in other genera, except *Cycas*, they are attached above the sporophyll stalk. In species of *Cycas* the ovules seem to be sunk into the margin of the sporophyll and in early stages in *Stangeria* they appear to be sunk into the base of the expanded sporophyll tip. In most cycad genera the micropylar end of the ovule faces inwards towards the central axis of the cone. The exception occurs in species of *Cycas* where the ovules point obliquely upwards and outwards. All cycad genera, again with the exception of *Cycas*, have the basic number of two ovules on each sporophyll (the presence of one or three is abnormal), whereas in *Cycas* four to ten is normal depending on the species.

Female cones of many cycads, especially species of *Zamia*, have an apical area which is sterile. The size and shape of this section of tissue has sometimes been used as a taxonomic character to distinguish between species. In some species the sterile apex can be substantial; for example in *Encephalartos ferox* it occupies about one-third of the cone. The size and shape of the megasporophylls of cycads can be a useful diagnostic feature, at least at generic level. The most obvious of these features in each genus are summarised below:

***Bowenia:*** Outer face of the megasporophyll flared into a hairy, flat, plate-like structure which lacks spines or any ornamentation.

***Ceratozamia:*** The apex of the megasporophyll has two horn-like structures, often with a small stalk between, which spread widely in young cones but are held obliquely erect at maturity.

***Chigua:*** The apex of the megasporophyll is hexagonal with raised mounds on the outer face.

***Cycas:*** The megasporophyll is flat, lobed or pinnate (*C. revoluta*) and somewhat leaf-like.

***Dioon:*** The megasporophyll has a long narrow stalk with an expanded, more or less domed apex which lacks any horns, spines or other ornamentation except dense hairs.

***Encephalartos:*** Thickened megasporophylls lacking spines, horns or other ornamentation and with the outer surface more or less three-lobed and developed into one or more facets.

***Lepidozamia:*** Megasporophylls shortly hairy, lacking any spines or ornamentation, the outer face prominently deflexed.

***Macrozamia:*** Megasporophylls have a thickened central area and a flat, terminal spine which is held vertically.

***Microcycas:*** Megasporophyll lacking ornamentation, much thickened and elongated with a distinctive apical cleft.

***Stangeria:*** The apex of the megasporophyll is widely flared and flat with a short, central stalk; this is the only cycad in which the young ovules are enclosed in a tissue-like membrane.

***Zamia:*** The apex of the megasporophyll is hexagonal and flat, lacking any raised mounds and ornamentation.

## Female Cone Development

The life of a female cycad cone is much longer than that of a male. Both sexes have a similar period of development up until pollination, but after that event life of the male cone is finished and it simply decays. The female cone on the other hand continues development, nurturing the expanding ovules until they are seeds of sufficient maturity to be released.

After pollination the female cone of a cycad enlarges and becomes a hormone sink, resulting in the transport of large levels of high-energy compounds and nutrients into the developing ovules. The increase in size of the cones and ovules is initially rapid then slows down. Considerable stress is placed on the female plant during this development period and if moisture and nutrient levels are insufficient, then the leaves of the female plant may suffer and often drop prematurely and no new leaves are produced that year.

Female cones take varying periods to mature their seeds depending on the species. Some species of *Encephalartos* may be as quick as six months (for example *E. ferox, E. hildebrandtii* and *E. manikensis*) whereas others such as *Dioon edule* and *D. spinulosum* may take from twelve to sixteen months and *Encephalartos transvenosus* takes up to eighteen months. Species of *Macrozamia* commonly take six to eight months. Strangely, cycads with the smallest cones, such as *Zamia pumila*, may still require nine to twelve months for their cones to mature.

At maturity the female cones begin to disintegrate from the apex downwards. Usually at this time the sporophylls fall off the central axis, often with the seeds still attached.

## Male Cones

The male cones of most cycads are much narrower than the female cones although in some species they may be longer. They also have more numerous sporophylls. At maturity male cones of *Cycas circinalis* may be seventy centimetres long whereas those of *Zamia pumila* may only be about eight centimetres long.

Each male sporophyll (microsporophyll) bears sporangia (microsporangia) on its lower surface. These are shortly stalked at the base and are arranged in groups, each of which may coincide with the sorus of a fern. Their function is the production of pollen (microspores) to achieve pollination and eventually fertilisation. Several hundred microsporangia may be crowded on the lower surface of each microsporophyll. Recent studies have shown that in most cycad genera the stalks of the microsporangia are fused into units and in some genera, such as *Macrozamia*, at least some of the sporangia themselves are fused. Such fused sporangia may be better interpreted as a synganium rather than a sorus.

## Motile Sperms

A unique feature of cycads, which in the seed plants is shared only with one other gymnosperm, *Gingko biloba*, is the production of motile sperm cells. Known as spermatozoids, the discovery of these structures in 1896 is regarded as one of the most exciting botanical discoveries of all time, since it provided a definite link between the gymnosperms and the ferns and fern allies, which also have spermatozoids. The discovery is made even more remarkable by the fact that two Japanese botanists, working independently in Tokyo, discovered the existence of spermatozoids in two different gymnosperms within a couple of months of each other. Those in *Gingko biloba* were discovered by Sakugoro Hirase early in 1896 and published in April of that year, about the time when Seiichuro Ikeno found spermatozoids in *Cycas revoluta*.

Spermatozoids of cycads are produced in the pollen tubes which have grown from the pollen grains deposited during the process of pollination. These pollen tubes, which may take up to two months to reach the vicinity of the ovules, burst when in close proximity to the egg, releasing the spermatozoids.

The sperms of cycads are quite large and are readily visible under the low magnification of a microscope. Those of *Dioon edule* are be about 230 microns in diameter (a micron is one-thousandth of a millimetre) and *Encephalartos altensteinii* 150–180 microns. Sizes range from about 80 microns in *Microcycas calocoma* to about 400 microns in *Zamia roezlii*. A spermatozoid can be seen to be more or less spherical in shape with two or more spiral bands which bear numerous cilia along their length. It is estimated that *Zamia roezlii* may have as many as 40,000 cilia. It is the pulsating beat of these cilia by which the spermatozoid swims towards the ovule to effect fertilisation.

CHAPTER 5

# The Economic Importance of Cycads

Cycads, although they have a multitude of uses, do not produce any products which are of major economic importance to humankind, nor do they contribute significantly to the economy of any country. As a food source they are of minor importance with their greatest usage being generally in times of shortage. Of increasing significance is their role in the ornamental plant industry, especially in South Africa, the United States and Australia. The poisonous properties of cycads attract considerable attention, especially in Australia, where their presence on grazing properties can be of considerable economic significance.

## EDIBLE CYCAD PRODUCTS

Cycads are used as a food source in many countries and by a wide range of cultures. Continuing research has shown that their exploitation by humans for food is not new and in fact some cultures may have been utilising them since prehistoric times. This evidence applies particularly to the Australian Aborigines and their use of species of *Cycas* and *Macrozamia*. Studies suggest that these remarkable people had developed the technology for the preparation of edible food from cycads at least thirteen thousand years ago. In Western Australia, empty *Macrozamia riedlei* seeds of about this antiquity have been found in cave deposits. Perhaps toxic cycads were one of the first dangerous plants to be tamed by humans.

In view of the presence of virulent toxins, the use of cycad parts by humans as food is quite extraordinary. Although the techniques of preparation are relatively simple (a discussion of which follows), the methods by which this is carried out must be precise since there is little room for error when the final product is being consumed as food. It is tempting to speculate on the hit or miss learning procedure which must have preceded the successful development of such methodology.

### Sago or Stem Starch

Sago, an edible, starchy material obtained from the central pith of the trunks of woody monocotyledons such as palms, is also present in the stems of cycads. Starch in plants is laid down as a storage reserve and is drawn on by the plant in times of high energy requirements such as occur at the onset of leaf production and coning. Thus the greatest concentration of starch in palms and cycads is to be found in the trunks just prior to a flush of growth or the emergence of flowers and cones.

Sago, which is also termed arrowroot in some countries, is a source of carbohydrate for people. Unlike palms, however, in which it is used as a primary source of food, the sago found in cycads seems to be mainly eaten in times of hardship or food shortage and then only after some method of treatment.

The earliest record of the use of cycad stem starch as a source of food appears in a paper by the Swedish botanist and plant collector Carl Peter Thunberg in 1775. While describing a new species of *Encephalartos* (*E. caffer*) he noted that the pith of the stem was eaten by local people. This material, later termed 'Broodboom' or 'Kaffir Bread', seems to have been mainly used in times of shortage or as emergency food. Treatment was variable depending on local customs but basically involved fermentation and leaching which were perhaps followed by drying. After treatment, the material was then kneaded to a paste with water and baked in hot coals as small loaves or cakes. In some areas a paste or porridge prepared by adding flour to boiling water may also have been eaten.

One species of cycad which seems to have been used as a staple part of a diet rather than as a supplementary ration is *Zamia integrifolia* from Florida. The rich starch, so abundant in its stems, was a regular article in the fare of the Seminole Indians. These people conferred upon it the name of 'Coontie', while the whites of the day, who also often ate the starch, preferred 'Florida Arrow-root'. As with other cycads used for food, the stems of this species were ground and leached prior to cooking.

Members of the genus *Cycas*, which are widely distributed in tropical regions, are more frequently used as a source of food than other cycads. Not only are these plants relatively fast growing but they are also often locally common and some species are widely distributed in many countries. Sago originating from species of *Cycas* has been used as a food source for centuries in many countries including India, Burma, Vietnam, Laos, Sri Lanka, China, Malaysia, the Philippines, Guam, Fiji, New Caledonia, Timor, the Solomon Islands and Madagascar.

Two common, widespread species (*C. circinalis* and *C. rumphii*) account for much of this usage but in some countries localised species such as *C. revoluta*, may be used in a similar manner. Whereas in normal times, cycad sago may be harvested and consumed at moderately low levels, the intake may be greatly increased during periods of shortage following natural disasters such as typhoons. The inhabitants of some Pacific Island communities occupied during World War II relied heavily on cycad sago as a source of carbohydrate. In the Ryukyu Islands starch (termed sotetsu) obtained from the stems of the locally common *Cycas revoluta*, sustained the local population during severe food shortages in the 1940s.

Collection of *Cycas* starch involves cutting the trunks at or near ground level and removing all the outer, woody material to expose the delicate inner cylinder of fleshy pith. This core is then sliced crosswise into thin discs which are dried in the sun until crisp and then pounded into flour. After mixing with water the starchy material separates as a sludge on the bottom and sides of the container. Following additional leaching, this material can be then used as a food source or prepared further into pellets by drying and partial steaming. Sago pellets, which may be known as pearl sago or bullet sago, are useful because they are readily transportable and can be stored for many years.

Analysis by Japanese chemists in 1918 of the stems and sago of *C. revoluta* revealed some interesting results. The starch content of the trunks of male plants over a year averaged about 50 per cent, whereas those of female trunks averaged 26 per cent. Seasonal variation was significant with starch in the male trunks ranging from 27 per cent in October to 61 per cent in June. Coning and seed production had a dramatic depleting effect on the starch levels of female plants, with the storage material not returning to normal levels until about twelve months after seed ripening.

Sago can also be obtained from the trunks of the Mexican cycads *Dioon edule* and *D. spinulosum* and these may have been utilised on a limited scale in times past. The fleshy subterranean stems of *Bowenia spectabilis* were eaten after cooking by various tribes of Aborigines in Queensland.

It would seem apparent that the stem pith of all cycads contains starch which can be used as a food source and many species have often been utilised in this way, if only on a localised basis. Many species however are too small or uncommon to be used to any significant degree.

### Commercial Starch Extraction

The abundance of *Zamia integrifolia* in south Florida and the richness of starch in its stems (up to 65 per cent), led to its exploitation as a commercial product. The first venture seems to have been set up in about 1845 and within about twenty years numerous mills had become established along the lower reaches of the Miami River near the Everglades and on coastal shores of local bays. At peak production, these mills produced a total of about fifteen tons of starch a day, most of which was used in the Key West and West Indian markets. This exploitation, in conjunction with clearing, draining and urbanisation, led to a dramatic reduction in the population of *Zamia integrifolia* and the mills gradually shut down, with the last closing in 1925.

Starch is present in the pith of the trunks of species of the Australian genus *Macrozamia*, but seems to have been utilised only on a limited scale as a food source by the Aborigines. Records indicate that material obtained from the pith of *M. riedlei* may have been consumed after considerable leaching and roasting. This latter species, from Western Australia, and *M. communis* of New South Wales attracted the attention of white settlers because of their trunk size and local abundance. As well as a source of food, starch from their trunks was used for the production of power alcohol, glucose, adhesive pastes and for laundry purposes. A company formed in New South Wales in 1921, harvested the trunks of *M. communis* (then confused with *M. spiralis*) from southern areas of the State for starch production. Preparation consisted of grinding the central pith in water and passing the mixture over silk sieves which separated the starch from any fibrous material. The starch was then purified by leaching and allowed to settle out at the final stage. The result was a high quality starch costing about four English pounds per ton which performed better than starches obtained from rice or corn. Another product from this starch, adhesive pastes, were in considerable demand in New South Wales during the 1930s. Although this commercial venture thrived for a few years it eventually closed due to technical problems.

## Seeds

Seeds are frequently a noticeable feature of cycads, particularly as a result of their often bright colours but also because they may be produced in abundance. The kernel, consisting mainly of a starchy megagametophyte makes up the bulk of the seed and it is probable that this part of all cycads can be used as food after treatment to remove any toxins.

Seeds of various species of *Cycas* are rich in starch (20–30 per cent of the megagametophye is starch) and are used for food in many countries. The widespread species *C. circinalis* and *C. rumphii* are most commonly used but localised species such as *C. media*, *C. revoluta* and *C. beddomei* may be of significance to the people who live where they grow naturally. In poorer areas this food source may be of major importance whereas in more affluent societies the seeds may be consumed only in times of shortage.

*Cycas* seeds contain toxins which must be neutralised or destroyed prior to their consumption and an array of preparative techniques have been developed by various cultures to overcome these toxins. Most of these techniques have a similar basic theme using heat and leaching, but often with some diversification.

After extraction from the seed, the kernel may be sliced (Sri Lanka), ground or crushed (Guam, Java), roasted (Andaman Islands) or boiled (Fiji), with the seeds treated by the latter two methods then being directly eaten. Seeds which are sliced or crushed are generally sun dried before pounding into a fine flour. Sometimes a short fermentation process may take place prior to drying in the sun. The powder obtained from crushing may be leached with water before the starch is allowed to settle and this material can then be readily separated from the liquid. The starch may then be stored for short periods or eaten immediately. It may be consumed as a porridge after first mixing with water or is more usually kneaded into a dough and baked in an oven or cooked in the ashes of a fire.

Cultures which appear to have made extensive use of *Cycas* seeds as a food source are to be found in the Andaman Islands, Guam, Comoro Islands, Ryukyu Islands and northern Australia. In parts of Australia the seeds of *C. media* and *C. armstrongii* were so commonly used by the Aborigines that researchers regarded them as a staple part of the diet of these people. Their significance was first observed by Captain James Cook in 1770 and has been studied by many workers since that time. Because of the antiquity of the Aboriginal culture and the fascinating insights into the developmental aspects which must have taken place over long periods of time it is worthwhile detailing the techniques used by these people.

The seeds, which are selected as being suitable (an inborn or inherent mechanism instinctively results in the rejection of those which are unsuitable), have their outer fleshy sarcotesta removed, are then broken and the kernels are roughly pounded. After sun drying for three or four hours the kernels are then packed in a fibrous bag and leached in water for seven to nine days. The first four or five days are in the running water of a stream and for the remainder of the period they are immersed in stagnant water. After immersion the kernels are ground between stones to a paste and are then baked in the hot ashes of a fire.

Countries in which *Cycas* are used as an occasional source of food include New Guinea, New Caledonia, Indonesia, Philippines, Fiji, Sri Lanka, Vietnam and the Nicobar Islands. In the Moluccas, sliced *Cycas* kernels are eaten as a delicacy after leaching in sea water, steaming and mixing with sugar and grated coconut.

The harvesting of *Cycas* seeds is more logical than the destructive harvesting of trunks for starch. Species of *Cycas* produce seeds regularly and the starch recovered from a single crop of seeds is nearly equivalent to that obtained from a trunk about one metre tall. A mature female plant of *C. media* may produce about five hundred seeds in a single crop whereas a large plant of *C. revoluta* can produce in excess of one thousand seeds.

Several accounts of poisoning recorded by early voyagers, settlers and explorers in Australia show clearly the toxic nature of *Macrozamia* seeds, and yet these items were an important food source to the local Aborigines. Seeds of various species of *Macrozamia* were treated by remarkably similar techniques in various parts of the country. The basic technique involved roasting the whole seeds for about thirty minutes in the hot coals of a fire, recovering the kernels after breaking the hard outer testa and then soaking these in a stream or pond (sometimes in the sea) for six to twenty days prior to consumption.

All species of *Dioon* have an edible kernel but only three species (*D. edule*, *D. spinulosum* and *D. mejiae*) are eaten regularly after roasting, boiling or being ground into meal and converted into tortillas. In fact *D. edule* obtained its name (which means 'edible') because of the local Mexicans fondness for making tortillas from a meal obtained from its seeds. The seeds of *D. mejiae* are so popular with local people that the plants are protected by the Honduran government.

The seeds of various species of *Encephalartos* are eaten by local African tribes after preparation techniques which are basically similar to those employed for species of *Macrozamia*. Species which are eaten include *E. altensteinii*, *E. cycadifolius*, *E. hildebrandtii* and *E. horridus*.

## Cycad Beverages

Alcoholic drinks are concocted from cycad products in some countries, although the scant information available on this subject suggests that these may be drunk on a limited scale and perhaps with some peril. It is recorded in Africa that some tribes prepare a beer by fermenting the central pith of some species of *Encephalartos*. The inhabitants of the Ryukyu Islands prepare a form of sake from *Cycas revoluta*. Both the kernel of the seed and the pith of the stem can be used for this purpose. This drink is almost as deadly as a game of Russian roulette, since it is slightly poisonous and occasionally a potent batch kills all who partake.

## TOXICITY OF CYCADS

All species of cycad contain virulent toxins which can cause severe debilitating symptoms or even death if ingested in sufficient quantities. Two of the major biochemical compounds involved, both of which are unique to cycads, are called cycasin and macrozamin. Cycasin and macrozamin, when ingested in sufficient quantities, are toxic to the liver and can cause cancer. They are also suspected of causing neurological disorders but the evidence to support this suspicion is equivocal. Both chemicals occur in all genera of cycads and it is speculated by research workers that the evolution of these toxins occurred in the Mesozoic Era (about 100–200 million years ago). It is known from fossil records that the morphology of cycads has changed little since that time, and these plants probably retained the toxins as a protective device against predation.

### Toxicity to Humans

Numerous instances are on record of cycad seeds causing poisoning in Europeans who did not known how to treat the seeds to render them safe. Probably the most famous incident involved the Boer Commander General Smuts who, along with other members of a commando, became violently ill after eating the seeds of *Encephalartos altensteinii*. Whole seeds must have been ingested by these soldiers since in this species the toxins occur in the kernels but are absent from the sarcotesta.

Many early visitors and explorers in Australia have been tempted to sample the attractive red seeds of *Macrozamia* species, but usually to their cost. It is recorded that as early as 1697 some Dutch sailors and the explorer de Vlamingh became violently ill after ingesting the colourful seeds of *M. riedlei* in the vicinity of present-day Perth. Members of at least three other exploratory expeditions also suffered distress, including violent vomiting and vertigo, after eating seeds of the same or related species. These included members of Matthew Flinders' expedition at Lucky Bay near Esperance (*Macrozamia dyeri*), Sir George Grey's men in Western Australia in 1839 (probably *M. riedlei*) and John McDouall Stuart's party in 1860 in the Macdonnell Ranges of Central Australia (*M. macdonnelli*). In eastern Australia some French sailors of the La Perouse expedition became violently ill in 1788 after eating seeds of *Macrozamia communis* at Botany Bay. It is known that the toxin macrozamin is present in the sarcotesta of two of these cycads (the others are yet to be tested). Thus even nibbling this attractive fleshy layer can constitute a health hazard. By contrast the sarcotesta of species of *Encephalartos* apparently lacks this toxin.

It is known that the kernels of *Cycas* seeds contain toxins since they have been employed as a fish poison (after crushing), and also have been used to concoct a poisonous drink for the elimination of unwanted children by nomadic tribes in the Celebes. Members of Captain James Cook's discovery expedition to Australia suffered violent purging after ingesting unprepared seeds of *Cycas media* at the Endeavour River in 1770.

## Toxicity to Livestock

Poisoning of domestic stock resulting from the ingestion of quantities of cycad seeds or leaves is a known hazard in some countries such as the Dominican Republic, New Guinea and Puerto Rico, but it is especially prevalent in Australia. Cattle and sheep are most commonly affected but other stock including horses, goats and pigs are susceptible. The disease, which is best known as 'zamia staggers' but may also be called 'rickets' or 'wobbles', can cause significant economic losses. Seeds and leaves can both be involved in the poisoning although the seeds may sometimes produce different symptoms than the leaves. Fresh young leaves are more toxic than mature leaves which in turn are more toxic than dry leaves. Poisoning is often severe in dry seasons since the cycads remain green and fresh when other vegetation has either been eaten or declined in visual appeal. Cycads are also one of the first plants to reshoot after fires and the young leaves might then be the only green food available to browsing stock.

In Australia the symptoms of zamia staggers can be induced by the consumption of various parts of species of all naturally occurring cycad genera (*Bowenia, Cycas, Lepidozamia* and *Macrozamia*). Most poisoning is attributable to the ingestion of *Macrozamia* leaves, probably because these are commonly available to browsing animals. The chief symptom of the disease is the loss of coordination and lack of control of the hind limbs. This includes staggers, dragging of the hind legs and complete collapse onto the haunches. Badly affected animals may become completely helpless for varying periods. Although these symptoms may be obvious for some period, the animals are not affected continually and the periods of sickness are interspersed with periods of apparent normality. The poisoning is cumulative and animals which continue to feed on cycads become progressively weaker and eventually starve to death. Zamia staggers is incurable and once the staggers become apparent the condition is irreversible and the symptoms will remain throughout its lifetime even if the beast is moved away from the source of the toxic food.

The toxic effects of cycads on livestock in Australia has undoubtedly given these unique plants a bad name with graziers and agricultural authorities and for a period they were declared as noxious and their eradication was encouraged by government agencies. The plants have obviously suffered from this backlash but there is no evidence that any species has been endangered or even reduced to rarity as a result of control measures taken.

The most drastic control measures involve eradication of cycads from areas where cattle graze. Commonly, poisons such as arsenic compounds were employed, these usually being applied to an axed notch in the trunk. Those species with a subterranean trunk were killed by splitting the trunk with an axe just prior to the onset of the wet season. The application of kerosene to the split ensured the plants' death. Some graziers these days choose to avoid such drastic measures and simply fence off areas where cycads are prevalent, thus preventing access by stock.

# HORTICULTURAL APPEAL

Cycads are a very popular group of plants with horticulturists and this appeal is increasing. The larger species make excellent subjects for landscape use and are often seen planted around public buildings and as lawn specimens. For this use they share a major advantage with palms in that their final size and shape are relatively predictable, certainly more so than in most other plants. Smaller cycads can be planted in the home garden with some of the lower, bushy species lending themselves well to margins and borders. Cycads are also appealing in containers and are excellent for the specialised art of bonsai. Some species are very adaptable and can be used for indoor decoration.

Cycads are mainly propagated and sold by specialist nurseries. Large specimens are quite expensive because these plants are generally slow growing. Even small plants, especially if they are difficult to obtain or rare, command higher prices than the general run of nursery plants. Popular species of cycads such as *Cycas revoluta* are regularly stocked by general nurseries and in recent years more species are becoming freely available to the general public. Even so there are many species with excellent ornamental qualities which are hardly known because of the scarcity of propagating material.

## ORNAMENTAL LEAVES

The leaves of many cycads are highly ornamental and may be used for indoor decoration where their long-lasting qualities are a decided advantage. In many countries the cut leaves are also used in religious ceremonies. Although these uses are mainly of local significance, in the Ryukyu Islands the leaves of *Cycas revoluta* are harvested and exported in significant quantities each year. The cut leaves are sorted into different classes, tied into bundles and hung in a shady place to dry for about two months. They are then packed in bales and shipped to Japan where they are exported to the United States, Germany and Switzerland. In the United States the leaves, which are often hand painted, are used in floral decorations and wreaths. International trade figures for 1988 suggest that about sixty thousand leaves of *Cycas revoluta* were traded. There is also a limited trade in the highly decorative leaves of *Bowenia serrulata* from Australia.

## MEDICINAL USES

Cycads have mucilage canals present in many organs but particularly in the petioles and rhachises of the leaves and the sporophylls. When the organ is damaged a mucilage or gum exudes from these canals which is initially transparent but later hardens and becomes light brown. These gums expand greatly in water (fifty to one hundred times) and become transparent. They have been used medicinally for the treatment of snake and insect bites as well as malignant ulcers and boils. A paste of *Cycas* seeds mixed with coconut oil is used in various Asian countries for the treatment of skin complaints, wounds, ulcers, sores and boils. In Sri Lanka *Cycas* seed flour is boiled and eaten as a remedy for bowel complaints and haemorrhoids. In New Guinea powdered *Cycas media* kernels are applied to wounds. In Mexico a decoction of the seeds of *Dioon edule* is used for the treatment of neuralgia.

## MISCELLANEOUS USES

The woolly hairs which occur at the base of the petioles of species of *Cycas* and *Macrozamia* have been used for stuffing pillows and mattresses. The leaves of some species may be used as a roof thatch, in the production of baskets, mats and hats or used as brooms. The stems of tall species of *Cycas* have been used as structural supports for dwellings. Whistles and toys may be made from the dry, hollowed seeds of some cycads such as *Dioon spinulosum* and *Cycas* species. Fancy matchboxes have been constructed from seeds of the Australian cycad *Lepidozamia peroffskyana* and in India snuff boxes have been created using the hollowed seeds of *Cycas circinalis*. Pipe bowls were made from the dried roots of Zamia integrifolia in Florida. In Japan cycad trunks may be carved into ornaments which are then exported. Trade figures show that some four thousand such carvings were exported from Japan in 1985.

CHAPTER 6

# The Biology of Cycads

Being a very ancient group of plants, cycads have retained some primitive characters, probably more than in any other living group of gymnosperms. Paradoxically they also have evolved some specialised features which belie their primitive nature. Features of their biology examined in this chapter include pollination, heat and odour production in cones, seed dispersal and seed predation, sex ratios, sex changes, algal symbiosis, root and stem contraction and the effects of fire.

## MATURITY TIME

Cycad plants produce cones when mature but the length of time from seed germination to maturity is largely unknown and seems to vary greatly with the species. Certainly cultivated plants which are regularly watered and fertilised grow much faster than plants in nature and mature much earlier. Many cycads may reach maturity within about fifteen years from seed. Some are much faster than this; for example some of the dwarf *Zamia* species, such as *Z. integrifolia* and *Z. fischeri*, may reach maturity within two years from germination. *Stangeria eriopus* has been recorded as coning at four years of age and *Bowenia* species often cone between three and five years of age. Even large cycads may take a surprisingly short time. A plant of *Lepidozamia hopei* growing under ideal conditions in Hawaii coned in seven years from seed; on the other hand a plant of this species at Fairchild Tropical Gardens took thirty years to produce cones. Large species of *Encephalartos* such as *E. altensteinii* and *E. natalensis* may take from twelve to fourteen years to cone.

While these figures are interesting it should be noted that there are no similar records for plants growing in the wild.

## CYCAD POLLINATION

There has been a commonly held belief among botanists and biologists that cycads are wind pollinated. Thus generations of students have graduated believing that in these plants, as in other primitive gymnosperms, the pollen released from the male cone is transported to a female cone by the agency of wind currents. Detailed studies on some cycads however have revealed a much more complex pollination biology involving insects as vectors.

The notion of wind pollination probably arose because of the long ancestry of these plants and their similarities and apparent relationships with conifers, the wind pollination system of which has been well studied. An obvious and significant supporting factor for the idea of wind pollination in cycads is the production of copious quantities of powdery pollen by the male cones of the larger species (about 100 cubic centimetres from a male cone of

*Cycas circinalis*). In conifers however the cone scales are widely separated at maturity exposing the ovules completely to air currents carrying pollen. By contrast the female cones of all cycads, except in species of *Cycas*, completely enclose their ovules and some or all of the cone scales are tightly closed at maturity. Wind tunnel tests on the cones of three species (*Cycas circinalis*, *Dioon edule* and *Zamia furfuracea*), have shown that while airborne pollen may settle on the cones of these species the site of deposition is usually too far from the ovules to result in fertilisation. It should be noted that the micropyle of the ovule in most cycads is separated by a distance of several centimetres from the outside of the cone. Impact sites for pollen are mostly on the outside of closed scales or on specialised landing platforms remote from the ovules. Thus in *Dioon edule* one or more whorls of basal scales, which open widely and are capable of trapping pollen, are sterile and are some distance from the fertile scales which remain tightly closed. Perhaps in some species, such as *Dioon edule*, wind is important to transport pollen onto the female cones and subsequent movement to the ovules is then effected by insects.

Further evidence has accumulated against wind pollination being the major agent of pollen transfer in cycads.

- Many species grow in closed forests where wind currents are much reduced and unpredictable.
- It is common for plants of many species of cycad to occur as scattered individuals which are often separated by considerable distances — in such cases insects could locate isolated plants by following scent trails whereas wind would be ineffective at such low plant densities.
- Some species such as *Zamia pumila* still produce fertile seeds from cones which are covered by forest debris and are therefore inaccessible to wind.

Cycad growers also know that exotic species rarely produce fertile seeds in collections, even when male and female plants of the same species are grown within touching distance of each other. For plants which are supposedly wind pollinated, it is amazing how such large quantities of its pollen end up in close proximity to the male cone, either covering adjacent leaves or as a dense coating on the ground.

Pioneering observations were made on the activities of beetles in the male and female cones of species of *Encephalartos* as early as 1913. The researcher, G. Rattray, thought that pollination may have been involved (see 'Notes on the Pollination of some South African Cycads', *Transactions of the Royal Society of South Africa* 3 (1913): 259–70). In relatively recent times, evidence has been accumulating to strongly support the contention that cycads are pollinated by insects, in particular beetles. Much of this evidence has arisen from controlled experiments which have been designed to exclude pollen borne either by insects or wind and often with a control to exclude pollen from any source. Pollination studies have been carried out in such diverse genera as *Cycas*, *Zamia*, *Lepidozamia*, *Macrozamia* and *Encephalartos*. The relatively few species of cycads which have been studied in detail show a surprising degree of specialisation which belies the traditional belief of their primitiveness. Some species, such as *Zamia furfuracea*, are known to have a specific insect pollinator (in this case the weevil *Rhopalotria mollis*), others have two vectors (*Zamia pumila*) whereas the Australian taxa *Macrozamia communis* and *Lepidozamia peroffskyana* are known to be both associated with the weevil *Tranes lyterioides*.

Similarities in the vectors and pollination procedure between species in different genera and on different continents suggests that these pollination systems were stabilised long before the break-up of the supercontinent of Gondwana. Members of the primitive beetle family Boganiidae provide solid evidence for this view. Insects of this family have links back to the Lower Cretaceous Period (about 100 million years ago when Australia and Africa were still joined) and are intimately related to cycads in both Africa and Australia. In South Africa, one insect of this family, *Metacucujus encephalarti*, is the probable pollinator of *Encephalartos lanatus*.

It may be inferred from their long ancestry that the pollination of cycads has evolutionary links with the very beginnings of insect–plant relationships. Some workers have suggested that such a close symbiosis could have its origins in the Palaeozoic Era (at least 260 million years ago). Support for such a theory lies both in the pollination syndrome adopted by the cycads and the primitive groups of insects which are involved as pollination vectors. The syndrome whereby the pollinating agents congregate, mate and lay eggs on the inflorescence of the plant after being attracted by a scent is believed to be the most primitive pollination system known. Most reports of cycad pollination involve beetles as vectors, particularly weevils (family Curculionidae), but less commonly members of other such ancestral beetle families as Tenebrionidae, Languriidae, Anthribidae, Boganiidae and Nitidulidae.

In *Cycas media* a species of small bee in the genus *Trigona* collects pollen from the male cones. This insect genus is the oldest known bee with fossil records reported from the Cretaceous Period (65–100 million years ago) and believed to exist even earlier. It seems highly likely that species of *Trigona* may have been collecting cycad pollen prior to the origin of flowering plants (about 140 million years ago). The fact that species of *Trigona* still exist today is proof of a successful adaptation to various changes and it is highly significant that an association between a species of *Trigona* and a *Cycas* (the most primitive cycad genus), has persisted to the present.

## The Pollination of *Zamia furfuracea*

The pollination of the Central American cycad *Zamia furfuracea* has been studied in more detail than any other cycad species, with the principal researcher involved being Knut Norstog, formerly of Fairchild Tropical Garden in Florida, USA. Several scientific papers have been written on the subject (see for example Knut Norstog, 'The Role of Beetles in the Pollination of *Zamia furfuracea* L. F. (Zamiaceae)', *Biotropica* 18, no.4 (1986):300–6 and K. Norstog-P. K. S. Fawcett, 'Insect Pollination of Cycads', *Encephalartos* 19 (1989): 38–41. By a chance occurrence the pollinating agent of *Z. furfuracea* was introduced into America along with live plants of the cycad and the vector has become well establised at the Fairchild Tropical Gardens. A summary of the pollination process is presented here.

*Z. furfuracea* is pollinated by the small snout weevil *Rhopalotria mollis*, the larvae of which feed on non-pollen-bearing tissues of the male cone. The adults, which emerge after boring through the cone scales (and pollen sacs), are coated with pollen from their activities and may then be attracted to the female cones. Feeding by the beetles takes place only in the male cones (which are rich in starch), not in the comparatively depauperate female cones.

The female cones of the cycad remain tightly closed at maturity, apart from narrow vertical cracks between the sporophylls. Experiments have shown conclusively that when wind-borne pollen is excluded, but not weevils, pollination rates are high but when weevils are excluded, virtually nil pollination is achieved. A strong attractant (probably an odour), lures the weevils to a mature female cone and to reach such cones they can fly over reasonably long distances.

In Florida the first weevils appear on the male cones in late June and feed on the sporophyll tips. After mating the females deposit eggs on these damaged sites. After hatching the legless larvae feed actively on the sporophyll tissue, but avoid the pollen sacs. Chance encounters between larvae result in cannabalism and eventually a single survivor (the fittest), pupates in each sporophyll. This cycle takes about ten days and is repeated several times during the coning period of the cycad. At this stage, adult weevils (carrying pollen) can be attracted to mature female cones.

As the coning period progresses, some larvae become extra fat and build thick, waxy pupa cases in which they remain in a suspended state known as diapause. These insects, which represent the carryover generation, can remain intact in the decaying male cone in their pupae cases at least until the following season when they will emerge as adults and begin the life cycle again.

### The Pollination of *Zamia integrifolia*

The pollination of this Florida species, studied by William Tang, closely parallels that of *Zamia furfuracea*. (see W. Tang, 'Insect Pollination in the Cycad *Zamia pumila* (Zamiaceae)', *American Journal of Botany* 74, no.1 (1987):90–9. Two vectors, both beetles, have been identified; one is a weevil *Rhopalotria slossoni* and the other a langurid beetle (*Pharaxontha zamiae*). The larvae and adults of both species feed on the tissues of the male cone and become covered with pollen. The adults of both species are attracted to the female cones. Experiments in which the beetles were excluded from mature female cones resulted in no viable seeds being produced.

### The Pollination of *Macrozamia communis*

The pollination of this common New South Wales cycad has been studied by C. E. Chadwick of the Australian Museum in Sydney. The pollination vector is the weevil *Tranes lyterioides*, a largely nocturnal species, which can be found in the stems and leaf bases of the plant throughout most of the year. Breeding takes place in the male cone and adult beetles congregate in large numbers on the male cones when pollen is produced. Examination of the faecal pellets of the weevils shows that they feed largely on pollen grains of the cycad. There is evidence that the beetles are attracted to mature female cones during night hours.

### The Pollination of *Lepidozamia peroffskyana*

The observations of Paul Kennedy in 'Cycad-Insect Relationships', *Encephalartos* 27 (1991):22–6 suggest strongly that *Lepidozamia peroffskyana* is also pollinated by the weevil *Tranes lyterioides* in a similar relationship to that of *Macrozamia communis*. The weevils swarm on the male cones in very large numbers during the day and apparently breed in the sporophylls of the male cone.

*Tranes lyterioides*

## Implications of Insect Pollination in Cycads

The evidence is very strong that cycads are basically insect pollinated. This knowledge impinges on two important aspects of cycads, their cultivation and conservation.

Cycads raised from seed in areas remote from their natural occurrence are unlikely to have their natural pollination vectors present. If seed is to be produced from such plants artificial pollination by hand must take place (see page 82 ).

The natural populations of many species of cycad have suffered greatly from overcollecting, with some species being reduced to great rarity or even becoming extinct in the wild. With the high degree of pollinator specificity, which we have just discussed, it is quite likely that populations of the insect pollinators have suffered a similar fate to that of the cycads. Programs to reintroduce cycad species into the wild or relocate them to new and safe localities will not be successful if the natural pollinating agent is not catered for in a similar way.

## Heat Production in Cycad cones

The production of heat by the inflorescences of plants is rare in the plant kingdom, although it is relatively common in some members of the family Araceae. It has been known for about one hundred years that heat production occurs in the male cones of some cycads. This phenomenon has recently been investigated and found to occur in members of all genera except *Stangeria* — see W. Tang, 'Heat Production in Cycad cones', *Botanical Gazette* 148, no.2 (1987):165–74 1987 and W. Tang, L. Sternberg, D. Price, 'Metabolic Aspects of Thermogenesis in Male Cones of Five Cycad Species', *American Journal of Botany* 74, no.10 (1987):1555–9 1987.

The studies, which used heat probes attached to a recording device, showed heat production to occur in male cones that are elongating and shedding their pollen and in mature female cones that are mature and ready for pollination. Temperatures measured in the cones ranged from 1°C to 17°C above ambient with the largest cones producing the most heat. The heat, which is produced by the breakdown of starch and lipids which are stored in the cone scales, coincides with cone maturity and peaks in the late afternoon and early evening, lasting from one to five hours. Heat production by cycads may facilitate the release of odours (see next heading), promote insect activity and aid in the shedding of pollen from the pollen sacs. Cones of *Stangeria eriopus* are exceptional in the Cycadales in that they do not show any significant heat production.

### Odour Production in Cycad Cones

It has often been noted that cycad cones at maturity release a distinctive odour. These odours, which occur in both sexes, have been variously categorised as musty (*Cycas circinalis, Dioon edule, Microcycas calocoma, Zamia furfuracea*), fruity (*Bowenia serrulata, Ceratozamia mexicana, Cycas media, Stangeria eriopus*) or resinous (*Encephalartos altensteinii, Lepidozamia peroffskyana, Macrozamia moorei*).

Odour production occurs in species of all genera including *Stangeria*, with the odours being strongest in late afternoon and early evening (coincident with maximum heat production in the cones). The production of these odours by cycad cones is probably related to the attraction of insects for pollination purposes.

## SEED DISPERSAL

When seeds are ripe it is important that they be dispersed to new sites where they can germinate and perhaps become established. Commonly, because of their bulk, most seeds end up in close proximity to the parent, but occasionally seeds are dispersed over some distance to a different region or habitat. In dioecious plants such as cycads this requires at least two such separate events if dispersal is to be successful. If conditions in the new site are suitable then the seedlings can become established and eventually a whole new colony of the cycad may develop. Cycad seeds seem to be dispersed by three main methods — by gravity, by water and by animals and birds.

Many species of cycads grow on ridges and slopes and because the seeds tend to be rounded, at least in one plane, they are capable of rolling and can be dispersed for short distances on the downhill side of the parent plant after the female cones breaks up. Some of this movement is probably enhanced by animals feeding on disintergrating cones or when fighting for feeding rights to the cone. By and large dispersal of seeds by this means is of limited importance to the distribution of the species. Most of the seedlings which germinate close to an established cycad plant lose the competition for survival and perish.

Three species of *Cycas* (*C. circinalis, C. rumphii* and *C. thouarsii*), are found mainly in coastal and near-coastal forests. These species tend to be widely distributed and are often common on islands. A study of the seeds of these three species shows that they have a layer of spongy tissue which is not found in other species of the genus. This layer confers buoyancy on the seeds and they are able to float for long distances in sea water and still retain their viability. Upon being deposited on shore, these seeds can germinate and become established if suitable growing conditions exist or else they simply perish if conditions are too harsh. It is noteworthy that the seeds of these three species of water-dispersed *Cycas* are much larger than all other species in the genus. It is also interesting that these seeds float when viable, whereas viable seeds of the other *Cycas* species sink and only inviable seeds float.

**Sections of Cycas seeds**

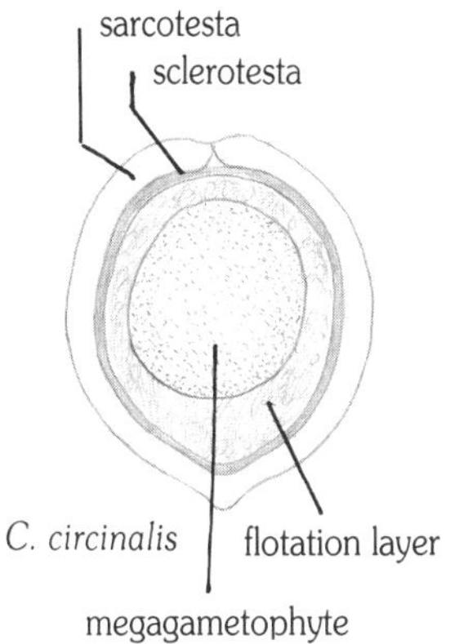


Most species of cycad tend to grow in regions distant from the coast and have relatively limited areas of distribution. These cycads are dependent on a range of animals for the dispersal of their seeds. To facilitate this dispersal it is important that the cycad advertises the availability of its seeds and provides some rewards to the visiting agents. Both roles are primarily carried out by the fleshy outer layer of the seed coat, which is known as the sarcotesta. This layer is brightly coloured to attract the eyes of passing animals and contains starch-rich cells which act as a food reward. It is generally believed that the sarcotesta lacks the toxic chemicals found in other parts of cycads, however recent studies on the Australian cycads, *Macrozamia communis* and *M. riedlei*, show that the sarcotesta contains higher concentrations of the toxin macrozamin than the kernel, and is thus more poisonous to eat. Why then is it eaten avidly by a range of animals?

It is important that a mature crop of cycad seeds should attract animals for their dispersal. Advertising of such a crop can take many forms. Seed colours are commonly red, scarlet, orange or yellow and they are often also glossy, adding to their display potential. A secondary feature which provides back-up to the seed display is the cone size and colour. Some species of *Encephalartos*, for example, have large cones which may be orange, red or yellow and these are well displayed against the dark green background of the foliage. Cones of *Macrozamia* species are green but provide an excellent contrast to the brilliant red colouration of the seeds. In *Bowenia* the sarcotesta of the seed has a distinctive, unpleasant smell.

A wide range of animals is attracted to the colourful seeds and feed actively on the sarcotesta.

Rodents (rats, mice and squirrels) gnaw the outer coat and in so doing often leave long grooves and scratches in the hard testa surrounding the seed. Refuge feeding, whereby an animal carries seeds to a secure area, such as a hollow log, rock ledge or dense covering shrub, is a common practice in rodents and results in aggregation of seeds. This feature is often noticeable in Australia within communities of *Macrozamia* and *Lepidozamia*.

Mammals such as bears, peccaries, baboons, monkeys and hyraxes are attracted to mature cones and ripe seeds. Monkeys and baboons may collect and carry away whole cones or sections of cones for their own consumption. Squabbles over feeding rights may result in the non-dominant animal making quick forays and carrying seeds for some distance before feeding. In Africa it has been reported that elephants eat the whole mature cones of *Encephalartos poggei* and the seeds are then voided whole in the dung a day or so later. In Mexico the large chestnut-like seeds of *Dioon edule* are a favourite item of diet for bears and peccaries.

Marsupials such as possums, kangaroos and wallabies feed on *Macrozamia* and *Lepidozamia* seeds in Australia. Possums are general poor dispersal agents but kangaroos and wallabies are far more effective. These animals may carry whole seed-bearing sporophylls in their mouths for some distance before stopping and eating the sarcotesta. These activities account for the puzzling, often uphill dispersal of these cycads and the odd plants which may be found well isolated from the main populations.

Large birds such as parrots, cockatoos, crows, emus, cassowaries, mockingbirds and hornbills also feed on the seeds of cycads. Most chew or peck on the sarcotesta and swallow the fragments but some species such as crows, emus and cassowaries may swallow the seed whole, to be voided later minus the digested sarcotesta. Cassowaries are particularly fond of the large scarlet seeds of *Lepidozamia hopei* and are important for its long distance dispersal. In Africa, hornbills may remain daily in the proximity of coning plants of *Encephalartos*. It has been recorded that large fruit bats also feed on cycad seeds but their method of feeding and effectiveness of dispersal is unknown.

The activities of man as a dispersal agent should not be overlooked. As mentioned in Chapter 5, cycad seeds have been used for food by various tribes and cultures over long periods of time. Transport by humans, discarding of unused seeds, inadvertent loss during travelling and trading must have influenced the spread of some species, although just how is largely a matter of conjecture.

## SEED PREDATION

Cycad seed kernels are surrounded by a very hard woody coat known as the sclerotesta. This protects the kernel which contains the embryo and its food supply, the megagametophyte, from being eaten by predators. As with most other parts of cycad plants, the seed kernel contains toxins and is believed to be toxic to vertebrates.

As with all rules there is an exception. In Australia the brown rat (*Rattus fuscipes*), and probably also some of its relatives, gnaw neat holes in one end of the sclerotesta of *Macrozamia* seeds. They then consume the contents, including the embryo, and discard the opened shell as telltale evidence of their activities. Although the rat is very active in colonies of *Macrozamia communis*, empty seed shells with distinctive gnawed holes have been observed by the author around female plants of a wide range of *Macrozamia* species. Another large Australian rodent, the white-tailed rat, *Uromys caudimaculatus* is very fond of the seeds of *Lepidozamia hopei*, eating both the sarcotesta and the kernel. A similar predation occurs in Mexico whereby rodents eat the kernels of *Dioon* species.

In Africa a group of weevils in the genus *Antliarhinus* have evolved an intimate life cycle centred around the seeds of various *Encephalartos* species. These insects have been studied by John Donaldson of the National Botanic Institute, Kirstenbosch, South Africa (see Abstracts of Papers, 7–8 *Cycad* 90 (1990).

Eggs of these weevils are laid in the developing ovules and the grubs feed on the kernel of the seed resulting in its destruction. Heavy infestations of these weevils can result in significant seed destruction and their activities may be important for rare species of *Encephalartos* (see also page 79).

The weevil *Antliarhinus zamiae* has a greatly elongated snout which it uses to drill through the sporophylls and into an ovule from the outside of the cone and it then uses a telescopic ovipositor to deposit a batch of eggs into the ovule. Its activities are limited by species which have particularly thick sporophylls. Thus seed predation is low in *Encephalartos longifolius* but is much higher in *E. altensteinii* which has relatively thin sporophylls.

The related weevil *Antliarhinus signatus* has a much shorter snout and is not capable of drilling through the cone from the outside. In this case the female weevil enters the cone and drills directly through the ovule before laying its eggs. The activities of this species are limited by the compactness of the cone with eggs only being laid in species where gaps in the sporophylls allow access by the adult female weevils.

## CONE PREDATION

In Africa, porcupines are very fond of the young tissues of developing cones of various species of *Encephalartos* and will destroy any cones which are accessible. Baboons may also destroy immature cones by using them as playthings. In lean years when coning is at a low frequency, rats may attack the immature cones of *Macrozamia* species in Australia, usually when the sarcotesta begins to show colour.

## ALGAL SYMBIOSIS AND NITROGEN FIXATION

In addition to their normal root system, all cycads have unusual specialised roots which fork repeatedly and form a dense, coral-like mass on the soil surface close to the crown of the plant. These roots, termed coralloid from their coral-like appearance, grow upwards against gravity (apogeotropic) and contain blue-green algae in some of their cortical cells. These algae, which are capable of fixing nitrogen from the atmosphere, show up as a distinct, dark green band when a coralloid root is sectioned. Coralloid roots are unique to cycads and are not found in any other group of plants.

Coralloid roots can be readily observed if the litter close to a cycad trunk is swept away. If the cycads are growing in sandy soil the coralloid root clusters may become exposed on the soil surface by normal erosion, for example *Macrozamia douglasii* on Fraser Island, Queensland, Australia. Coralloid roots appear as irregular clusters of roots with rounded tips and have a general appearance similar to some fungi. On large species of *Cycas*, clusters of these roots may be up to thirty centimetres across. They are often prominent in potted plants, frequently protruding above the surface of the potting mix.

The relationship between the cycad and the blue-green algae in the coralloid root is an example of a true symbiosis with both organisms obtaining benefit from the arrangement. Measurements on populations of *Macrozamia riedlei* showed that each year this species is capable of fixing between nineteen and thirty-five kilograms per hectare of nitrogen from the atmosphere. Although small by agricultural standards this amount of nitrogen is of major significance in nutrient-tight plant communities growing on impoverished soils.

Cycads are the only gymnosperm known to form an association with a nitrogen-fixing organism. In all, three genera of blue-green algae are known to be involved with cycads. These are *Nostoc* with *Bowenia, Cycas, Dioon, Encephalartos, Macrozamia, Stangeria* and *Zamia*; *Anabaena* with *Cycas revoluta*; both *Anabaena* and *Nostoc* in *Macrozamia communis* and *Calothrix* in *Encephalartos hildebrandtii*.

## SEX RATIOS

Cycads are dioecious having separate male and female plants. It would be expected that nearly equal numbers of male and female plants would be present in any population of a species, however in practice this is not always the case. Studies by Robert Ornduff on the Western Australian cycad *Macrozamia riedlei* showed that in seven of the nine populations of coning plants he sampled, two-thirds or more of the plants were male and in the other two populations male and female plants were present in equal proportions. (See R. Ornduff,

'Male-biased sex ratios in the cycad *Macrozamia riedlei* (Zamiaceae)', *Bulletin of the Torrey Botanical Club* 112 (1985):393–7.) Similar results have been obtained for *Zamia integrifolia*.

The disproportionate number of males in these populations may be a reflection of the relative energy requirements for cone production in both sexes. Female cones, being much more massive than the males, deplete the starch reserves of the plant to such an extent that several years of restoration are required before the next coning event. The male plants on the other hand use far less stored reserves in producing smaller, shorter lived cones and are thus able to cone more frequently. Perhaps the seasonal availability of water is also a factor. Evidence from a study of *Encephalartos transvenosus* in the Transvaal Province of South Africa shows that the availability of water may influence the proportion of plants bearing male or female cones in a given year.

## SEX CHANGES

It is well known that some animals, such as fish, are capable of changing their sex, however sex changes in unisexual plants are rare occurrences indeed. It should be noted however that a number of cases of spontaneous sex change have been reported in cycads. Most of these reports appear to be of a serious nature and involve plants of known sex that have been under relatively close observation for a number of years Many of these reports indicate that the changes occurred after some period of stress such as transplanting, physical damage, drought, heavy frost or freak freezing spell. Various genera are involved in these sex changes, including *Cycas*, *Encephalartos*, *Stangeria* and *Zamia*.

These reports of apparent sex changes in cycads have been treated with scepticism by some geneticists who point out that the gender of the plants is controlled by sex chromosomes and is irreversible. Thirteen cases of sex change however have been documented (see R. Osborne, 'Two New Reports of Cycad Sex Changes', *Encephalartos* 23(1990):18–20), and the evidence from these cases indicates that a different mechanism of sex determination may exist in cycads.

Biologists and horticulturists have been quick to seize on the possibility of inducing sex change in *Encephalartos woodii*. This species, which exists in many of the world's botanical gardens and is known only from male plants, is propagated solely by vegetative means. If a sex change could be induced in *E. woodii*, then for the first time there would be the possibility of pollination and seed production taking place within this species.

## SEXUAL DIMORPHISM

It is often contended that male and female cycad plants can be distinguished by features other than their cone structure. Proponents of this theory put forward such vegetative characters as trunk height and thickness and leaf size and number. Careful studies by researchers on species in various cycad genera indicate that no such differences exist between male and female plants. Thus the only way to determine the sex of a cycad plant is when it produces a cone at maturity.

## ROOT AND STEM CONTRACTION

Many species of cycad face adversity in regions where they grow, from hostile factors such as heat, cold, seasonal aridity, fires or poor soils. Over the eons they have developed mechanisms to aid with their survival, one of the most interesting of which is to have subterranean trunks. This is achieved by their main roots and/or stems having the ability to contract. Such contractable roots and stems, which are characterised by a wrinkled appearance (the wrinkles often extend around the organ), actually pull the plant underground. This phenomenon is achieved by certain cells in the cortex and pith of the stem or root collapsing and the length of the whole organ is then reduced by contraction. Root contraction is known from other plants (e.g. bulbs), but stem contraction is apparently unique to cycads. Measurements indicate that in some species of cycad, stem contractions may decrease stem length by as much as 30 per cent. The greater the contraction the greater is the degree of wrinkling on the surface of the organ involved. With the stems being underground, a degree of protection is conferred to the growing point from factors such as fire, aridity and cold. If however the soil is too hard or rocky the contractions of the roots and/or stems will have a minimal effect and the trunk will probably remain emergent.

Contraction of the tap root would appear to be of major significance for the successful establishment of young seedlings of all cycads, including those with emergent trunks. All species studied exhibit this feature. In their early stages, seedlings of cycads are succulent and are susceptible to predation, fire and drying out. The contractile primary root pulls the susceptible parts underground where they are protected by the soil cover until the plants grow sufficiently large to develop protective leaf bases.

## THE EFFECTS OF FIRE

Many species of cycad occur naturally in areas where fires are of frequent occurrence, often occurring annually. These fires may start naturally, such as following lightning strikes, but are more usually lit by humans. The common result of such fires is the above-ground destruction of living plant parts and plants have evolved various mechanisms to survive these events.

Cycads generally cope very well with fire. Whereas any leaves, cones, exposed seeds and some seedlings are destroyed, the leaf bases which persist on the trunks of most species protect the important internal growth apex from damage. Those species with subterranean stems are well insulated by the soil. Soon after the fire new leaves are produced and the cycad plants quickly recover. For large cycads and in some species of *Cycas*, the new leaves emerge in a flush which adds greatly to their ornamental appearance.

Cycads are an important food source for the northern tribes of the Australian Aborigines and there is some evidence that burning on a regular basis may increase the seed productivity of *Cycas* species which occur in the area. A comparison between two groves of cycads, one burnt and the other unburnt, showed a dramatic increase in production, with plants in the burnt area producing more than seven times the number of seeds than those in the unburnt grove.

In a study of the Western Australian cycad *Macrozamia riedlei*, researchers found that fire stimulates leaf growth and coralloid root activity. It has also been reported that cone production in this species is stimulated by fire. It is a generally held belief that all species of *Macrozamia* respond to fire by coning but in the author's experience this is fallacious. This is not so in the savannahs of South Africa, however, where regular fires are an important requisite to induce coning in a number of species of *Encephalartos*.

CHAPTER 7

# The Cultivation of Cycads

One species or another of cycad can be grown as a garden plant in most areas which have a moderate climate. Some species are better suited to the tropics, others to subtropical and temperate regions and there are even some which can tolerate short periods of extreme cold and dryness. Extremes of climate, such as intense cold, intense heat and long periods of dryness, are unsuitable for their cultivation unless modifications can be made, such as by the construction of greenhouses..

Cycads are generally very easy plants to grow if the basic requirements of unimpeded soil drainage, good soil, warmth and plenty of water are met. Some species react adversely to very acid soils and prefer those of a neutral to slightly alkaline pH. This probably arises because of the symbiotic nitrogen-fixing cyanobacteria which prefer alkaline conditions and supply the cycad with its nitrogen. Most cycads are sun-loving plants, but some species, especially those originating in rainforests, may need a position protected from hot sun, certainly when they are young. Frost is a major limiting factor to the cultivation of many species in temperate regions.

## LANDSCAPING WITH CYCADS

Cycads share a major advantage with palms when used as plants in landscaping. Their growth habits and dimensions are well documented and entirely predictable, thus a cycad can be chosen with confidence to fill a particular niche. Apart from palms and some other woody monocotyledons, few other plants can be chosen accordingly.

Cycads like company and look appealing when planted in groups. Interesting effects can be obtained by group planting a single species, closely related species or those which have a similar growth habit. Leaf colours are a highly noticeable feature of cycads and some interesting effects can be obtained by group planting species which have complementary leaf colouration. A collection of mixed species of cycad will provide continuing interest but some thought must be given to their placement. Smaller growing species are best sited towards the front and larger species towards the centre of a garden bed. Low growing, bushy species, especially those which have a suckering habit such as *Zamia integrifolia* and *Z. furfuracea*, can be ideal subjects for the margins of a garden bed or for lining footpaths and roads.

When planting cycads in groups, consideration should be given to the spacing of plants. Large-growing species must be given sufficient room to develop and without impinging on smaller types. Maintenance considerations are important and sufficient space must be left between individuals so that dead fronds can be removed without difficulty.

The larger growing cycads look impressive when grown as lawn specimens. Different effects are achieved depending on whether the chosen cycad develops a solitary trunk or is more bulky because of a suckering, clumping habit. Only those cycads which can tolerate direct hot sun should be chosen for this type of use. Cycads used as lawn specimens can add an interesting feature to an expanse of lawn and have minimal interference with the growth of the grass which will grow right up to the base of the cycad trunk. Cycad plants cannot be ringbarked and close mowing will cause minimal plant damage.

Many of the larger growing cycads have sharp to pungent leaf tips which can readily penetrate the skin and can pose a major health hazard to sensitive areas such as the eyes. For this reason it is best to distance any maintenance operations such as mowing, by digging a small garden bed around lawn specimens. For similar reasons such plants should not be sited close to paths or other areas where they are in close proximity to people. For massive species such as *Macrozamia moorei*, an adequate safety zone will need to be quite large.

Cycads which form a solitary trunk have a unique, primitive appearance and a well-grown subject, if not the source of admiration or envy, certainly will arouse comment and interest. Such plants if well sited and healthy, can make excellent specimens, and the effects they create will differ with their growth habit. Thus species of *Cycas* with their widely spreading fronds have a different impact to the rounded crowns of *Microcycas calocoma* or some species of *Encephalartos*. Cycads with a slender trunk such as *Cycas media* or *Zamia obliqua* impart a very different impression to those which develop a massive trunk (for example *Macrozamia moorei*).

Large cycads can create an excellent effect when spaced at regular intervals along paths or driveways (with due consideration to pungent leaf tips). Although they are not as successful as many palms for avenue planting, some of the larger, non-suckering species such as *Macrozamia johnsonii* and *M. moorei* can still create an impressive impact when used in this manner. An entirely different effect is obtained if large suckering species such as *Encephalartos altensteinii, Cycas circinalis, C. rumphii* or *C. revoluta* are chosen.

Cycads generally mingle well with other garden plants particularly low shrubs. Small-growing species mix well with low shrubs, ferns and garden perennials and do not appear out of place in such an assortment. Thus most gardens could successfully incorporate a few cycads in the general planting. It should be borne in mind, however, that cycads resent overcrowding and sufficient space should be maintained around each individual to ensure adequate air movement. Cycads also survive successfully under large, established trees. In fact they are one of the best groups of plants to choose for such a situation where root competition for water and nutrients can be critical.

## Cone and Seed Colour

The cones and ripe seeds of many cycads are extremely colourful and can be a highly ornamental feature. Species of *Cycas* are not generally noted for their colourful seeds however those of *C. revoluta*, and particularly *C. taiwaniana*, are an exception being bright orange to red. By contrast to the females the male cones of many *Cycas* species are very prominent at maturity ranging from bright yellow to orange or brick red. Species of *Macrozamia* and *Lepidozamia* have red to scarlet seeds and in the former group these seeds contrast with green to lime green cones. Interesting variants having yellow seeds are sometimes encountered in *Macrozamia communis* and *Lepidozamia peroffskyana*, and these provide a contrast to the usual form.

Many species of *Encephalartos* have female cones which are highly coloured at maturity and in some the male cones may also be colourful (*E. woodii*). *E. ferox* is one of the most noticeable with its reddish cones and in many others such as *E. lebomboensis, E. villosus, E. poggei* and *E. tegulaneus* the cones are in yellow tones. Bright reds and scarlets predominate in the sarcotesta of *Encephalartos* seeds (*E. arenarius, E. lehmannii, E. longifolius, E. natalensis, E. villosus*) while others may be in tones of orange (*E. dolomiticus,*

*E. heenanii, E. laevifolius*), brown (*E. dyerianus, E. eugene-maraisii, E. middelburgensis*), or yellow (*E. bubalinus, E. cupidus, E. humilis, E. lanatus*). Frequently, as in *Macrozamia*, the red to scarlet seeds contrast with green cones (*E. arenarius, E. caffer, E. inopinus, E. latifrons, E. lehmannii, E. longifolius, E. ngoyanus, E. princeps, E. trispinosus*).

Species of *Zamia, Microcycas* and *Ceratozamia* are generally dull coloured being in shades of green or brown, although in some species velvety hairs may be prominent. Hairs are prominent on the pendulous female cones of *Dioon edule* and *D. spinulosum* and the erect cones of *Stangeria eriopus*. In all three species the outer surfaces of the sporophylls are coated with appressed, silky hairs of a distinct silvery white colouration.

## SOIL

A wide range of soil types from sands and gravels to clay loams can be suitable for the cultivation of cycads. Cycads are generally not exacting in their soil requirements. As a general rule the better the soil the better will be the growth of individual cycads and the greater the variety that can be grown.

Drainage is of major significance since no cycad will tolerate waterlogged soils for any length of time. While it is true that a few species of cycad occur naturally in soils which may be wet for some period, this should not be used as an indication that these plants will tolerate waterlogged soils in the garden. In natural systems a host of factors ensure that water does not stagnate; such a complex interchange is less commonly present in a garden. In an ideal garden situation, water should drain quickly through the larger soil pores and then be replaced by an atmosphere rich in oxygen. Such a system ensures strong, healthy root growth and the plants are then better able to resist root pathogens. Cycads are generally sensitive to the attacks of strong root-rotting fungi such as species of *Phytophthora, Rhizoctonia* and *Pythium*, and the effects of such attacks are magnified many times in poorly drained soils.

Most cycads seem to prefer an acid to neutral soil with a pH between 6 and 7, but a few species, such as *Encephalartos dolomiticus, E. inopinus* and *Cycas calcicola* may respond better to soils that are slightly alkaline (pH 7.5). Where soils are more acid than pH 6, some reduction in acidity may be necessary. This can be achieved by applying quantities of ground limestone or dolomite. After a suitable period the pH of the soil should be retested.

Although soil texture does not exert a major influence on cycad growth a few notes on techniques to optimise growth in different soil types are pertinent.

**Technique of planting in a heavy clay soil**

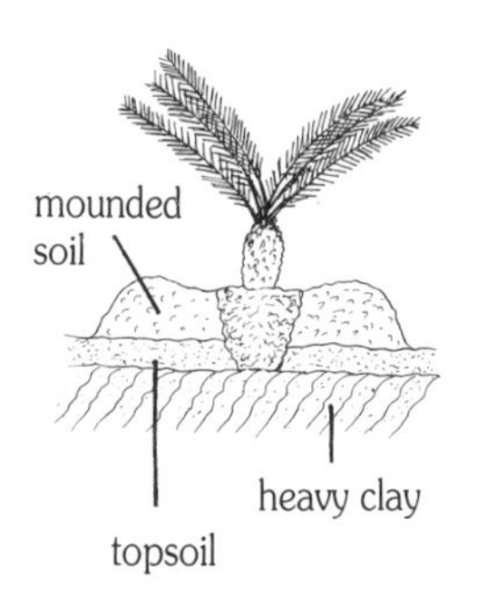


### Clay Soils

Cycads can be grown in heavy clays or even exposed subsoil but the plants are often difficult to establish or grow poorly unless the clay is improved in some way. Excessive moisture should be removed by drainage and the structure further improved by the addition of gypsum and organic matter. The gypsum causes aggregation of the soil particles thus improving water penetration and aeration. This process is enhanced by the addition of organic matter particularly well-rotted materials. These ameliorants should be worked into the top fifteen centimetres of the clay until its friability is improved. The greater the effort that is put into this improvement, the better the results that will be obtained. Surface compaction, cracking and surface drying are common problems encountered in clays, but these can be overcome by regular mulching with organic matter. Planting into clay soils may require special techniques (see left).

## Sandy and Gravelly Soils

Cycads usually grow well in these type of soils because drainage and aeration are excellent, the soils are often deep and they warm up quickly after cold weather. Such soils however, are usually deficient in organic matter, lack much in the way of surface litter and are often poor nutritionally. Some sands with very fine particles may also become water repellent when very dry. Heavy applications of organic matter to the surface of such soils is very beneficial. Organic mulches of this type should be regularly topped up and supplementary watering and fertilising will further promote healthy plant growth.

## Loams

These soils are generally well-structured, drain freely, have adequate nutrition and do not dry out too rapidly. They are ideal for the cultivation of a wide range of cycads but still require some basic assistance such as mulching, fertilising and watering during dry periods.

## Alkaline Soils

Soils which contain an excess of calcium salts in the profile are termed alkaline or calcareous soils. Such soils often suffer from nutritional problems, particularly involving the elements iron, magnesium, zinc and manganese. Whereas hardy cycads may grow in such soils, there are some species (e.g. many *Zamia* species) which are very sensitive to the nutritional problems and will not thrive if indeed they survive. Even hardy species, such as *Encephalartos altensteinii, Macrozamia communis* and *M. moorei*, may suffer from nutritional problems in strongly calcareous soils and while some growth is achieved the plants often appear unhealthy. The health of plants in calcareous soils can be improved by heavy mulching with organic material and possibly also acidification with sulphur.

## Acid Soils

Some soils, which may have a pH as low as 4.5, are excessively acid and cause poor plant growth. Many species of cycad are intolerant of such soils and grow very poorly as a result. Nutritional problems are common in acid soils, and these include a shortage of some elements such as calcium and magnesium and an excess of others such as aluminium and boron. Acid soils can be greatly improved by heavy applications of garden lime (calcium carbonate) and organic matter.

## Saline Soils

Soils with an excess of salt or sodium chloride in their profile are known as salted or saline soils. Such soils pack tight, are poorly aerated because they have lost their structure and are nutritionally difficult. Often the conditions in such soils are exacerbated by poor water, usually originating from bores. Cycads are generally intolerant of such conditions. Saline soils can be improved by drainage so that the salt is leached through the profile and removed in the drainage water.

## PLANTING

Planting cycads is not significantly different from planting palms or any other container-grown plant, although if carried out with care the plant's establishment will be enhanced.

The planting hole should be sufficiently wide and deep to accommodate the root system of the plant. If the soil is dry the planting hole should be filled with water and allowed to drain prior to planting. Some fertiliser (either inorganic, organic or slow release) or well-rotted animal manure should be thoroughly worked into the soil at the bottom of the hole before planting. Animal manures and compost are of particular benefit in sandy soils. Planting in clay soils should be carried out with care; in particular avoid forming a sump by digging too deep into the clay (see page 68).

The cycads should be thoroughly soaked in their containers before planting. If the plants are very dry or the pots filled with masses of roots, this pre-planting soak may be best achieved by immersion of the container in water.

At planting time the cycad plant should be tipped out of the container, the plant placed in position with the top of the container soil just below that of the garden soil and the soil firmed around the roots and watered thoroughly. If necessary a small basin can be created around the plant to direct and retain the water in the vicinity of the root system. The area around the plant should then be mulched (see the later section on 'Mulching').

## WATERING

Although cycad plants (especially once established), are tolerant of short to moderately long dry periods, they benefit from supplementary watering applied during these times. Plants watered during such periods of stress not only have a better appearance but are also healthier and better able to withstand attacks by pests and diseases.

Species originating from rainforests and other mesic habitats, together with those from high rainfall climates, are less able to cope with dry situations than are cycads from open harsher habitats or those which grow naturally in climates having low or irregular rainfall. Thus plants of the former group require more frequent watering than those originating from the latter type. It is also important to realise that established plants have a more extensive root system than do those recently planted. The root system of established plants means they are able to tap a large reserve of soil for their water requirements and the plants can last for longer intervals between waterings than those which have been planted recently.

Sufficient water should be applied to thoroughly soak the root zone of the plant. Provided this need is met, the method of water application (i.e. whether it be by hose, sprinkler or drip irrigation) is not of major significance to the cycads. For efficiency, water is best applied during cool periods in the evening or early morning. Mulches are a feature which greatly improves water penetration and retention and their use is strongly recommended.

## FERTILISERS AND MANURES

Cycads have similar nutrition needs to those of other plants, however they are often treated by growers as being different. This folklore, which has probably arisen because of the often enforced slow growth rate of these plants, is reinforced by their primitive nature, their ability to withstand neglect and their symbiotic association with specific blue-green algae which occurs in specialised coralloid roots. It should be realised, however, that cycads are typical plants and as such they respond in a normal way to the application of fertilisers and manures. In fact under optimum conditions these plants can be very fast growing (see page 72).

Inorganic fertilisers are the cheapest form of fertiliser available and because they are water soluble the nutrients quickly become available to the plants. A balanced complete fertiliser should be applied annually. The benefits of inorganic fertilisers are maximised if they are applied in conjunction with organic mulches.

The rates of inorganic fertiliser application to cycads need not be as high as are used for strong growing plants such as palms, although some growers do advocate moderately high levels. A reasonably large specimen would benefit from about one kilogram of complete fertiliser at each time in two or three dressings per year. Smaller plants benefit from lesser amounts applied in similar split dressings or even lighter applications at regular intervals throughout the year.

Fertilisers are best applied during the warm months when the plants are in active growth. Spring is an excellent time to begin fertiliser application in temperate or subtropical regions as growth at this time is just beginning and the nutrients can help the plants recover from the hardships of winter. For similar reasons fertilisers should be withheld in autumn so that plants can become hardened to endure the cold winter. Fertilisers are best applied to moist soil and watered in. If this is not possible, they should be applied just before or during rain. Application should take place over the surface of the ground, with the main concentration being scattered within the drip line of the leaf canopy. Slow release fertilisers can be used in a similar way but are especially valuable on young plants at the time of planting. Organic fertilisers such as blood and bone, bone meal or hoof and horn can be successfully used for cycads. Well-rotted animal manures are also excellent and are especially valuable when used on sandy or gravelly soils.

## PRUNING

Cycads require some trimming of untidy, damaged, protruding or dead leaves but in general very little pruning is required for these plants. Some species retain dead leaves which may appear unsightly to some people; other species shed the leaves as they die. A skirt of hanging dead leaves may add interest to a species, for example *Encephalartos latifrons*, *E. longifolius*, but it may also leave the plant susceptible to vandalism by fire and able to harbour vermin such as rats and some birds. A few species, such as *E. altensteinii*, may shed the leaflets from dead fronds but retain the rhachis and petiole.

Many cycads have stiff, pungent tips to their leaflets and these can easily penetrate skin or result in smarting and irritation after contact. In public areas, cycads planted near paths or in areas accessible to the public should be observed regularly and any hazardous leaves should be removed.

Cones are a significant feature of interest in cycads and in many species these structures are highly decorative, especially when ripe. The colourful, attractive but toxic seeds may be regarded as a hazard to children, however, and if this is a problem the cones are best removed early in their development.

## MULCHING

It is noticeable in many public gardens which feature cycads, that these plants are not mulched. Cycads generally have deep roots but an important part of their root system, which is associated with their algal symbiosis, develops near the root surface. In mulched areas these specialised coralloid roots are very noticeable when the mulch is scraped aside and it would seem that mulches are highly beneficial to cycads and their use is to be recommended. Mulches have a number of other benefits including improvement of water penetration, reduction in temperature fluctuation of the soil surface and the reduction or elimination of weed competition.

Mulches should be applied thickly as soon as possible after planting to minimise drying of the soil surface and weed germination. After application the mulch should be watered heavily to compact the surface and reduce dispersal by wind. Mulches should be topped up at least annually. Organic materials make the best mulch and extra fertilisers will be required to compensate for nutrient use during their breakdown. Suitable materials include chipped bark, wood chips, shavings, sawdust, peanut shells and hay.

## OPTIMISING CYCAD GROWTH

The reputation which cycads have for slow growth is largely undeserved. A number of cycad enthusiasts and professional nurserymen have shown conclusively that these plants can respond with rapid growth if provided with optimum growing conditions. In fact cycad plants in garden situations not only grow more rapidly than they do in the wild, but often also grow larger and have more numerous, longer and broader leaves. Thus in the garden or nursery situation these plants are achieving a greater potential than is possible in the wild, where competition and numerous environmental factors are continuously imposing limitations.

Amazing growth rates can be achieved with artificially grown cycads especially in tropical regions where temperature is rarely a limiting factor. If all of the soil conditions outlined previously are attended to, mulches applied, pests and diseases controlled and the plants regularly watered and fertilised, these plants can grow at an incredible rate and reach maturity in much shorter periods than in the wild ( see also page 36 ). Instead of a single flush of leaves, plants so treated may produce four or five flushes of leaves in a calendar year and the stems may enlarge at about five or six times their normal rate. Trunks can increase in height quickly and a substantial plant can develop in a relatively short period of time.

## TRANSPLANTING CYCADS

Cycads are generally very easy to transplant (even from the wild) and even very large specimen plants can be moved successfully providing a few basic horticultural procedures are observed. Although cycads generally transplant readily the response is variable with the species. For example, species of *Cycas* generally grow away quickly after transplanting, whereas some species of *Encephalartos* may take twelve months or longer to produce new leaves.

The best time to transplant established cycads is before a flush of new leaves. In temperate regions this is usually spring but in tropical regions it may be just before the onset of the wet season. If irrigation is not available then transplanting should not be undertaken during dry periods and should be postponed until rain has fallen. In warm temperatures and with adequate soil moisture available, transplanted cycads will establish relatively quickly.

Cycads are generally tolerant of damage to their root system although if a substantial proportion is retained at transplanting the better will be the results. The retention of a solid root ball will ensure a quick transition to the new position but frequently cycads are transplanted successfully with less care than this. As a general rule the more care taken during transplanting the better will be the result obtained. For very large cycad specimens mechanical equipment such as backhoes, cranes and trucks will be necessary for the task.

When transplanting a large cycad the leaves are first removed and the roots are severed in a circle around the trunk by cutting with a sharp spade and a trench is excavated to reach the deep roots. Once the roots are cut right through the root ball can be surrounded by hessian or similar material to hold in the soil and reduce drying of the root system until the plant is repositioned. The plant can then be transported to its new position where it is placed in a well-prepared hole of suitable dimensions to accomodate its root system.

The aftercare of such a transplanted cycad is extremely important. Following planting the soil should be firmed thoroughly around the root ball and a temporary reservoir created for the retention of water around the root zone. This reservoir should be filled several times until the entire profile and root ball is wet. This watering should be undertaken as soon as practical after transplanting and should be continued regularly (especially during dry periods) until the plant is established.

If the transplanted cycad has a tall trunk, staking may be required to prevent movement and consequent root damage. The most effective system of staking is to use three stakes arranged like the points of a triangle and guy wires to support the trunk.

New growth should be regularly observed for any pest problems as the first crop of new leaves will be vital for the establishment of the plant. Regular watering during dry periods, mulching and light fertiliser applications will aid with the establishment of the transplant.

CHAPTER 8

# Pests and Diseases

## PESTS

Cycads are attacked by a range of pests, some of which can cause serious damage whereas others are more of a nuisance and only cause minor damage. Most pests are seasonal although it is noticeable with cycads that stage of growth is of significance, with young leaves which are just hardening off being often a primary source of attack. Most of the pests detailed below have life cycles adapted closely to the growth cycle of various cycads and may be very restricted in their distribution.

### Cycad Weevils

A few weevils are highly significant pests of cycads and severe infestations by them can result in debilitation of growth and even plant death. Although information is scanty it seems likely that various species of pestiferous weevils may occur wherever cycads are to be found naturally. These insects seem to spend all of their life cycle in association with a cycad and mostly their activities are out of sight and plant damage is difficult to detect.

It is a relatively common practice for established cycad plants to be transplanted to new sites and the trunks of these plants may contain adults or larvae of a pestiferous weevil. The transport of these pests to a new site where a wide range of cycads is present (e.g. a botanic garden) may allow them to flourish. In these new circumstances, where natural predators may be absent, the activities of these weevils will be much more devastating than in their natural state.

It should be noted by growers that these pests can survive long distance transport and may then become established in a new region with devastating results. At least two species have become transported internationally in this way, and other instances of regional relocation within countries are known. There is evidence also to suggest that these weevils are able to survive methyl bromide treatment applied under vacuum. The significance of this should not be lost on cycad growers and nurserymen, as methyl bromide is a standard treatment used by quarantine authorities in most countries for the elimination of pests in imported plant material.

### Macrozamia Weevil (*Tranes internatus*)

A very destructive pest which is native to Australia and has become naturalised in California, probably being introduced in the stems of imported plants. It was also recorded in 1886 from the stem of a cycad growing in Europe. In its native country this weevil usually attacks species of *Macrozamia*, but may also be found in *Cycas* and possibly also in *Lepidozamia*. In gardens in which it has become naturalised it is known to be very destructive of *Encephalartos* plants and may well attack a whole range of cycads.

Adult weevils, which are very dark blackish, are usually about twelve millimetres long and have numerous ribbed lines along their wing cases. The larvae, which grow to about one centimetre long, are legless and cream in colour with a brown or blackish head. They burrow in cycad stems and leaf bases creating a network of cylindrical tunnels. Usually there is very little evidence of their activities visible from the outside of an affected plant. Sometimes an odd leaf will yellow, wither and die (especially in *Macrozamia*), but usually the first and final symptom is a sudden and complete collapse of the plant. On inspection, the stem of collapsed plant is usually found to be riddled with tunnels and larvae and adults of the weevil can often by found if the debris is sifted through carefully. Seedlings, as well as adult plants, may be attacked with similar results. Because all life stages of this insect are spent within the stem of a cycad plant, control has proved to be extremely difficult.

## Zamia Weevil (*Phacecorynus zamiae*)

This weevil apparently attacks *Zamia* species and has been recorded from plants cultivated in Europe. The adults are black and have red spots on the wing covers and a velvety black oval spot on the front of the thorax.

## Encephalartos Weevil (*Phacecorynus funerarius*)

This weevil apparently attacks *Encephalartos* species and was recorded in Europe on trunks imported from the interior of South Africa prior to 1870. Several species of *Encephalartos* were found to be infested with this pest, which caused similar damage to that detailed above for the *Macrozamia* weevil. The black adults, which are unmarked, are about twelve millimetres long and the yellowish white larvae are wrinkled and have a light brown head. An apparently similar weevil has been found in South Africa infesting the trunks of *Encephalartos* species which were originally imported from Zimbabwe.

Another weevil, *Calandra sommeri*, has been reported to attack plants of *Encephalartos altensteinii* in the Cape Province of South Africa. The brownish adults of this species, which are about two centimetres long, have a series of dotted lines on their wing covers and black markings on the thorax and abdomen. As with the other weevils, the larvae of this species tunnel extensively in the stem and around the leafbases, causing the latter to be shed prematurely.

## Leopard Moth (*Zeronopsis leopardina*)

This destructive pest is prominent in the Natal Province of South Africa. The moths, which are bright orange with distinctive large black spots on their wings, are active in summer and lay batches of tiny yellow eggs on cycad leaves. These eggs hatch within twenty-four hours and the young caterpillars begin feeding on the leaflets, quickly increasing in size and developing a voracious appetite. Newly emerged leaves are the primary target and severe attacks may completely devour all new leaves and debilitate the plant. Large caterpillars may also damage leaf rhachises, petioles and young parts of the stem. The caterpillars, which are fleshy and nearly hairless, have two prominent dark marks on each segment. Successive broods of leopard moths are produced over summer with the pupation period being as short as ten days. This highly destructive pest poses a major problem for cycad growers in South Africa and in some cases rare species have been severely damaged. Attacks may be sporadic in some years and severe in others.

Control of leopard moth is by regular vigilance and the application of a contact insecticide or a stomach poison, such as carbaryl or the spores of *Bacillus thuringiensis*, applied as recommended by the manufacturer. Regular spraying may be necessary throughout the summer.

## Zamia Butterfly (*Eumaeus atala florida*)

This is a rare butterfly found in Florida, the caterpillars of which feed on the young leaflets of *Zamia integrifolia*. Attacks are usually sporadic and of a minor nature and control is rarely required.

An interesting side issue was revealed by studies on this insect. Toxins such as cycasin and macrozamin are taken in by the caterpillars and impart chemical protection to them and also to the later developmental stages of pupae and adults. Thus the caterpillars are not only immune to the plant toxins but can also use these chemicals for their own protection. This characteristic may occur in other species of insect which feed actively on cycads. It is certainly the case for the larvae of another lycaenid butterfly, *Eumaeus minyas*, which feeds on *Zamia* species in Costa Rica, particularly *Z. skinneri*. Newly hatched larvae of this insect eat soft new leaves and developing cones of both sexes, whereas older caterpillars can eat the hardened leaflets and even the stem.

## Emperor Moth (*Bunaea alcinoe*)

The large, spiky larvae of this South African species normally feed on *Cussonia*, but sporadic attacks may also occur on *Cycas thouarsii*. As a pest this species is of minor significance and control is readily attained by squashing.

## Miskins Blue Butterfly (*Theclinesthes miskini*)

The larvae of this small, blue butterfly feed on young leaves of species of *Cycas* and *Macrozamia* in eastern Australia (from Cairns to Moruya). The larvae, which range in colour from green to brown, have a dark dorsal band and are attended by small black ants which defend the larvae in return for food exudates. Plant tissue damaged by the feeding of the larvae turns papery and withers. This insect is present in most years, with sporadic increases in abundance in certain seasons. Control is rarely necessary.

## Palm and Cycad Beetle (*Anadastus* species)

This insect, a native of tropical Australia, can cause severe damage to palms and cycads. Attacks appear to be very sporadic but in some years the insect appears in large numbers and can devastate cycad collections. Soft new leaves and those which have recently matured are favoured and this pest can cause immense destruction in a short time. Seedlings are also susceptible and are readily killed by its ravenous feeding. Control measures, such as spraying with a contact insecticide, should begin immediately the pest is noticed.

## Cycas Leaf Beetle (*Lilioceris nigripes*)

This robust beetle is common on *Cycas* in Queensland, Australia, particularly *Cycas media*. The adults, which are about one centimetre long with metallic markings, together with their fleshy caterpillars, feed actively on young and recently mature leaves, destroying all of the upper leaf surface and sometimes skeletonising the whole leaf. Affected tissue turns white and papery and frequently all of the leaves in a new flush of growth may be destroyed. This insect can damage cultivated *Cycas* and affected plants should be sprayed with a contact insecticide immediately damage is noticed.

Recently Gary Wilson, who is studying cycad-insect interactions in Australia, has noted *Lilioceris nigripes* to be active on leaves of *Bowenia* species 'Tinaroo' in the wild. The beetle was found to be present on all new growth of some four hundred plants surveyed and massive damage was noted. It is interesting that systematic surveys carried out by Gary Wilson on *Bowenia serrulata* and *B. spectabilis* have indicated that *Lilioceris nigripes* does not feed on these species.

## Gall Midge Fly (Family Cecidomyiidae)

The larvae of these tiny flies attack the developing leaves of species of *Encephalartos* in certain parts of South Africa. Afflicted leaves develop strongly distorted or misshapen leaflets which may turn black and die. The larvae is a yellow maggot about two millimetres long. This insect can migrate to new leaves as they develop and may spread rapidly through cycad collections. Control should be achieved by spraying with a systemic insecticide such as dimethoate.

## Mealy Bugs (Family Pseudococcidae)

Various species of these soft, plump insects attack cycads in different countries. The insects congregate in colonies and suck sap from young growth particularly the emerging leaves and leaflets. Often the site of attack becomes discoloured and white waxy secretions and powder may build up. Moist sugary secretions from the insects may encourage the growth of certain fungi such as sooty mould which disfigures the site of attack even further. Mealy bugs tend to congregate in dry protected areas and their feeding may result in distorted leaflets or yellow patches of tissue. Weakened plants are more susceptible to attack and a vicious cycle of interaction between susceptible plants and the pest may take place. Ants often collect the sugary exudates from these insects and may even move the mealy bugs about on the plants.

Mealy bugs are not easy to control, often because they are protected by waxy secretions or surrounding plant material. Small infestations can be eradicated by dabbing with methylated spirits but on larger plants spraying may be necessary. Exposing the pest to the elements is a good start and ensures better success with sprays. Systemic sprays such as dimethoate or contact sprays such as maldison can be effective but successive applications may be required.

Dennis Stevenson has noticed that *Dioon* species seem to be immune to mealy bug attack and if *Dioon* plants are mixed with *Zamia* plants, the incidence of mealy bugs on the latter is reduced.

## Macrozamia Mealy Bug (*Pseudococcus* species)

This Australian pest is common on the emerging leaves of various species of *Macrozamia*. It usually congregates at the base of leaflets and causes localised yellowing.

## Aphids

Aphids, also called plant lice, greenfly or blackfly, may congregate on the newly emerging leaves of cycads. They feed by sucking the plant's sap and their activities can result in leaf and leaflet distortion. Aphid colonies are readily disrupted by a strong jet of water from a hose, with many drowning as a result. If spraying is necessary a suitable contact insecticide, such as pyrethrum, is usually successful.

## Scale Insects (Families Coccidae and Diaspididae)

These pests are readily recognised by their shells which may be waxy, leathery or cottony. The insect itself shelters beneath this shell and feeds by sucking the sap of the plant to which it is attached. Scale insects usually cluster in colonies and cause growth distortion and localised yellowing of affected tissue. Fungi such as sooty mould often grow on the sugary exudates secreted by these insects and ants commonly attend scales to collect this sugary material.

Significant infestations of scale insects should be controlled by spraying with successive applications of white oil. This material may damage plants particularly at high temperatures. White oil should not be applied at temperatures above 25°C. The addition of the contact insecticide maldison to the white oil often enhances control.

Various species of scale insects feed on cycads in different countries, with most infestations being of minor significance. Some however are deserving of special mention. Growers have noted that species of *Macrozamia* seem to be particularly susceptible to attack by scale insects particularly coconut scale.

### Flat Brown Scale (*Eucalymnatus tessellatus*)

This is a relatively minor pest which attacks native and cultivated cycads in Australia. It is usually found in small groups and causes localised leaf yellowing. The oval adults are dark brown (to five milimetres long) whereas the juveniles are almost transparent and are appressed flat to the leaf.

### Oyster Scale (*Quadrospidiotus ostreaformis*)

This European scale has a flat, yellowish to cream, oyster-like waxy shell. It is known to attack cycads and causes leaflet distortion and yellowing.

### Nigra Scale (*Parasaissetia nigra*)

This is a minor pest found on some cycads in Australia and some other countries. The adults form a raised, leathery, black, waxy covering to five millimetres long and are usually surrounded by numerous juveniles.

### Coconut Scale (*Pinnaspis aspidistrae*)

This is a persistent pest which is common on small-growing species of *Macrozamia*, particularly those which have divided leaflets. Infestations resemble shredded coconut sprinkled on the leaflets. Numbers of this scale increase greatly on unthrifty or weakened plants and worsen the condition of the plant. Control should begin as soon as this pest is noticed.

## Termites

Termites are abundant in tropical regions and their feeding activities may damage cycad plants. Usually roots are damaged first and the affected plants may appear unthrifty or else odd leaves die sporadically. Termites are generally very difficult to control and if damage is suspected the plant should be excavated and any eaten areas treated with an insecticide.

## Spider Mites

These pests, which are not insects because they have eight legs, feed by sucking sap and their activities can severely debilitate cycad plants, especially seedlings. They form colonies often on the undersides of leaflets. The adults are very tiny and move about on fine, delicate webs. Reproduction is very rapid, particularly during warm dry weather, and mites can quickly build up into severe infestations unless controlled at an early stage.

Leaves afflicted by mites have a dry appearance often accompanied by bronze or silvery tonings. These leaves may yellow and die prematurely. Mites favour dry conditions and are found on plants growing in adverse conditions such as those weakened by planting in a dry position, as in the lee of a wall or under eaves where rain does not penetrate. Cycads used for indoor decoration, particularly those which are neglected or held in very dry atmospheres may also suffer from mite attack. Regular moistening or hosing of their leaves will reduce the mite build-up somewhat but spraying is usually necessary for their control. Several treatments of a miticide such as difocol, dimethoate or tetradifon may be required for adequate control.

## Cycad Seed Borers

Insect larvae are frequently found feeding on the storage tissue of cycad seeds. This situation seems to occur in most cycad genera and is common to the various countries where they occur naturally. If damage is confined to the storage tissue then the seeds can still germinate but not if the embryo is destroyed.

Most damage to cycad seeds seems to result from the activities of weevil larvae. These are usually noticeable as small, fat, white legless grubs. The adults are typical weevils often with long, curved snouts.

Few studies have been carried out on the activities of cycad seed weevils but research in South Africa has shown that at least fifteen species of *Encephalartos* suffer seed loss in this way. Two species, *E. altensteinii* and *E. villosus*, may lose all of their seeds in some areas to the activities of two cycad weevils, *Antliarhinus zamiae* and *A. signatus*. Others such as *Encephalartos horridus*, *E. lehmannii* and *E. princeps* are also commonly damaged. Both weevils feed exclusively within the seeds of several species of *Encephalartos*, although their level of damage varies with the species involved (often as low as 10 per cent).

A significant point which emerged from this research has profound implications for cycad growers. Seeds enclosing such pests often show no external damage and can be an unwitting form of dispersal of these weevils. It is known that the two weevils mentioned above can infect a range of cycad species from which they have never been collected. Thus cones or seeds should never be transported without first checking for seed weevils or until they have been treated with a suitable contact insecticide. All seeds found to be infected should be destroyed by burning. The adults of these weevils are able to fly and once in a new region can quickly move from cycad to cycad. The effects on indigenous cycads, if these pests were to be introduced into other countries such as Australia, could be devastating.

# DISEASES

Few diseases affect cycads and most appear to be of a sporadic nature. Some leaf diseases may be a problem when plants originating in relatively dry climates are grown in gardens where they are watered excessively or the prevailing climate is wetter. Often such diseases may also result when plants are grown in unsuitable sites (e.g. too shady) or become overgrown by adjacent shrubs.

## Root-rotting Fungi

As a general rule cycads require well-drained soil for successful growth and it would seem that most species are sensitive to the attacks of root-rotting fungi such as species of *Pythium*, *Phytophthora* and *Rhizoctonia*. These fungi generally become more virulent in poorly drained or waterlogged soils or in heavy clays which waterlog readily after heavy rain or excessive watering. Cycads grown in such soils will suffer from repeated attacks by root-rotting fungi and may become stunted or quickly succumb and die.

Species of *Macrozamia* have proved to be highly sensitive to attacks by the virulent fungus *Phytophthora cinnamomii* with large, mature plants succumbing quickly after infection. This fungus is now common in Australian gardens. In parts of Australia, particularly in the jarrah forests of south-west Western Australia, the same fungus has become established in natural areas resulting in drastic changes to the flora. Natural stands of *Macrozamia riedlei*, a native component of these forests, die out quickly following spread of the infection.

Control of these fungi is mainly by improving soil drainage through techniques such as installing subterranean drains, raising garden beds and mulching with organic material.

## Sooty Mould

This is a common disease which disfigures a wide range of plants including cycads. Afflicted parts of the plant become covered with a black sooty growth. The fungus grows on the secretions of sucking insects such as scales and mealy bugs. Affected plants suffer very little damage from this fungus but look unsightly. Control is achieved by eliminating the sucking pest and the fungus usually disappears within a couple of weeks of their removal.

# OTHER DAMAGING FACTORS

## Frost Burn

Cycads originating in temperate regions are generally tolerant of light to moderate frosts, however, tropical species are highly sensitive to cold weather and especially frost. Whereas such species may survive occasional light to moderate frosts, repeated heavy frosts will usually cause their death. Frost damage shows up as brown patches on the leaflets and these eventually turn white and papery. Developing leaves may blacken and collapse completely. Plants of sensitive species collapse and die.

Protection from mild frosts can be obtained by planting under the protective canopies of established trees or close to buildings. Cold sensitive species should be grown in glasshouses.

## Sun Burn

Cycads are generally very hardy to long periods of hot sun with the only obvious problem being a bleaching of the leaflets. Those species which grow naturally in sheltered habitats, such as species of *Zamia* which grow in rainforests, are more sensitive to sun damage than hardy types which originate from open habitats. Unexpected exposure to hot sun can, however, result in sunburn even of hardy species. This shows up as white or brown papery patches in the leaflets. Sunburn of this type can occur when protective shrubs or trees are suddenly removed or when plants are taken out of glasshouses or bushhouses without being hardened sufficiently. It can also show up on the underside of leaflets when leaves become twisted out of their normal plane, for example following strong winds. Young cycads in general need to be protected from excessive hot sun for their first two or three years of life and then gradually hardened to the effects of the sun.

## Nitrogen Deficiency

Because this element is important for the formation of the green leaf pigment chlorophyll, its deficiency results in pale yellow leaves. The symptoms appear on the older leaves first — these become a uniform pale green, whereas the young leaves may be a normal green colour. In severe cases deficient leaves become severely bleached and may develop necrotic patches. Nitrogen deficiency can be readily corrected by the application of fertilisers rich in nitrogen such as ammonium sulphate, ammonium nitrate or calcium nitrate. The best answer to nitrogen deficiency is good coralloid root development. If a plant is suspected to be deficient in blue-green algae, adding algal material collected from a glasshouse floor or pots, to the soil surface should alleviate the problem.

## Iron Deficiency

In contrast with nitrogen deficiency, a lack of iron shows up in the new growth. Initially this is pale green but severe chlorosis can develop with the veins remaining green and providing a dramatic contrast to the rest of the leaf. Iron deficiency is common in plants growing in calcareous soils and can be corrected by the application of iron chelate or iron sulphate to the root zone.

## Magnesium Deficiency

Magnesium is used in the manufacture of chlorophyll and a deficiency shows up first in the older leaves. In cycads whole leaflets may become yellow or sometimes green areas are retained towards the centre of the leaflet. The effect can be quite dramatic with a number of leaflets on a frond changing colour in contrast with the remainder. Correction is by applying magnesium sulphate to the root zone.

CHAPTER 9

# The Propagation of Cycads

As with most plants, cycads can be propagated sexually or by a range of vegetative techniques, including from suckers, from trunk offsets and hopefully in the near future from tissue culture. Most cycads can be raised readily from seed and the equipment needed and techniques used are well within the scope of amateurs and enthusiasts. Thousands of seedlings of the more popular horticultural species are raised each year by professional nurserymen and the interest in less common and rare species by enthusiasts and collectors remains strong. Vegetative techniques of propagation are of much less significance to the nursery industry but are of major importance to gardeners and cycad enthusiasts. The contribution of such techniques to the survival of *Encephalartos woodii* cannot be questioned. Tissue culture, although still a technique requiring a high degree of specialisation, offers much potential for the future and perhaps an avenue of survival to species threatened and endangered by humans.

## SEED PROPAGATION

Untreated seeds of all species of cycad are slow to germinate and may take periods ranging from six months to two years with, in extreme cases, some seeds taking as long as five years. The germination process of cycad seeds has been relatively well studied and a number of steps can be taken which will greatly reduce the time lag for seed germination and enhance the results overall. The theory controlling cycad seed germination and various practical aspects are dealt with in the following paragraphs.

### Hand Pollination

As pointed out elsewhere it is rare for cultivated cycads to set seed even if male and female plants of the same species are growing side by side and their cones reach maturity together. Thus if cycad growers wish to raise seedlings it is imperative that hand pollination be carried out. This procedure is relatively straightforward. Many growers simply cut a ripe male cone and dust the pollen over mature female cones. This may be necessary two or three times a day for several days or as long as the male cone is producing fresh pollen. Nurserymen and society members may be more sophisticated than this and actually store collected pollen.

#### Pollen Collection

Just prior to shedding pollen the male cone begins elongating. This process is readily observed as the extension is quite rapid; some species increase in length by one-quarter in a day. Also the cone scales begin to noticeably separate and often there is a lightening in colour, especially a yellowing if the male cones are green. Odours may also be detectable and the cone temperature will rise above ambient as heat is produced.

Pollen shedding begins as the elongation process slows down. At this stage the male cone should be cut off carefully and placed on a smooth piece of paper. If placed indoors in a warm, dry place the pollen will shed rapidly over a couple of days. This pollen can then either be dusted direct onto mature female cones or stored.

### Pollen Storage

Unless the pollen is stored under proper conditions it will deteriorate within a few days. Small plastic containers can be suitable but it is preferable to use small paper envelopes which are then sealed inside an airtight jar containing a desiccating agent such as silica gel. These can then be stored in a refrigerator where the pollen will remain viable for two or three weeks. Remember to label the envelope with the cycad species and its date and place of collection.

Recent studies using *Encephalartos* pollen have shown that it can be successfully stored in liquid nitrogen (−196°C) without loss of viability. This is a major breakthrough for it means that it may be possible for cycad pollen to be stored over long periods, perhaps indefinitely.

### Pollinating Female Cones

At maturity, when the female cone becomes receptive, at least some of the cone scales will usually part to reveal gaps somewhere on the cone. This feature is variable with the genus; for example in *Dioon* only a few of the basal cone scales separate, whereas in *Zamia* often only a few scales at the top will open, sometimes along with a few vertical cracks. In *Ceratozamia* and *Macrozamia* cracks appear between most of the scales and in *Stangeria* the overlapping scales separate on the upper side to form channels. In *Cycas*, regular close examinations will reveal tiny drops of liquid at the micropylar end of the ovules.

Despite these changes which are supposed to occur in female cones at maturity, in practice it is often difficult to decide just when the female cone is receptive. Some growers simply assume that when the male cone begins shedding pollen any female cones of the same species will also be at the correct stage. If male cones are absent then close inspection of the female cones will be necessary.

The easiest way to introduce pollen into a female cone is by puffing it into the cracks using an eye dropper with a rubber bulb attached or a kitchen syringe. Some growers may cut off one or more of the upper cone scales and puff the pollen directly into the gap. Pollination should continue for several days to ensure a complete coverage of the ovules or to coincide with the peak period of receptivity.

## Seed Collection

As cycad cones mature their colour intensifies and the maturing seeds become obvious as they expand and force the sporophylls apart. Upon reaching maturity the female cone begins to disintegrate from the apex downwards and the seeds fall to the ground, often while still attached to a sporophyll. These seeds are now mature and can be collected for sowing. Any cone which has started to disintegrate can also be collected as all of the seeds will be mature enough to be sown.

Sometimes it is necessary to collect immature seed and if the seed is about two-thirds mature there is a chance some may germinate. Immature seeds usually shrivel readily, often quickly after collection and they should be stored in a plastic bag of moist peatmoss or else sown straight away. At the Fairchild Tropical Garden in Florida, USA, immature seeds of *Microcycas calocoma* were successfully germinated even though they were only about one-half developed and with a green sarcotesta.

## Seed Toxins

Because of the possible presence of virulent toxins in the sarcotesta of cycad seeds, precautions should always be taken when handling them. Gloves should be worn during seed collection and when carrying out operations such as removal of the sarcotesta prior to sowing. Seed collectors and commercial growers who handle large quantities of seed wear heavy duty gloves, protective goggles and face masks to minimise skin contact.

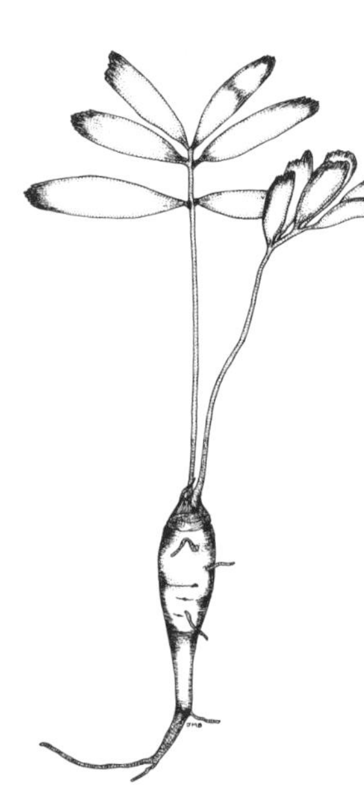

*Zamia integrifolia* seedling

## Viability Tests

The most reliable method to test the viability of cycad seeds is to shake them vigorously. Any which rattle can be discarded for it means that the endosperm has shrunken away from the sclerotesta and such seeds will not germinate or will do so slowly and with difficulty. Many cycad growers employ a flotation test to separate viable and inviable seeds and this is especially useful for large quantities of seed. Basically infertile seeds float whereas viable seeds sink. This method is not however 100 per cent accurate and it must be realised that the seeds of *Cycas circinalis*, *C. rumphii* and *C. thouarsii* (which are dispersed by ocean currents), float when viable.

Slicing a sample of seeds longitudinally may save some later heartache. Seeds with discolouration in the embryo will usually rot. The presence of a small to large hollow in the megagametophyte usually indicates the absence of an embryo and such seeds will not germinate no matter how healthy the tissue looks.

## After-Ripening of Seed

Studies on ripe seeds of cycads have shown that in most species the embryo is immature at this stage and is incapable of germination. If the seed is kept moist after harvest the embryo slowly continues developing until it is mature and capable of germination. This latter length of time, which is known as the after-ripening period, is variable for each species, however generalisations can be made for the various cycad genera. *Ceratozamia*, *Chigua*, *Dioon*, *Microcycas* and *Zamia* have the shortest after-ripening period with the embryo maturing after about a month, while the remaining genera (*Bowenia*, *Cycas*, *Encephalartos*, *Lepidozamia*, *Macrozamia* and *Stangeria*) may take from six to twelve months for this process to occur.

As with all rules, exceptions do occur, and it is obvious that more observations and studies are required on this subject. Growers often report that *Zamia* seeds will germinate quickly after sowing and may respond rapidly if the sclerotesta is removed. Seeds of *Microcycas calocoma* often have a well-developed embryo at maturity and occasional seeds will be found germinating in the cone as it disintegrates. Similarly the embryo in species of *Encephalartos* (such as *E. transvenosus* and *E. manikensis*), where the female cone has a long developmental period (nine to eighteen months), may be mature by the time of seed dispersal.

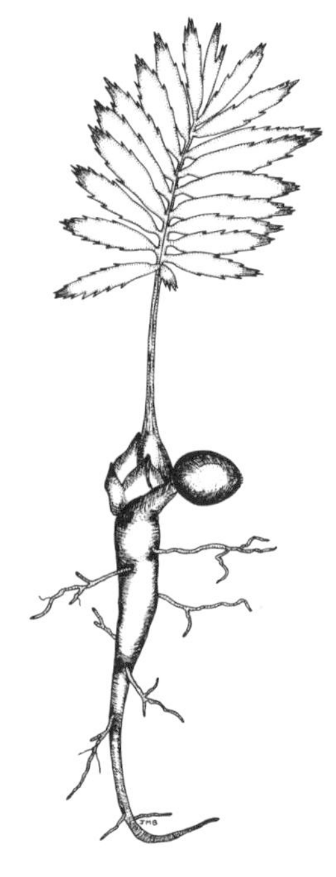

*Dioon spinulosum* seedling

## Sarcotesta Inhibitors

The fleshy sarcotesta which surrounds cycad seeds contains chemical inhibitors which delay seed germination. This layer also, however, may assist in maintaining the moisture content of the megagametophyte by reducing water loss. In nature the sarcotesta is often eaten by animals and so any inhibitory or protective effects it confers are often removed quickly after the seed matures.

Experiments have shown conclusively that removal of the sarcotesta speeds up the germination rate of cycad seeds, and most growers carry out this process as routine. Since the sarcotesta may contain toxic materials (it certainly does in *Macrozamia* species), it is strongly recommended that gloves be worn during this procedure.

## Seed Scarification

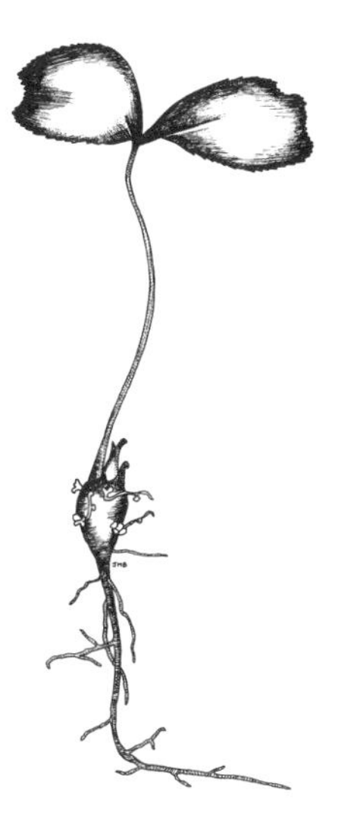

*Zamia furfuracacea* seedling

Scarification consists of mechanically gouging, scratching or cutting the hard outer coat of a seed in order to enhance germination. In cycads, the seed is surrounded by a hard woody coat known as the sclerotesta. This coat appears to be a mechanical barrier to germination, however it is known from experiments carried out on *Encephalartos* seeds that water can be imbibed through the sclerotesta. There is a possibility that chemical inhibitors may be present in the sclerotesta although this has not been proven.

Slicing through the sclerotesta to the surface of the endosperm beneath greatly increases the germination rate of cycad seeds. In *Zamia integrifolia* for example treated seeds may germinate in one or two weeks whereas unscarified seeds take in excess of six months to germinate. Various techniques for cutting the sclerotesta can be employed including the use of a sharp knife or file. Cutting at one end of the seed seems to be the most effective at inducing germination, particularly the micropylar end. Care must be exercised during these operations to avoid excessive damage to the megagametophyte and particular care must be taken to avoid the embryo.

Researchers have shown that softening the seed coat of *Zamia integrifolia* by soaking in concentrated sulphuric acid greatly improves germination of this species. Best results were obtained with seeds soaked for periods of thirty to sixty minutes in the acid.

It is apparent from research into the stage of embryo maturity at seed ripening time (see previous page), that not all genera of cycads will respond successfully to scarification. Thus species of *Lepidozamia* and *Macrozamia* may require months of after-ripening before the embryos are developed and ready for germination. Scarification on its own would be of limited value for these species where the embryos are immature, but would still be of value following the after-ripening period.

## Seed Stratification

Stratification is the exposure of seeds to low temperatures in the presence of moisture. It is commonly employed as a germination technique for the seeds of species of plants which would naturally be subjected to very low temperatures for periods in their life cycle.

Tests with seeds of *Encephalartos natalensis* show that periods of stratification at 5°C will not produce any significant promotion of germination and long periods of exposure are inhibitory. Similar results are likely to occur for most cycads, especially those of tropical origin. Stratification tests on those *Encephalartos* species which occur naturally at high altitudes (e.g. *E. friderici-guilielmi, E. ghellinckii*) may provide interesting results, as these species are exposed naturally to periods of intense cold.

## Pre-Soaking Seeds

Water can be imbibed through the sclerotesta and some growers advocate that soaking the seeds for a couple of days prior to sowing improves the rate of germination.

## Germination Requirements

Apart from physical and chemical barriers inherent within the seeds, cycad seeds for germination require warmth, moisture, moderate humidity, an abundant supply of oxygen (especially in the propagation medium) and freedom from plant pathogens. The latter condition is extremely important for those cycad species which take a long time to germinate.

Cycad seeds seem to have an optimum germination temperature of about 30°C. In tropical areas such temperatures can be provided without special facilities, however, in temperate regions heated glasshouses or propagation units fitted with bottom heat cables will be needed for successful germination. Response to temperature is probably cumulative and fluctuations such as occur with overnight cooling are not detrimental and do not negate the high temperatures of the day. Temperatures as low as 20°C can still be satisfactory for cycad seed germination (although the time taken for seeds to germinate will be correspondingly longer), however temperatures lower than about 10°C may be damaging, especially to those species of tropical origin. Temperatures in excess of 38°C may also cause damage, and exposure over long periods to such high temperatures may be lethal.

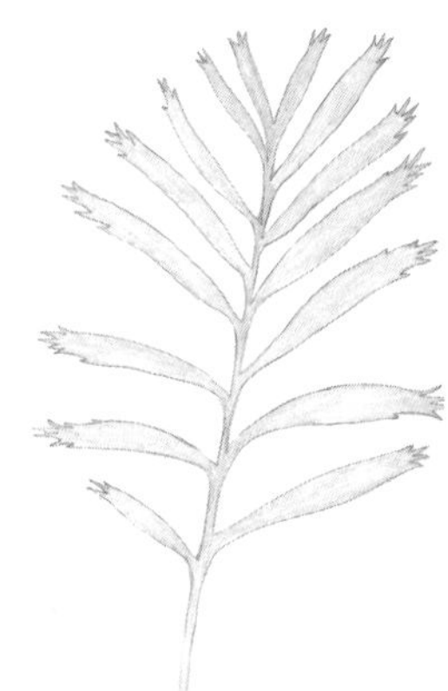

*Macrozamia lucida* seedling

Despite its hardness and relative thickness, the sclerotesta is not impervious to water, and the embryo of fertile seeds can imbibe water that is present in its immediate surroundings. Fertile seeds which have the sclerotesta still attached have a rapid initial water uptake when placed on moist propagation medium. Tests on such seeds show that about a 20 per cent increase in moisture content of the embryo and megagametophyte occurs within one or two days of sowing. Subsequent water uptake is very slow. Humidity at moderate to high levels is important for successful seed germination and is readily achieved by watering the propagation mixture, facilities and the immediate surroundings on a regular basis.

Oxygen is very important for seed germination and it is vital that any medium used for seed propagation be of coarse, irregular particles thus allowing rapid passage of water and adequate aeration.

Cycad seeds can be damaged or destroyed by pathogenic fungi and bacteria. The effects of these pathogens are exacerbated in immature, poorly harvested, poorly stored or damaged seeds or if the germinating conditions are unsuitable. For example root-rotting fungi will be a problem after germination if the drainage and aeration of the propagation mix are inadequate. As a general rule, healthy, undamaged seeds sown within a couple of days of harvest under good conditions will not suffer greatly from the activities of pathogenic fungi and bacteria. If in doubt the seeds can be dipped in a solution of a broad spectrum fungicide prior to sowing.

## Rapid Seed Germination

It is a general rule with seed propagation that the faster the germination process takes place the better will be the final result. As can be seen from discussions in previous paragraphs, cycads, for a variety of reasons, tend to take a long time to germinate with periods of up to five years being recorded.

Some growers are content to let nature take its course but research horticulturists have studied the cycad germination process in detail, and have introduced some techniques which are successful in reducing the time period involved. Most of these are dealt with elsewhere in this chapter and not all of the techniques are equally applicable to all cycads.

The germination of *Zamia integrifolia* has been studied in some detail and a technique for its improved germination has been published (see B. Dehgan and C. R. Johnson, 'Improved Seed Germination of *Zamia floridana* (*sensulato*) with $H_2SO_4$ and $GA_3$', *Scientia*

*Horticulturae* 19 (1983):357–61). The results obtained are summarised here.

The sarcotesta was removed from freshly collected seeds using sharp sand and a long-stemmed wire brush attached to a mounted variable speed drill. Soaking the seeds in concentrated sulphuric acid for thirty to sixty minutes greatly improved germination especially if the treated seeds were placed in intermittent mist. Soaking the seeds in gibberellic acid ($GA_3$) at a concentration of one gram per litre, for twenty-four to forty-eight hours also improved the seed germination. The combination which resulted in the best germination of the cycad was soaking in concentrated sulphuric acid for sixty minutes followed by rinsing in water, soaking in gibberellic acid for forty-eight hours and sowing under intermittent mist. Nearly 100 per cent of seeds treated in this way germinated within six weeks of sowing.

Using these results as a basis, an improved germination could be expected in species of *Ceratozamia, Chigua, Dioon, Microcycas* and *Zamia* following similar treatment. Results may well be different in the remaining genera, however, because in these cycads the embryo has a long after-ripening period before it reaches maturity (see page 84).

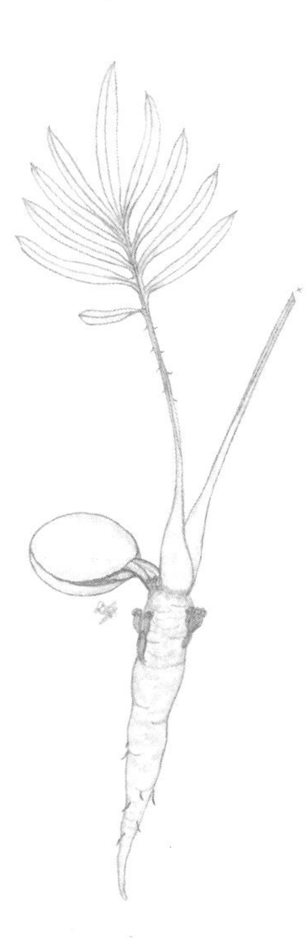

*Cycas media* seedling

## Propagating Mixes

For best germination, cycad seeds should be sown in coarse, well-drained materials which drain freely after watering, yet retain some moisture for germination, and provide excellent aeration. Cycad seeds can be sown in a well-structured garden soil and will often germinate satisfactorily. Soils however are not recommended for cycad seed germination, especially for large quantities of seed, because of often poor aeration, and the possible presence of disease organisms, weeds and pests. Most cycad enthusiasts and commercial growers use materials such as coarse sand, peatmoss, vermiculite, perlite and pine bark.

**Coarse Sand** Usually obtained from alluvial deposits. It must be washed thoroughly to remove dirt and weed seeds. Drainage is excellent but it dries out rapidly after watering and is best mixed with a water-retentive organic material.

**Peat Moss** Rotted organic material which has reached a stable point of decay. It is very acid (pH 4.5), sterile and absorbs many times its volume in water. It has excellent aerating properties and mixes well with other materials such as coarse sand. It can be used on its own to germinate cycad seeds.

**Vermiculite** A naturally occurring mica which is expanded by being subjected to temperatures in excess of 1000°C. It is a very light material with a high water-holding capacity and excellent aeration. It mixes well with other materials but must not be overwatered or compacted as it readily becomes clogged. It is sterile because of the high temperatures used in its production.

**Perlite** A naturally occurring silicate material which is treated with temperatures in excess of 700°C. It forms grey, spongy particles which are very light and is best mixed with a water-retentive organic material.

**Pine Bark** Finely ground pine bark is often called pine peat because of its similarity in appearance and in some properties to peat moss. When fresh it contains toxins and must be stored moist for six to eight weeks before use. It is best mixed with coarse sand or perlite.

**Sawdust** Like pine bark, fresh sawdust contains plant toxins and hence must be stored moist in a heap for six to eight weeks before use. It mixes well with sand or perlite and has good moisture retention and aeration properties.

Peatmoss and vermiculite can be used by themselves as a germination medium for cycad seeds but it is more common to use a mixture. Very successful mixtures can be made by combining two parts coarse sand or perlite with one part peatmoss, vermiculite, pine bark or sawdust. The mixture must be moistened before sowing as materials such as peatmoss and sawdust are difficult to wet when dried out. If weeds, pests or diseases cause problems during germination it may be necessary to sterilise the sowing mixture.

## Sowing Techniques

There is some variation between growers in the way cycad seeds are placed in the propagating medium, but all agree that about half of the seed's surface should remain uncovered and be exposed to the atmosphere. Thus cycad seeds are sown on the surface of the propagation medium and are not buried as are the seeds of many other plants.

Most cycad enthusiasts place the seed flat on the surface of the propagation medium so that its long axis is parallel to that of the medium. Many professional nursery growers however claim to obtain better results if the seeds are planted vertically with about half their length buried. Most growers plant the micropylar end downwards but some prefer to leave this end exposed and bury the chalazar end. Perhaps all of these variations are an indication that the orientation of the cycad seed is not critical for successful germination. Certainly if the micropylar end is left exposed then the emergence of the first root is readily observed. It also means that the micropylar area is kept relatively dry and this reduces the chance of pathogenic fungi and bacteria penetrating to the kernel via the micropylar opening.

Cycad seeds do not mind being crowded and can be sown close together. Crowded seeds, however, should be observed regularly for germination, with the germinated seeds being potted quickly to avoid suffering from competition.

Suitable containers for sowing cycad seeds must be fairly deep as a root appears first and grows rapidly downwards before the first leaf is produced. A container with a depth of fifteen centimetres will give ample room for the initial root development of young seedlings. Cycad enthusiasts mainly sow into pots whereas professional growers may use pots, deep trays or specially constructed sowing beds. These latter structures, when furnished with systems of bottom heating and controlled intermittent mist, provide optimum conditions for the germination of cycad seeds. Scaled down versions of these propagating systems are frequently available from firms specialising in horticultural hardware and are ideal for small scale propagation. Fortunately cycads are amenable to variation, germinating under a range of conditions and thus allowing enthusiasts living in different regions to be successful with this fascinating group of plants.

### Bottom Heat

Although already mentioned briefly, the use of bottom heat for the germination of cycad seed is deserving of greater consideration. In such systems heat is provided in the propagating medium below the seeds while the atmosphere above the seeds remains at ambient temperature. Electricity at low voltage run through high resistance wire cables, hot water or steam pipes are the most commonly used forms of heating. The bottom heat unit is set at 25–30°C and the cycad seeds are place on the surface of the propagating mixture just above the heating elements. The use of bottom heat not only reduces the time taken for cycad seeds to germinate but also often promotes a more even germination so that uniform batches of plants can be obtained. Seeds must be observed regularly and any which have germinated should be removed from the propagation unit and potted.

### Intermittent Mist

Placing seeds under intermittent mist results in an improvement of cycad seed germination. This may be due to the leaching of chemical inhibitors from the seed coat or as a result of the softening of the sclerotesta and increased water uptake.

## Seedling Establishment

The care of cycad seedlings is of paramount importance if they are to grow and become successfully established as plants. Cycad seedlings are capable of rapid growth and respond well to tender loving care. If neglected at this critical stage of their life, cycad seedlings may linger and become unthrifty or succumb to diseases. Whereas cycad seedlings in the wild have a very high mortality rate, it is possible for most seedlings in a batch to survive in cultivation if given good conditions.

Successful growers ensure that seedlings are potted soon after germination is noticed into a suitable well-drained mixture. If then provided with warmth, bright light, humidity, air movement, abundant water and regular fertilising, the seedlings are capable of amazing growth rates. Under optimum conditions a seedling may double in size every two or three months for several years.

It should be noted that small seeds (such as species of *Zamia*) generally have proportionately reduced nutrient reserves. Seedlings of such species should always be potted in preference to seedlings of species which have large seeds. The latter type not only have sufficient reserves to maintain a seedling for a longer period but also produce larger seedlings.

## Causes of Poor Germination

The most common causes of poor seed germination in cycads are the result of sowing seeds which are either immature or overmature or have deteriorated in storage. Misunderstanding the after-ripening process that occurs in the embryos of most genera can also be a contributory factor (see page 84).

Considerable time and anguish can be saved by examining a small sample of seeds from a batch prior to sowing. The seeds will appear normal from the outside but when the sclerotesta is cut may show shrunken, discoloured or rotted kernels. Such seeds will rarely germinate satisfactorily. Cultivated specimens of *Encephalartos* sometimes produce seeds which are apparently viable (although often a little smaller than normal), but lack any embryo and cannot germinate. Such seeds, which have not been fertilised, may have resulted from parthenocarpy, with the apparently normal development occurring because of high hormone levels in parts of the cone.

Seeds which have a cracked sclerotesta often germinate poorly, probably due to the entry of pathogenic fungi or bacteria through the cracks. Treatment with a fungicide may improve the survival rate. Similarly the complete removal of the sclerotesta and soaking in gibberellic acid prior to sowing may be of benefit. The results of such treatment will probably be more successful with species of *Ceratozamia, Chigua, Dioon, Microcycas* and *Zamia* (where the embryo is fully developed by shedding time), than in other genera in which the embryo has a long post-shedding development period.

In seeds which are overmature or have become desiccated, water loss occurs from the megagametophyte resulting in a shrunken kernel. Such seeds will rattle when shaken and are often discarded by growers because they germinate poorly. If however the sclerotesta is removed and the seeds soaked successively in solutions of a fungicide followed by gibberellic acid prior to sowing, some germination can sometimes be obtained.

In some species such as *Lepidozamia peroffskyana* and *Macrozamia moorei*, the primary root may have difficulty in emerging from the micropyle. Subsequent development occurs within the sclerotesta resulting in a crowded, convoluted root that has nowhere to go. Such seeds will not germinate unless the sclerotesta is removed.

## Seed Storage

If kept moist and cool, cycad seeds will retain their viability for many months. In fact stored seeds of some species may germinate quicker than those sown fresh because of the after-ripening period required by the embryo (see page 84).

A significant deterioration in the viability of cycad seeds occurs as the result of moisture loss through the sclerotesta. This results in the endosperm shrinking away from the inner wall of the sclerotesta. When about 20 per cent water loss has occurred the process seems irreversible. If such seeds are cut in half, the megagametophyte will be seen to be greatly shrunken, discoloured and with a dry texture.

Moisture loss from cycad seeds is influenced by the prevailing temperature, humidity of the atmosphere around the seed and time. Thus in high temperatures and a dry atmosphere the seeds will lose their viability much faster than in moderate temperature and high humidity. Low temperatures may cause damage to the embryo. A successful method of storage is to mix the seeds with moderately moist peatmoss, seal in a plastic bag and store at 10–15°C.

## Transport of Seeds

If collected fresh, despatched within a short time of collection and packed properly, cycad seed will survive transport between most countries of the world. Air transport is certainly not a problem but may be prohibitively expensive. If sealed in a plastic bag containing moist peatmoss, fresh cycad seeds will withstand the rigours of five to seven months travel by surface transport and still be viable at the end of their journey.

# VEGETATIVE PROPAGATION

Vegetative techniques useful for the propagation of cycads are fairly limited. Most of the procedures used involve the removal of growths or offsets with the specialised technique of tissue culture still proving to be difficult and not able to achieve its potential.

## Propagation from Basal Offsets

Basal offsets or suckers are commonly produced by many species of cycads and these growths are often particularly prominent on cultivated plants. Suckers of this type are genetically identical to the parent plant and represent an important source of plant material since they can be readily used for propagation. In fact if it was not for its suckering habit, *Encephalartos woodii* would be much rarer than it is today.

The removal of suckers is part of a technique of vegetative propagation known as division. Division is a simple technique of propagation but if it is to be successfully carried out, care is required both during the removal process and in the establishment phase of the new plant. Unfortunately many growers handle the process of sucker removal roughly or casually, often resulting in death of the sucker and loss of a potentially valuable resource.

The tools required for the excavation and removal of a sucker, such as spades, knives, saws and chisels, must be clean and sharp. The soil should be dug away carefully so as to expose the base of the sucker where it joins the parent trunk. If the soil does not fall away readily, the base of the trunk and sucker should be hosed to expose the point of attachment.

The sucker should then be severed using a strong sharp knife or large chisel. Damage to the tissues should be kept to a minimum and any ragged edges should be trimmed. All cut surfaces should then be sealed by rubbing in garden lime or powdered sulphur.

Aftercare consists of removing all leaves (to reduce dehydration) and immersing the whole sucker in a solution of a broad spectrum fungicide for one or two hours. The sucker is then held in a cool, dry place for one to three weeks until the wound seals, after which it can be potted up or planted in the ground. Placement of the sucker in a bottom-heat unit at this stage will usually result in a strong root system and quick establishment of the new plant.

Even suckers which lack roots of their own can be removed and established successfully, however, they may require plenty of tender loving care. Potted suckers should be held in a semi-shaded, protected situation until sufficiently established to be planted. At no stage should the new plant be allowed to dry out.

The optimum time for removing basal offsets from cycads is during spring and early summer when the climate is warm and congenial for the quick establishment of the new plant.

## Propagation from Trunk Offsets

Many cycads, particularly species of *Cycas*, are capable of producing new growth or offsets from aerial parts of their trunks. Such growth may appear spontaneously but often they are noticeable following some form of mechanical damage.

In species of *Cycas* trunk offsets often begin as small, bulbous swellings which are usually leafless but may develop cataphylls at the apex. As these swellings enlarge, odd small leaves may be produced from the apex. In other genera, particularly *Encephalartos*, a small tuft of leaves develops from an early age.

When the trunk offsets are well developed and preferably with a few small roots at the base, they can be removed from the parent trunk. This is best achieved by cutting with a clean, sharp knife or chisel at the point where the offset emerges from the trunk. After removal, all ragged edges on both the parent and the offset should be trimmed and sealed by rubbing in lime or powdered sulphur. The leaves are then removed from the offset and it is then treated in the same way as outlined previously for basal offsets or suckers.

Although the emphasis is on the removal of established offsets, it should be pointed out even small, bulbous swellings on the trunks of *Cycas* species can be removed and established successfully. This is particularly true of *Cycas circinalis* and *C. revoluta*.

## Regeneration from Trunk Sections

Cycads have proved to be remarkably resilient plants and stories of them surviving long periods out of the ground, adversity and high levels of damage are relatively common. Thus it is perhaps not surprising to learn that if the trunks of *Cycas circinalis* and *C. revoluta* are cut into sections about fifty centimetres long and placed upright in a bottom heat propagating unit they respond in a similar way to cuttings. New roots are produced strongly from the basal area and new leaves emerge from the apex of those pieces with an intact crown. In those trunk sections which lack a crown a series of buds develop from among the axils of the apical scales. Although this method of propagation is rather drastic, it serves to illustrate a survival technique of these plants and suggests that cycad trunk material should never be discarded without first testing to see if it will regenerate in a similar way. Instances of offset production have also been recorded in species of *Encephalartos* and *Lepidozamia* following mechanical damage or partial rotting of the trunk.

## Stem Cuttings

The much-branched stems of *Stangeria eriopus* can be cut into pieces and if placed into a bottom-heat unit will develop as separate plants. Some success has also been obtained using this method on plants of *Bowenia spectabilis*.

## Leaf/Stem Cuttings

This is a novel propagation technique used by some enthusiasts. A healthy leaf is trimmed off at about one-third of its length and a small section of the stem is removed together with the sheathing leaf base. The basal area is then buried in sand and peatmoss in a bottom-heat propagating unit and treated as a cutting. Small outgrowths often arise from the stem region and if nurtured carefully can be grown as separate plants.

## Cone Outgrowths

On rare occasions the cones of a cycad may sprout a couple of leaves from the apex. Usually this phenomenon is of passing interest, but occasionally the region around the base of the leaves swells and a plantlet or bulbil forms. When sufficiently developed, the whole cone, together with the apical growth, can be removed and treated as an offset.

## Tissue Culture

Tissue culture, or meristem culture, is a modern technique of laboratory propagation which uses very small pieces of plant tissue which are cultured on nutrient media under sterile conditions. These tiny pieces of tissue (termed explants), are first sterilised then induced to form a mass of undifferentiated callus cells by growing them on a balanced nutrient medium. After sufficient increase in quantity, clumps of callus are then placed on specialised nutrient media containing a critical balance of hormones, and differentiation into organs such as leaves and roots takes place. This propagation technique must be carried out in a laboratory and requires the use of expensive and specialised equipment.

Tissue culture has obvious advantages, notably the use of small amounts of plant material and the potential for the production of large numbers of new plants. Such a technique of micropropagation is of major significance for the survival of rare and endangered species. It has been estimated that about seventy species of cycad are currently assessed by conservation authorities as being rare, endangered or vulnerable, hence a successful program of tissue culture could greatly enhance their prospects for survival.

Detailed experimentation on the tissue culture of cycads has been carried out by scientists in many countries. The results show clearly that callus growth can be obtained readily for a range of species but considerable difficulty has been encountered in getting the callus to become organised into plantlets. Occasionally odd leaves, buds or roots have been produced in the cultures but this has not yet resulted in the formation of viable cycad plants, despite the ease of production of vigorous cultures of callus cells. Some encouraging results have been obtained in *Stangeria eriopus* using explants from the root. Further research is proceeding as a successful conclusion to this propagation technique is of major importance for the future survival of many cycads.

CHAPTER 10

# Cycads for Containers, Indoors and Out

Cycads are excellent plants for indoor decoration, and some species, especially of the genus *Cycas*, have been used in this manner over long periods of time. Early this century it was almost as fashionable to have a plant or two of *Cycas circinalis* or *C. revoluta* for indoor decoration, as it was to have a kentia palm. They were especially popular in Europe and today they are still used to grace the large rooms and hallways of stately homes and mansions.

Although a large number of cycad species make excellent foliage plants for indoor decoration, they are still only used on a fairly limited scale for this purpose. Cycads compete with a large number of other plants, particularly palms and aroids, for this niche and it is difficult to displace species which already have a proven track record.

Some cycads perform very well indoors, others are less successful. Light seems to be a significant factor in their performance and if this is adequate then a wide range of species can be grown. More species succeed in tropical regions not only because of the extra warmth but, often as a result of the open, airy design of the houses. Larger numbers of cycads can find a congenial place in the modern home with its courtyards, skylights, bright bathrooms and large windows, than would have been possible in the older style homes of the past.

## Suitable Species

In general the cycads which succeed best indoors are those which originate in sheltered habitats, such as occur on protected slopes, in gullies and under the shady canopies of forests. These seem to be able to tolerate fairly dark conditions (although they do better with good light) and some degree of neglect. Species of *Zamia* have performed particularly well indoors, especially *Z. furfuracea*. Both species of *Lepidozamia* are also excellent as indoor plants. For a full list of recommended species see table on page 95 .

A word of caution may be timely here. A number of cycads have spiny leaflet tips which can inflict a painful wound if contacted with any degree of force. Such species are generally not considered for indoor use even if they succeed in the conditions.

### Pot Size

It has been noted by growers that cycads which usually have a subterranean trunk are more vigorous and grow better when the trunk is buried rather than being exposed. Such species normally have contractile roots and stems which regulate the depth of the trunk in the soil. Pots which are sufficiently deep should be used when potting species of cycad which have this growth feature.

## Choice of a Plant

Selection of a plant for indoor use should be straightforward. Never feel sorry for the weaklings for they are probably the genetic runts of the batch and will more than likely perish. Choose a sturdy, healthy, actively growing specimen with unblemished dark green leaves. Avoid those with a dull sheen to the leaves as they may have been neglected or held too long in unsuitable conditions in the store or nursery. Plants with dry, sporadically mottled leaflets may have an active infestation of mites. Plants with extremely soft or lush growth may have just been taken out of a glasshouse and could be expected to deteriorate quickly when placed indoors. Do not be afraid to tip a plant out of its container to check the health of its root system. If concerned about the mess ask the sales attendant to tip it out. Do not forget when inspecting the root system of a cycad that the much-branched coral-like growths present are a normal part of a cycad's root system.

Buyers should be wary of plants which have pests already established. A few eaten leaflets resulting from the feeding activities of caterpillars should cause no anguish but colonies of scales, mites and mealy bugs can be expected to proliferate in the indoor environment. These pests, which are difficult enough to eradicate outdoors, can cause major problems indoors and it is best to start with clean stock.

## Indoor Conditions

### Light

This factor seems to be of major significance for the successful performance of many cycads indoors, although there is a variation in response between various groups of cycads. As a general rule species of *Zamia* perform best in conditions which are moderately bright to very bright, whereas *Lepidozamia* species and the large-growing *Cycas* will tolerate much darker conditions.

All cycads when grown indoors need a position where they will receive some natural light or else require supplementation from artificial lights. Direct sun for part of the day can promote excellent growth in all seasons excepting summer when long exposure to hot sun will cause bleaching or burning. Direct exposure to the morning sun is preferable to later in the day. Sun filtered through the foliage of shrubs and trees can provide ideal lighting for cycads. Bright light through coloured or frosted glass provides an attractive background for foliage and may not be harmful to their growth. Glass or plastic of this type however may transmit heat and the plants will require more regular watering and misting than would be the case in other areas. Solar films applied to windows to reduce heat and glare are generally detrimental to indoor plant growth.

### Humidity and Air Movement

The atmosphere inside offices and homes is generally quite different from that occurring in the natural habitats where cycads grow. Humidity is generally low and air movement is usually either stagnant or in the form of draughts. Humidity may also fluctuate greatly, especially when heaters are used or air conditioning. Often it is difficult to attain a balance between air movement and humidity. Gentle air movement is necessary to replenish oxygen supplies and humidity is important for reducing stress and maintaining health in the plants. In low humidities or atmospheres which fluctuate greatly, sensitive plants quickly lose their lustre, appearing dull and are then often severely attacked by pests such as spider mite and mealy bug.

The humidity in the atmosphere around a plant can be enhanced in a couple of simple ways. Watering is one but the plant must not be overwatered or the results will be even more disastrous. If the leaves of an indoor cycad lose lustre and appear dry, but the potting mix is still moist, then the problem lies in the atmosphere and excess watering of the potting mix will only cause death of the roots. Humidity around indoor plants can be boosted by simple techniques such as misting the plants at regular intervals with a fine spray, grouping plants together so that each contributes to the atmosphere around the other and growing more than one plant in each pot. Standing pots in a large saucer of wet, evaporative material such as scoria or coconut fibre is also a useful technique.

### Temperature

Cycads which are of tropical origin generally dislike low temperatures and are extremely sensitive to frosts. Such conditions rarely cause problems indoors, however leaves in contact with windows may suffer cold damage. In highland districts and temperate zones the winter temperatures experienced indoors can drop to levels which may cause damage to sensitive species. Plants which are dormant when these conditions prevail are less likely to suffer damage than those in active growth.

Internal house heaters, especially those with forced draughts or ducts, often create excessively dry atmospheres which may result in excessive water loss and desiccation of indoor plants.

## Bonsai Culture

Some species of cycads are admirable subjects for bonsai culture. Species such as *Cycas revoluta* are commonly used for this purpose in Japan and Korea and venerable old specimens show that they have been valued in this way for decades and even centuries. More recently growers have tested other species such as *Lepidozamia peroffskyana*, *Macrozamia heteromera*, *Zamia pumila* and *Z. furfuracea* with considerable success. It may be that because of their structure and growth habit, most species of cycad have potential as bonsai subjects.

Some pointers for their culture in this manner are pertinent. The root systems should be restricted by growing the plants in small containers and about one-third of the root system is trimmed back annually. Further restrictions on plant size are imposed by using pure sand as a potting medium, applying no fertilisers either in the potting medium or as side dressings and by withholding water at strategic periods in their growth cycle. The latter technique, applied just as new leaves are opening, results in reduced or miniature leaves and leaflets. Conversely new growth may be forced by regular watering in conjunction with removal of the old leaves.

# Cycads Suitable for Indoors

| *Species* | Light Tolerance | Comments |
|---|---|---|
| *Bowenia serrulata* | Dull–bright | Excellent, highly decorative |
| *Bowenia spectabilis* | Dull–bright | Excellent, highly decorative |
| *Cycas chamberlainii* | Bright | Good, attractive foliage |
| *Cycas circinalis* | Bright | Good, needs regular spelling and a large container |

| | | |
|---|---|---|
| *Cycas media* | Bright | Fair, needs regular spelling |
| *Cycas pectinata* | Bright | Good |
| *Cycas revoluta* | Dull–bright | Excellent, tolerates neglect |
| *Cycas rumphii* | Bright | Good, needs regular spelling and large container |
| *Cycas siamensis* | Bright | Good |
| *Cycas thouarsii* | Bright | Good, needs regular spelling and large container |
| *Lepidozamia hopei* | Dull–bright | Excellent, tolerates neglect |
| *Lepidozamia peroffskyana* | Dull–bright | Excellent, tolerates neglect |
| *Macrozamia lucida* | Dull–bright | Excellent |
| *Microcycas calocoma* | Bright | Good |
| *Zamia acuminata* | Dull–bright | Good |
| *Zamia chigua* | Dull–bright | Good, needs room to spread |
| *Zamia fairchildiana* | Dull–bright | Good |
| *Zamia fischeri* | Bright | Good |
| *Zamia furfuracea* | Dull–bright | Excellent |
| *Zamia integrifolia* | Bright | Good |
| *Zamia loddigesii* | Dull–bright | Good |
| *Zamia pygmaea* | Bright | Excellent |
| *Zamia roezlii* | Bright | Good |
| *Zamia skinneri* | Dull–bright | Good |
| *Zamia splendens* | Bright | Good |
| *Zamia standleyi* | Bright | Good |
| *Zamia vazquezii* | Bright | Good |

## Care of Indoor Cycads

### Watering

Cycads are generally tolerant of some degree of neglect in the watering department but even so for best appearance and optimum growth they should be watered at sufficiently regular intervals to meet their requirements. Cycads that are kept too dry take on a dry appearance because their leaves lose the healthy sheen and the plants generally look unthrifty. Cycads that are kept too wet may have brown areas in the leaflets, especially towards the apex and if the roots are examined closely rotting of the root tips is often prominent. If cycads are kept too wet for long periods then frequently whole leaves may die and in severe cases the whole plant collapses.

The watering of cycad plants which are used for indoor decoration is basically common sense. Vigorous, actively growing plants will need more water at each watering and will need to be watered more frequently than those plants which are growing slowly or not at all. In any group of indoor plants some specimens are going to require more regular watering than others. It is a temptation to water all of the plants at the same time but this should be avoided and the needs of individuals should be catered for separately. If, for example, all of the plants are watered each time the most vigorous specimen dries out, then the least vigorous plants will receive too much water.

Because of the higher prevailing temperatures, cycads indoors will need to be watered more regularly in summer than in winter, and plants growing in brightly lit areas will dry out more quickly than those in dim positions. In summer, plants can be safely watered on a daily basis whereas in winter their needs are much less.

Some potting mixes retain less water than others and plants growing in them will need more frequent watering to maintain growth. Potbound plants, in which the pot is full of roots, may need special attention with regard to watering and would probably benefit from repotting if room is available to accommodate the larger pot size.

The ideal watering regime keeps the potting mixture sufficiently moist to supply adequate water and oxygen to the plant's roots for healthy growth. Small quantities of water supplied at regular intervals may serve to satisfactorily top up moisture levels in the potting mix, but heavier applications which result in water flowing out of the drainage holes are also important. Such a flow ensures a thorough wetting of the root system, eliminates dry pockets in the potting mix and also leaches out salts which may have accumulated from the breakdown of fertilisers. Heavy leaching is best performed out of doors or in a bath or sink. If difficulty is experienced in wetting a potting mix, then the pot should be stood to about half its depth in a container of water and left until moisture is obvious on the surface of the potting mix.

## Recuperation

All indoor plants, including cycads, appreciate a recuperation period whereby they are removed to congenial conditions outdoors. If this is carried out regularly at relatively short intervals the healthy appearance of the plants is readily maintained. More usually however, recuperation consists of reinvigorating plants which have lost much of their health and vigour because they have been kept too long in often unsuitable indoor conditions. The worse the condition of the plant the longer will be the recuperation period needed.

Indoor cycads can be rested by moving them to a shady position in the garden or bush house. A recuperation period of two or three weeks after a two or three months spell indoors is usually highly beneficial to the health and appearance of the plant. The leaves should be hosed down to remove dust, discourage mite infestation and refreshen the plants generally. While recuperating, cycad plants should be well watered and fertilised, and repotted if necessary. They should be generally encouraged to develop new growth. Once spelled, the reinvigorated plants can be moved back indoors. With some forethought, a series of cycads can be cycled in this way and those which are indoors can always be at their peak.

## Fertilisers

Cycads respond to the use of fertilisers and those plants used for indoor decoration are no exception. Fertilisers should only be applied during the warm months of the year when plant growth is active. Fertilisers are best applied in small doses at regular intervals and watered in to the potting mix. Correct fertiliser usage will ensure the maintenance of healthy growth but incorrect application may be highly detrimental.

Best results for indoor cycads are obtained by using supplementary materials such as slow-release fertilisers, plant pills or liquid preparations. Liquid fertilisers are usually safe (except where the plant is suffering from waterlogging), easy to use and generally beneficial to plants. Some kinds, such as those extracted from seaweed or fish meal, may be smelly and are best used on plants held outdoors. Commercial liquid preparations have the dilutions to be used printed on the packets and these recommendations should be adhered to. Some nutrients can be applied through the leaves in a process known as foliar feeding. This is generally a much less satisfactory and more expensive means of boosting growth than applications through the root system.

Quick-release fertilisers should be used with care and should never be applied to newly potted cycads or those with weakened roots as burning can result. Fertilisers are also incorporated in the potting mix to encourage growth after repotting. Complete fertiliser mixes are usually used in the potting mixes and these may be rapidly soluble or of the slow-release type. Organic fertilisers and well-rotted animal manures can be very beneficial but some such as blood and bone have the drawback of being smelly and attractive to dogs and vermin. A suitable potting mixture, including fertilisers, is presented on page 99.

### Pests

Indoor plants may suffer from a range of pests which find the conditions indoors to their liking. The three most serious pests of container-grown cycads are mealy bugs, spider mites and scale insects (see also Chapter 8). Spider mites revel in dry conditions and their numbers can build-up very rapidly during warm weather. Often their build-up can be reduced by frequent misting or hosing of the foliage. Mealy bugs and scale insects often become significant pests on plants weakened from growing in unsuitable conditions or debilitated from neglect. The maxim to remember is that healthy cycads resist pests far better than weakened ones.

## CYCADS AS OUTDOOR CONTAINER PLANTS

Cycads make excellent container plants for the decoration of outdoor areas such as verandahs, patios, terraces and entertainment sites around barbecues and pools. Providing they can tolerate the climatic regime of the area, a wide range of cycads can be grown in this manner. In fact all cycads will make suitable container subjects for a while but vigorous species will soon outgrow containers and will then need to be planted out. Those species which have prickly leaflets or an abundance of sharp spines should be placed with care.

Cycads suitable as outdoor container plants are summarised in the table on p. 100.

## Outdoor Conditions

### Sun

As a general rule most cycads are very tolerant of exposure to sun, especially when their root systems are kept regularly moist. Exceptions are to be found in those species which originate in rainforests, particularly species of *Chigua* and some *Zamia*. Sun-hardy species of cycad will tolerate hot sun from an early age whereas sensitive species such as *Zamia manicata* and *Z. wallisii* may need continual protection even when mature. It is useful when purchasing a new cycad from a nursery to take note of its surroundings, in particular how much direct sun it is receiving. If it has been grown in a glasshouse or shadehouse it will need to be hardened off before being place in full sun. If the prevailing humidity is also regularly high then the sudden removal to a dry atmosphere may cause some desiccation. Hardening consists of increasing the exposure to sun and/or lower humidity that the plant receives each day while keeping it regularly moist. A sudden prolonged exposure, especially to hot summer sun, will lead to severe burning by ultraviolet rays.

## Wind

Attention should be paid to the placement of container-grown cycads in respect to exposure to wind. Sun-hardy species of cycad are generally tolerant of wind but those species which originate in sheltered forests may suffer from wind damage. Draughtly positions should be avoided. Cold winds may cause deformed growth or stunting whereas hot winds result in severe drying and desiccation of the leaflets. Hardy cycads will generally adapt well to a windy position, often producing short sturdy leaves as a result.

## Frost

Frost is a major limiting factor against the successful cultivation of many cycads. It is dealt with in more detail in Chapter 7. Cycads in containers have the major advantage that they can be moved to more congenial surroundings if severe frosts are imminent.

# Care of Outdoor Container Cycads

## Watering

If container-grown cycads are to be maintained in a healthy condition and with good appearance they must be watered on a regular basis and at no time should the root system be allowed to dry out completely. In the hot conditions experienced over summer, watering on a daily basis will be necessary whereas in winter once or twice a week may be sufficient. Windy weather and periods of low humidity result in drying conditions and extra watering may be needed. Heavy rain will water plants but light rain will not bring about the same result.

In addition to normal waterings it is a wise policy to thoroughly soak the potting mix every two or three weeks. This prevents dry patches developing in the potting mix and also leaches out excessive salts which have resulted from the breakdown of fertilisers. Hosing down the foliage on a regular basis is also useful because it reduces the build-up of dust and discourages the activities of pests such as spider mites.

## Potting Mixes

A suitable potting mix is of major importance to the successful growth of a container cycad whether it be held indoors or outside. A potting mix must supply anchorage for the roots and encourage their growth by ensuring adequate aeration, moisture and nutrient supply. Two suitable potting mixes are described.

**Mix 1**
3 shovels coarse sand
4 shovels milled pine bark or peatmoss
3 shovels friable loam
45 g Osmocote (3–4 months formulation)
130 g Osmocote (8–9 months formulation)
40 g dolomite
3 g iron sulphate

**Mix 2**
5 shovels coarse sand
3 shovels peat moss
10 shovels milled pine bark
70 g Osmocote (3–4 months formulation)
240 g Osmocote (8–9 months formulation)
60 g dolomite
5 g magnesium sulphate
45 g iron sulphate
40 g trace element mix

## Fertilisers

Some of the fertilisers incorporated into the potting mixes detailed above are slow release and will maintain growth for up to nine months. After this period treatment with additional slow-release fertilisers will be necessary or liquid fertilisers can be used.

## Repotting

Cycads grown in containers eventually develop a congested root system and the potting mix becomes exhausted of nutrients. At this stage or preferably before this situation is reached, repotting becomes necessary to maintain the health, growth and appearance of the plant.

Repotting is generally necessary every one to two years, although vigorous plants may need moving on at shorter intervals. Strong-growing plants are generally potted into the next size container and weaker plants can be put back into the same container after removal of the old potting mixture and perhaps also some of the root system. Care should be taken to avoid excessive damage to the coralloid roots which are generally brittle and readily damaged. Repotting is best carried out in spring or early summer so as to allow the maximum growing season for the plants to stabilise. The plants should be thorougly watered after repotting.

## Pests

Pests and diseases are dealt with in detail in Chapter 8.

## General Hints

Prickly cycads, particularly those species of *Encephalartos* in which the basal leaflets are reduced to spines (e.g. *E. villosus*), can inflict nasty wounds and should be placed where there will be minimum human contact. Such plants can also be used to advantage, however, if placed where they will be a barrier to unwanted guests or animals such as dogs.

Some species of cycad, especially the vigorous growers, develop a large root system that quickly fills a container. If these cycads are not repotted regularly their roots are quite capable of bursting the container, whether it be of plastic, terracotta or concrete.

Containers used for growing cycads should not be placed directly on the ground, but should be supported above the surface on concrete or bricks. This discourages entry via the drainage holes into the potting mix of pests, including grubs and worms. It also prevents the cycad roots from growing through the drainage holes and becoming entrenched in the soil.

# Cycads Suitable for Outside Containers

| *Species* | Position | Climatic Conditions | Notes |
|---|---|---|---|
| *Bowenia serrulata* | Semi-shade | T–WT | Highly ornamental |
| *Bowenia spectabilis* | Semi-shade | T–WT | Highly ornamental |
| *Ceratozamia mexicana* | Sun | T–ST | Vigorous |
| *Ceratozamia robusta* | Sun | T–ST | Very vigorous |
| *Cycas beddomei* | Sun | T–ST | Neat habit |
| *Cycas calcicola* | Sun | T | Deciduous |
| *Cycas chamberlainii* | Sun–Semi-shade | T–ST | Ornamental |
| *Cycas circinalis* | Sun | T–ST | Excellent, large container |
| *Cycas media* | Sun | T–ST | Excellent |
| *Cycas pectinata* | Sun–Semi-shade | T–ST | Attractive |

| | | | |
|---|---|---|---|
| *Cycas revoluta* | Sun | T–WT | Excellent, very hardy |
| *Cycas rumphii* | Sun | T–ST | Excellent, large container |
| *Cycas siamensis* | Sun–Semi-shade | T–ST | Attractive |
| *Cycas taiwaniana* | Sun–Semi-shade | T–WT | Hardy and adaptable |
| *Cycas thouarsii* | Sun–Semi-shade | T–ST | Excellent, large container |
| *Dioon edule* | Sun | T–ST | Highly ornamental |
| *Dioon mejiae* | Sun | T–ST | Highly ornamental |
| *Dioon spinulosum* | Sun | T–ST | Robust |
| *Encephalartos concinnus* | Sun | ST–TE | Excellent |
| *Encephalartos cupidus* | Sun | ST–TE | Excellent |
| *Encephalartos friderici-guilielmi* | Sun | ST–TE | Slow Growing |
| *Encephalartos horridus* | Sun | ST–TE | Very Spiny |
| *Encephalartos humilis* | Sun | ST–TE | Dwarf |
| *Encephalartos lehmanni* | Sun | ST–TE | Blue leaves |
| *Encephalartos ngoyanus* | Sun | T–STE | Attractive |
| *Encephalartos trispinosus* | Sun | ST–TE | Spiny |
| *Encephalartos umbeluziensis* | Sun | T–ST | Attractive |
| *Lepidozamia hopei* | Sun–Shade | T–STE | Highly ornamental |
| *Lepidozamia peroffskyana* | Sun–Shade | ST–TE | Highly ornamental |
| *Macrozamia communis* | Sun–Semi-shade | ST–TE | Large container |
| *Macrozamia fawcettii* | Semi-shade | ST–TE | Attractive leaves |
| *Macrozamia lucida* | Semi-shade | ST–TE | Excellent |
| *Macrozamia miquelii* | Sun–Semi-shade | T–ST | Large container |
| *Macrozamia moorei* | Sun | T–WT | Large container |
| *Macrozamia riedlei* | Sun | ST–TE | Large container |
| *Microcycas calocoma* | Sun | T–ST | ttractive |
| *Stangeria eriopus* | Sun–Semi-shade | ST–WT | Attractive |
| *Zamia acuminata* | Semi-shade | T–ST | Shiny leaves |
| *Zamia angustifolia* | Sun | ST–TE | Hardy |
| *Zamia chigua* | Semi-shade | T-St | Attractive |
| *Zamia fischeri* | Semi-shade | T–ST | Neat habit |
| *Zamia furfuracea* | Sun–Semi-shade | T–WT | Adaptable |
| *Zamia loddigesii* | Semi-shade | T–ST | Attractive |
| *Zamia muricata* | Shade | T–ST | Grows well |
| *Zamia pygmaea* | Sun | ST–WT | Dwarf |
| *Zamia splendens* | Semi-shade | T–ST | Shiny leaves |
| *Zamia vazquezii* | Semi-shade | T–ST | Neat habit |

Abbreviations: T = tropical; ST = subtropical; WT = warm temperature; TE = temperate.

Colony of *Cycas pruinosa* in habitat.

K. HILL

PART TWO

# The Living Cycads

# Key to Cycad Genera

1 Leaves bipinnate ........ ***Bowenia***
Leaves pinnate ........ **2**

2 Leaflets with a prominent midrib ........ **3**
Leaflets lacking a prominent midrib ........ **5**

3 Leaflet margins entire, lateral veins absent ........ ***Cycas***
Leaflet margins irregular or toothed, lateral veins present ........ **4**

4 Lateral veins arising at right angles to midrib ........ ***Stangeria***
Lateral veins arising at acute angle to midrib ........ ***Chigua***

5 Leaflets attached to the upper surface of the rhachis ........ ***Lepidozamia***
Leaflets attached to the margins of the rhachis ........ **6**

6 Swollen, often colourful small patch of callous tissue at base of upper leaflet margin ........ ***Macrozamia***
Colourful callous tissue absent from the base of the upper leaflet margin ........ **7**

7 Leaflets jointed to the rhachis, cataphylls absent on cone stalk ........ **9**
Leaflets not jointed to the rhachis, cataphylls present on cone stalk ........ **8**

8 Sporophyll ends faceted, seeds attached directly to sporophyll, leaflets weakly decurrent on rhachis ........ ***Encephalartos***
Sporophyll ends not facted, seeds attached to sporophylls via a basal stalk, leaflets strongly decurrent on rhachis ........ ***Dioon***

9 Leaflets reflexed, prickles absent on petiole and rhachis ........ ***Microcycas***
Leaflets spreading, prickles present or absent on petiole and rhachis ........ **10**

10 Sporophylls with two horns, leaflets entire ........ ***Ceratozamia***
Sporophylls lacking horns, leaflets with teeth or entire ........ ***Zamia***

GENUS

# *Bowenia*

The genus *Bowenia* consists of five species (two are fossils), all being restricted to Australia where the living species are found in tropical Queensland (for fossil species see page 26).

**Derivation:** The genus *Bowenia* honours the first governor of Queensland, Sir George Ferguson Bowen, later governor of Victoria. The genus, together with the type species *B. spectabilis*, was described by the English botanist Joseph Dalton Hooker in 1863 after the names had been first proposed (but not published) by his father William Jackson Hooker.

**Generic Description:** Terrestrial cycads with a fleshy, branching, subterranean, naked, tuberous stem, a tuberous taproot and separate leaf- and cone-bearing branches. New leaves produced singly, coiled like watchsprings. Young parts covered with hairs. Cataphylls irregular and when present short, flat, scattered among the leaf bases. Leaves erect, arching, bipinnate, one to seven on each rootstock branch, with long, slender petioles, swollen and hairy at the base; pinnae slender, spreading widely; pinnules (ultimate leaflets or leaf segments) opposite to alternate on each secondary rhachis which itself ends in a terminal pinnule, thin textured, decurrent and stalked at the base, not articulate. Cones terminal on short branches of the rootstock, shortly stalked. Sporophylls peltate, the outer face shortly hairy and lacking any spine or ornamentation. Seeds radiospermic, the sarcotesta cream to pinkish or purplish.

**Notable Generic Characters:** Species of *Bowenia* are the only cycads to have bipinnate leaves and this obvious feature provides a ready means of recognising members of this genus. Other useful characters include:

- a subterranean, much-branched, fleshy, naked rootstock;
- irregular cataphyll production;
- leaf bases not retained at senescence;
- young parts bearing short, curved, coloured hairs;
- vascularised stipules present on leaf base;
- young leaves coiled like watchsprings;
- long, slender petioles;
- non-articulate pinnules with a midrib (the rhachilla);
- cones terminal, borne singly on short branches of the subterranean, tuberous stem;
- stomata occur on both surfaces of the pinnules, those on the upper surface are scattered, those on the lower are in bands between the veins;
- sporophylls peltate, hairy, lacking spines, appendages or ornamentation; and
- seeds attached below the sporophyll stalk.

**Recent Studies:** The most recent studies on the genus *Bowenia* were by L. A. S. Johnson in 'The Families of Cycads and the Zamiaceae of Australia', *Proceedings of the Linnaean Society of New South Wales* 84 (1959): 109–13.

**Habitat:** Species of *Bowenia* grow in sheltered forested situations such as in rainforests, on the fringes of rainforests or on protected slopes in open forest and close to streams. Whereas these cycads are often prominent in lowland situations they may also extend to moderate altitudes on the ranges and tablelands. The soils where they occur are generally well-structured loams. Fires are generally uncommon in areas where these cycads grow.

**Miscellaneous:** Leaves of *Bowenia* have been used for decorative purposes such as backing material for cut flowers and wreaths. For many years fronds were collected from the wild and sold commercially. They are long lasting (even out of water) and popular for their fresh green appearance.

All species of *Bowenia* contain the toxic compound macrozamin and are potentially toxic to stock (and humans) if the leaves or seeds are ingested. It is interesting to note that the sarcotesta of *Bowenia* seeds gives off an unpleasant fragrance.

Two fossil *Bowenia* species have been described. The morphology of the pinnules of these fossils is very similar to that of living species and suggests that leaflets in this genus have changed little over 45 million years.

**Cultivation:** Species of *Bowenia* are popular subjects for cultivation and all species grow readily if given their basic requirements of warmth, humidity, air movement and moisture. Humus-rich, acid loams are ideal and mulches are extremely beneficial. Slow-release fertilisers or regular, light fertiliser applications produce strong growth. All species of *Bowenia* are decorative subjects for containers and can be used for beautification in glasshouses, conservatories and indoors. Generally these plants are not tolerant of frosts but provenances originating from highland situations are more cold-hardy than those from the lowlands.

**Propagation:** Species of *Bowenia* are usually propagated from seed which for best results should be sown fresh. The seeds have an after-ripening period of about six to twelve months and consequently germination takes about twelve to eighteen months (see also page 84). The tuberous stems of these cycads can be divided successfully providing considerable care is taken, all cut surfaces are treated with lime or sulphur and the divisions are not too small.

## DISTRIBUTION OF *BOWENIA*

## KEY TO SPECIES

**1** Leaflets with entire margins .................. ***B. spectabilis***
Leaflets with toothed margins ...................................... **2**

**2** Rootstock much-branched, leaves 5–30 per plant.......... ................................................................ ***B. serrulata***
Rootstock sparsely branched, leaves 1–6 per plant ........ ................................................ ***B.* species 'Tinaroo'**

# *Bowenia serrulata*

(W. Bull) Chamberlain
(with finely toothed leaflet margins)

**Description:** A small cycad with a subterranean, nearly round stem, to 30 cm across, much branched into five to twenty crowns, the taproot tuberous. **Young leaves** bright green, sparsely hairy. **Mature leaves** 1–2 m x 0.5–1 m, five to thirty-five on each plant, erect with spreading branches; **petiole** 0.5–1 m long, 0.2–0.8 cm across, slender, wiry, hairy at the base 'imary rhachises with one or two channels on the upper surface; pinnules five to thirty

Leaves and leaflets of *Bowenia spectabilis*.

Leaf and leaflets of *Bowenia serrulata*.

on each pinna, 8–15 cm x 1.5–4 cm, lanceolate, falcate, thin textured, dark green and shiny, the margins with numerous pungent teeth to 0.3 cm long (sometimes the margins lacerated), numerous veins fairly prominent, apex acute to acuminate. Cones dissimilar. **Male cones** 4–6 cm x 2–3 cm, ovoid, brownish, shortly hairy; **sporophylls** about 2 cm x 1.2 cm, wedge-shaped, outer face hairy; **peduncle** short. **Female cones** 8–12 cm x 7–10 cm, barrel-shaped to ellipsoid, brownish green; **sporophylls** 3–5 cm x 2 cm, outer face shortly hairy; **peduncle** short. **Seeds** 2.5–3.5 cm x 1.5–2.5 cm, oblong to ovoid, cream to pale pinkish or purplish.

**Distribution and Habitat:** Endemic to Australia where it is restricted to the Byfield area in central-eastern Queensland. This species grows in sheltered, moist areas in open forest, sometimes extending into moderately dry situations. The climate is tropical with summer day temperatures to 30°C and night temperatures to 23°C. Winter day temperatures average 21°C and night temperatures 11°C. Frosts are unknown. Rainfall is about 1500 mm per annum with most rain falling in summer.

**Notes:** *B. serrulata* was first described as a variety of *B. spectabilis* in 1878 and then raised to specific rank in 1912. It is similar in many respects to *B. spectabilis* but is more vigorous, forming larger clumps with a much-branched rootstock and more numerous leaves, the pinnules of which have prominent, rather pungent marginal teeth. This cycad is locally common in the vicinity of Byfield, north of Rockhampton, where it forms fairly dense and extensive stands. It is known locally by the inappropriate common name of 'Byfield fern'.

A population of a *Bowenia* from the Tinaroo Hills on the Atherton Tableland was formerly assigned to this species but recent studies have shown it to be distinctive (see *Bowenia* species 'Tinaroo')

**Cultivation:** Suited to tropical, subtropical and warm temperate regions. This species is popular for its attractive leaves and is grown as a garden plant and in containers in glasshouses and conservatories. Best in filtered sun in warm, humid conditions. Mulches, watering during dry periods and light fertiliser applications are all beneficial practices. Tolerates light frosts only.

**Propagation:** From seed and by division of the tuberous stem.

*Bowenia spectabilis*
From *Botanical Magazine*, vol. 98, tab.6008 (1872)

# *Bowenia spectabilis*

Hook. ex J. D. Hook.
(spectacular, showy)

**Description:** A small cycad with a subterranean, carrot-shaped stem, to 12 cm across, sparsely branched into two to five crowns, the taproot tuberous. **Young leaves** bright green, sparsely hairy. **Mature leaves** 1–2 m x 0.4–1 m, one to seven on each plant, erect with spreading branches; **petiole** 0.5–1 m long, 0.2–0.8 cm across, slender, wiry, hairy at the base primary and secondary rhachises with one or two channels on the upper surface; pinnules seven to thirty on each pinna, 7–15 cm x 1.5–4 cm, lanceolate, falcate, thin-textured, dark green and shiny, the margins entire or with a few coarse teeth, numerous veins fairly prominent, apex acute to acuminate. Cones dissimilar. **Male cones** 3–6 cm x 2–3 cm, ovoid, brownish; **sporophylls** about 1.8 cm x 1.2 cm, wedge-shaped, outer face hexagonal, shortly hairy; **peduncle** short. **Female cones** 10–12 cm x 7–10 cm, barrel-shaped to ellipsoid, brownish green; **sporophylls** 3–5 cm x 2cm, outer octagonal face shortly hairy; **peduncle** short. **Seeds** 2.5–3.5 cm x 1.5–2 cm, oblong, cream to pale pinkish or purplish.

**Distribution and Habitat:** Endemic to Australia, where it occurs in north-eastern Queensland from the McIlwraith Range on Cape York Peninsula south to near Tully. This species grows in or close to rainforest, close to streams and on sheltered slopes in wet sclerophyll forest, mostly in lowland areas but extending to 700 m altitude on the Atherton Tableland. The climate is tropical with wet summer days to 35°C and nights to 22°C. Winter day temperatures vary from 29° to 24°C with night temperatures down to 14°C. Frosts are unknown. Rainfall ranges from 1500 mm to 2500 mm per annum with most rain falling in summer.

**Notes:** *B. spectabilis* was described in 1863 from a living plant grown at Kew Gardens and originally collected from near Rockingham Bay by Walter Hill. Plants of this species are locally common in some areas and may be a distinctive feature along the margins of tracks and roads.

**Cultivation:** Suited to tropical and subtropical regions but also widely grown as a glasshouse plant and is also successful for indoor decoration. Best appearance is achieved in filtered sun although plants will tolerate considerable exposure to sun in humid climates. Requires good drainage and regular moisture. Responds well to mulches and regular light fertiliser applications. Tolerant of light frosts only. Large specimens transplant readily.

**Propagation:** From seed and by division of the tuberous stem.

# *Bowenia* sp. 'Tinaroo'

**Description:** A small cycad with a subterranean, carrot-shaped stem, to 10 cm across, sparsely branched into two to six crowns, the taproot tuberous. **Young leaves** bright green, sparsely hairy. **Mature leaves** 0.5–1.5 m x 0.3–0.8 m, one to five on each plant, erect with spreading branches; **petiole** 0.3–0.6 m long, 0.2–0.6 cm across, slender, wiry, hairy at the base, **primary and secondary rhachises** channelled on the upper surface; **pinnules** five to twenty on each pinna, 5–15 cm x 2–4 cm, lanceolate, falcate, thin textured, dark green and shiny, the margins with several teeth to 0.15 cm long, apex acuminate. **Cones** dissimilar. **Male cones** 4–6 cm x 2–3 cm, ovoid, brownish; **sporophylls** about 1.8 cm x 1.2 cm, wedge-shaped, outer face hexagonal, shortly hairy; **peduncle** short. **Female cones** 8–10 cm x 6–8 cm, barrel-shaped,

brownish green; **sporophylls** 3–5 cm x 2 cm, outer face hairy; **peduncle** short. **Seeds** 2.5–3.5 cm x 1.5–2 cm, oblong, cream to pale pinkish.
**Distribution and Habitat:** Endemic to Queensland where it is restricted to the Tinaroo Hills on the Atherton Tableland in north-eastern Queensland at about 700 m altitude. This species grows in rainforest and moist, sheltered slopes in open forest. The climate is tropical with wet summer days to 30°C and nights to 20°C. The drier winter days rise to 23°C and the nights drop to 7°C. Frosts are very rare.
Rainfall is approximately 1000 mm per annum falling mainly in summer, with long periods of light, misty rain.
**Notes:** In many respects this species is intermediate between *B. serrulata* and *B. spectabilis* but it has not originated as a hybrid. Whereas the plants have small, sparingly branched rootstocks and few leaves as in *B. spectabilis*, the leaflets are toothed as in *B. serrulata*, although these teeth are shorter. Gary Wilson has noted that a chrysomelid beetle (*Lilioceris nigripes*) causes severe damage to the developing leaves of this cycad, but apparently does not feed on the other species of *Bowenia*.
**Cultivation:** Suited to highland tropical regions as well as subtropical and temperate regions. Plants attain their best appearance in semi-shade or filtered sun. An excellent container plant which can be used for indoor decoration. Grows well in humus-rich loamy soils. Responds well to mulches, regular light fertiliser applications and watering during dry periods. Tolerates moderate frosts.
**Propagation:** From seed and by division of the tuberous stem.

GENUS

# *Ceratozamia*

The genus *Ceratozamia* consists of eleven living species which have a centre of development in the mountains of Mexico, most species being endemic there and with a couple of widely distributed species extending into Belize and Guatemala. Distribution of the species is in a relatively narrow band in a north–south arrangement which more or less parallels the mountain systems of the region. Species of *Ceratozamia* have suffered at the hands of poachers and all are included on Appendix 1 of CITES. For a fossil species of *Ceratozamia* see page 27.

**Derivation:** The genus *Ceratozamia* was described by Adolphe Theodore Brongniart in 1846. The generic name refers to the paired, horn-like projections which are found on the male and female sporophylls of all species (Greek *ceras*, 'horn'; *Zamia*, the name of another genus).

**Generic Description:** Terrestrial cycads with a relatively slender, ovoid or cylindrical, rarely branched trunk which may be partly subterranean and partly emergent or wholly emergent. Leaf bases falling free at senescence. New leaves emerging in flushes, green or copper coloured, glabrous or hairy. Mature leaves pinnate, oblong or lanceolate in outline, straight, mostly flat in cross-section. Petioles swollen at the base, hairy, bearing prickles although in some species these are extremely sparse or absent. Rhachis straight or twisted, bearing prickles or unarmed. Leaflets articulate at the base, opposite to nearly opposite, evenly spaced or in clusters, straight or falcate, entire, margins flat, involute or revolute; veins prominent or immersed and obscure. Male cones one or two, cylindrical, erect, hairy, pedunculate; sporophylls usually with two prominent spine-like horns, those on the female cones longer and stouter than the male. Female cones generally solitary, ovoid, shortly hairy; sporophylls with two prominent spine-like horns. Seeds radiospermic, ovoid to subglobose, the sarcotesta cream to whitish.

**Notable Generic Features:** Species of *Ceratozamia* share many characters with the genus *Zamia* but all species of *Ceratozamia* can be immediately distinguished by the paired, horn-like projections on the peltate sporophylls. Other useful generic features include:

- stipules present, at least in seedlings;
- pinnate leaves borne in whorls;
- leaf bases deciduous;
- leaflets articulate;
- trichomes with an oblong basal cell; and
- spinose prickles on the petiole and rhachis (in some species these prickles may be much reduced or are extremely sparse).

**Infrageneric Features:** Although at this stage there is no formal infrageneric classification, studies by researchers at the New York Botanical Garden, USA, University of Naples, Italy, and the University of Veracruz, Mexico, have shown the existence of two groups within the genus.

It should be noted that the leaflets of seedlings of all species in both groups are similar in shape and these resemble the mature leaflets of Group 1.

Group 1: This includes *C. euryphyllidia, C. hildae, C. latifolia, C. microstrobila* and *C. miqueliana*. Members of this group have small cones and relatively broad, thin-textured to almost papery leaflets which are developed in a strongly asymmetrical manner, taper gradually to the base and abruptly to the acuminate apex.

Group 2: This includes *C. kuesteriana, C. matudae, C. mexicana, C. norstogii, C. robusta* and *C. zaragozae*. Members of this group have small to large cones and the leaflets of mature plants are narrow, thin-textured to thick and leathery, are symmetrical or almost symmetrical, do not taper noticeably to the base and the apex may be drawn out or acute.

**Recent Studies:** The genus *Ceratozamia* has been well studied in recent times and no major changes in taxonomy are expected. For a recent detailed treatment see Stevenson, Sabato and Vazquez-Torres in *Brittonia* 38 (1986): 17–26.

**Habitat:** Species of *Ceratozamia* are primarily found in mountainous districts at approximate altitudes of between 800 m and 1800 m, but some widespread species extend to the lowlands. At high elevations these cycads grow in forests where clouds and mists are frequent. Vegetation types range from the almost constantly wet tropical rainforests to less wet broad-leaved forests and to drier types which contain a mixture of pines and oaks and have a pronounced seasonal wet-dry regime. Researchers have noted strong correlations between species in factors such as leaflet width and texture and the wetness of the habitat (broad, thin-textured leaves in wet habitats versus narrow or inrolled, leathery leaves in drier habitats).

**Cultivation:** Species of *Ceratozamia* are easily grown in containers and in the ground provided some basic parameters are met. As a group they grow in conditions of light shade to heavy shade or with filtered sunlight or dappled sunlight. Excessive hot sun, especially when combined with low humidity, is detrimental and can cause leaf damage. Seedlings and larger plants, when suddenly exposed to direct hot sun, are more sensitive to damage than others. Being forest plants these cycads revel in atmospheres with moderate to high humidities. In cultivation, a buoyant atmosphere with high humidity and unimpeded gentle air movement produces excellent growth. Those species from high altitudes will tolerate some exposure to frosts (especially if the plants are not in active growth) but some degree of frost protection for all species is advisable.

Soils should be well drained with humus-rich loams producing good growth. Most species seem to prefer slightly acid soils but a couple (e.g. *C. hildae* and *C. latifolia*) are

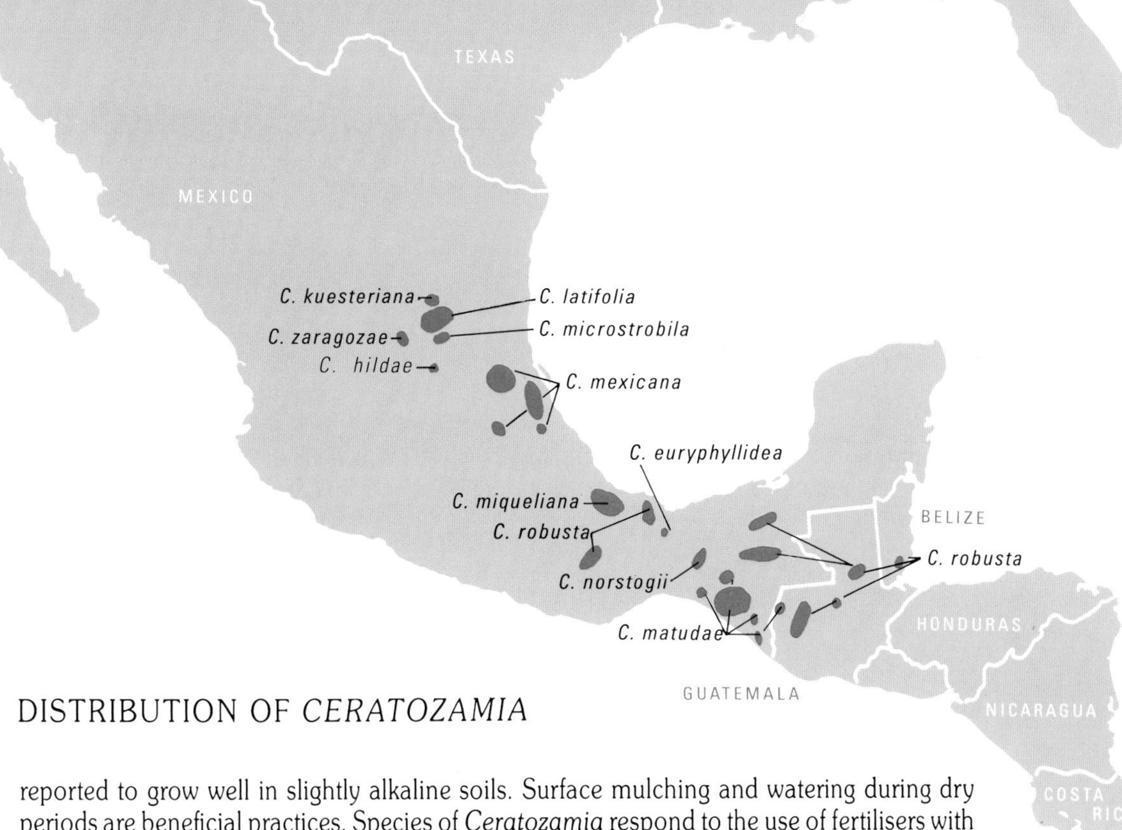


DISTRIBUTION OF *CERATOZAMIA*

reported to grow well in slightly alkaline soils. Surface mulching and watering during dry periods are beneficial practices. Species of *Ceratozamia* respond to the use of fertilisers with regular light dressings of slow-release types being preferred over spring and summer.

**Propagation:** Species of *Ceratozamia* are mainly propagated from seed although a few (such as *C. latifolia*) produce basal suckers and their numbers can be increased by division. Seeds have a very short period of viability and should be collected as they attain peak colouration and the cone begins to disintegrate. Seed-boring weevils may be very destructive in members of this genus. After collection the seeds should be sown within a short time or stored in moist peatmoss to prevent dehydration. Seeds of *Ceratozamia* have an after-ripening period of one to two months and seed germination for most species takes about six to twelve months.

## KEY TO SPECIES

Adapted from Stevenson, Sabato and Vazquez-Torres, *Brittonia* 38 (1986):17–26.

**1** Rhachis spirally twisted .......... **2**
Rhachis not spirally twisted .......... **3**

**2** Leaves to 2 m long, leaflets to 50 cm long .......... ***C. norstogii***
Leaves to 1 m long, leaflets to 25 cm long .......... ***C. zaragozae***

**3** Leaflets arranged in clusters .......... ***C. hildae***
Leaflets arranged evenly along the rhachis .......... **4**

**4** Leaflets widest above the middle .......... **5**
Leaflets widest at or below the middle .......... **8**

**5** Leaflets more than 5 cm wide .......... **7**
Leaflets less than 5 cm wide .......... **6**

**6** Leaflets to 4 cm wide .......... ***C. latifolia***
Leaflets to 3 cm wide .......... ***C. microstrobila***

**7** Leaflets 5–8 cm wide, papery, margins flat .......... ***C. miqueliana***
Leaflets 9–16 cm wide, translucent, margins wavy .......... ***C. euryphyllidea***

**8** Leaflets more than 1.5 cm wide, margins flat .......... **9**
Leaflets less than 1.5 cm wide, margins revolute .......... **10**

**9** Leaflets less than 2.5 cm wide, straight to slightly falcate .......... ***C. mexicana***
Leaflets more than 2.5 cm wide, strongly falcate .......... ***C. robusta***

**10** Petiole and rhachis smooth or sparsely prickly .......... ***C. kuesteriana***
Petiole and rhachis with numerous stout prickles .......... ***C. matudae***

P. VORSTER

Male cone of *Ceratozamia hildae*.

Leaves of *Ceratozamia hildae*.

## *Ceratozamia euryphyllidia*

Vazquez-Torres, Sabato & D. Stevenson
(with broad leaflets)

**Description:** A small cycad which in nature develops a slender trunk to about 20 cm tall and 10 cm across. **Young leaves** light green, translucent, sparsely hairy. **Mature leaves** less than ten in a crown, 2–3.2 m long, obliquely erect to spreading, dark green, glabrous; **petiole** 60–90 cm long, expanded at the base, armed with robust spines; **leaflets** twelve to twenty-six on each leaf, 18–30 cm x 9–16 cm, obovate to oblanceolate, thin-textured and almost membranous to translucent, dark green above, blue-green beneath, with numerous, prominent rigid veins, the distal margins wavy, apex asymmetrically acuminate. **Male cones** 20–28 cm x 2–3 cm, cylindrical, greenish red when young, grey when mature, usually solitary; **sporophylls** 7–10 mm x 6–9 mm, with two horns 1–2 mm long; **peduncle** 6–8 cm long, woolly. **Female cones** 15–20 cm x 4–5 cm, cylindrical, wine red when young, brown when mature, usually solitary; **sporophylls** 1.5–3 cm x 1.5–2 cm, with two stout horns separated by many fine ridges; **peduncle** to 12 cm long. **Seeds** 2.3–2.7 cm x 1.8–2 cm, ovoid, smooth, cream to white.
**Distribution and Habitat:** Endemic to Veracruz, Mexico, near the Oaxaca border at about 120 m altitude, growing in wet, tropical evergreen rainforest.
**Notes:** This species, described in 1986, is known from only about thirty plants, and precise locality details were not presented in the original publication in an attempt to protect the species in the wild from extinction by collectors. This distinctive species is readily recognised by its large leaves and broad, thin-textured to almost papery or translucent leaflets which taper to the base and have wavy margins.
**Cultivation:** Suited to tropical and subtropical regions. This species is rare in cultivation and experience is limited. With its large attractive leaves this is one of the most distinctive species of *Ceratozamia*. An excellent subject for containers or in the ground.
**Propagation:** From fresh seed.

## *Ceratozamia hildae*

Landry & M. Wilson
(after Hilda Guerra Walker)

**Description:** A small cycad which in nature develops a slender trunk to about 15 cm tall and 12 cm across. **Young leaves** hairy. **Mature leaves** five to twenty in a crown, 1–1.5 m long, dark green, smooth, glabrous; **petiole** 20–30 cm long, woolly at the base, armed with prickles; **leaflets** twenty to fifty on each leaf, 7–22 cm x 1–5 cm, lanceolate, spreading to recurved, clustered along the rhachis in groups of three, thin-textured, somewhat

papery, light green above, silvery beneath, flat, apex acute. **Male cones** 18–25 cm x 2–3 cm, cylindrical, yellowish brown; **sporophylls** with two horns to 3 mm long; **peduncle** to 3.5 cm long, slightly woolly. **Female cones** 10–14 cm x 3–5 cm, nearly cylindrical, olive green; **sporophylls** with two large horns, separated by a prominent oval ridge; **peduncle** to 9 cm long. **Seeds** 1.8–2 cm x 1.5 cm, ovoid, smooth.
**Distribution and Habitat:** Endemic to San Luis Potosi and Quiretaro in eastern Mexico, growing at 900–1200 m altitude. This species, which is known from only two localities, occurs in deciduous cloud forests which are dominated by species of *Quercus*.
**Notes:** Although first collected in the 1950s by a commercial collector, Luciano Guerra, and named after his daughter, this species was not described until 1979. Plants cultivated in Lousiana, USA, originally collected from San Luis Potosi, Mexico, were used to provide the type specimens. Unfortunately the species has become virtually extinct in the wild due to overcollecting. It belongs to the group of small species within the genus and can be distinguished by its thin-textured, somewhat papery leaflets which are arranged in clusters along the rhachis. It should be noted that in some plants of this species, occasional leaves have a normal, non-clustered arrangement of leaflets, and rarely some plants of *C. latifolia* may produce an occasional leaf which has clustered leaflets. Additionally some leaves may be devoid of prickles while others are densely prickly even on the same plant.
**Cultivation:** Suited to subtropical and warm temperate regions. This species is uncommon in cultivation but grows readily in sheltered conditions and germinates readily from seed. Grows well and quickly reaches adult size taking about five years under ideal conditions to produce cones. Some growers advocate slightly alkaline soils for its culture. Makes an attractive subject for containers.
**Propagation:** From fresh seed.

## *Ceratozamia kuesteriana*

Regel
(after Baron von Kuster)

**Description:** A small cycad which in nature develops a somewhat globose trunk to about 30 cm tall and 12 cm across. **Young leaves** bronze-coloured, lightly hairy. **Mature leaves** a few in a spreading crown, 1–1.8 m long, dark green, glabrous; **petiole** 30–60 cm long, very sparsely prickly and appearing smooth; **leaflets** seventy to one hundred and twenty on each leaf, 10–22 cm x 0.6–1.5 cm, linear-lanceolate, dark green above, paler beneath, sessile, the margins inrolled, veins prominent beneath, apex long acuminate. **Male cones** 20–30 cm x 5–7 cm, usually solitary, cylindrical, brown; **sporophylls** with two small horns; **peduncle** to 15 cm long, woolly. **Female cones** 15–20 cm x 8–10 cm, cylindrical, usually solitary, dark grey-brown; **sporophylls** with two stout horns. **Seed** details not recorded.

*Ceratozamia kuesteriana.*

**Distribution and Habitat:** Endemic to the Sierra Madre Oriental Range in southern Tamaulipas, Mexico, where it grows in cloud forests dominated by species of *Pinus* and *Quercus* at between 1000 m and 1800 m altitude.
**Notes:** This species was described in 1857 from plants growing in the Leningrad Botanic Garden. It was subsequently lost to science until its rediscovery was reported in 1982. Plants have been cultivated since at least the 1970s, but under the horticultural name of *C. angustifolia*.
**Cultivation:** Suited to subtropical and warm temperate regions. A very attractive species which merits wider planting. Ornamental features include attractive bronze new leaves and a pleasant crown of spreading leaves with narrow leaflets atop a small, slender trunk.
**Propagation:** From fresh seed.

# Ceratozamia latifolia

Miq.
(with broad leaves)

**Description:** A small cycad which in nature develops a slender, light brown trunk to about 20 cm tall and 10 cm across. **Young leaves** pale green to bronze, lightly hairy. **Mature leaves** two to four per plant, 0.5–1 m long, dark olive green, smooth, glabrous; **petiole** 15–20 cm long, swollen and woolly at the base, with very sparse prickles and appearing smooth; **leaflets** thirty to eighty on each leaf, 15–24 cm x 2–4 cm, broadly lanceolate, falcate, sessile,.not crowded, thin textured and papery, dark olive green and shiny above, brownish beneath, the margins slightly inrolled, apex acute. **Male cones** 15–17 cm x 2–2.3 cm, usually solitary, cylindrical, brown; **sporophylls** with two small horns; **peduncle** about 5 cm long, woolly. **Female cones** 5–6 cm x 4–4.5 cm, usually solitary, broadly cylindrical, greenish brown, woolly; **sporophylls** peltate, woolly on the edges, with two small horns. **Seeds** 1.8–1.9 cm x 1.4 cm, ovoid, smooth.
**Distribution and Habitat:** Endemic to San Luis Potosi and Hildago in eastern Mexico growing at about 850 m altitude. This species occurs in cloud forests which are dominated by species of *Quercus*.
**Notes:** This species was described in 1848 and since none of the original type material could be located a new type was designated in 1986 from material collected in San Luis Potosi. *C. latifolia* belongs to the group of small species within the genus. It is closely related to *C. hildae* but can be distinguished by its leaflets being evenly arranged along the fronds. It should be noted however that rarely some plants of *C. latifolia* may produce an occasional leaf which has clustered leaflets.
**Cultivation:** Suited to cool tropical and subtropical regions. An attractive species for a lightly shaded position in the ground or as a container subject. Plants can sucker freely under suitable conditions. Established plants are tolerant of some neglect.
**Propagation:** From fresh seed or by division of suckers. Plants produce cones from an early age and seeds germinate readily.

# Ceratozamia matudae

Lundell
(after Eizi Matuda, the original collector)

**Description:** A small cycad which in nature develops a slender trunk to 50 cm tall and 20 cm across. **Mature leaves** 0.7–1.2 m long, green to somewhat yellowish green, smooth, glabrous; **petiole** 20–35 cm long, swollen and densely woolly at the base, with few to numerous short, stout prickles; **leaflets** forty-six to eighty-eight on each leaf, 20–40 cm x 0.6–1.5 cm, linear-lanceolate, sessile, leathery, green to yellowish green and shiny above, paler beneath, apex long acuminate. **Male cones** 8–16 cm x 3–4.5 cm, cylindrical; **sporophylls** 1–2.1 cm x 8–11 mm, bearing two horns to 4 mm long; **peduncle** to 11 cm long, covered with small red scales and woolly at the base. **Female cones** 12–15 cm x 8–9 cm, ellipsoid; **sporophylls** 3–3.5 cm x 1.5–1.8 cm, with two prominent spreading horns 1.5–4 mm long; **peduncle** to 22 cm long, hairy. **Seeds** 2.5–3 cm x 2–2.3 cm, obovoid, smooth.
**Distribution and Habitat:** Endemic to south-western Chiapas and Oaxaca in Mexico and eastern Guatemala, growing at about 1000 m altitude. This species grows on sheltered slopes in cloud forests of broad-leaved trees.
**Notes:** *C. matudae* was described in 1939 from material collected on Mt Ovando in Chiapas, Mexico. The species is similar to *C. kuesteriana* but can be distinguished by the numerous, stout prickles on the petiole and rhachis. The spelling of the specific epithet used in the original description (*matudai*) is incorrect according to botanical rules.
**Cultivation:** Suited to subtropical and warm temperate regions. An attractive species for a sheltered position or as a container subject.
**Propagation:** From fresh seed.

# Ceratozamia mexicana

Brongn.
(from Mexico)

**Description:** A medium-sized cycad which in nature develops a trunk to 0.5 m tall and 20 cm across. **Young leaves** light green and hairy. **Mature leaves** up to twelve in a graceful crown, 1–1.5 m long, dark green, glabrous; **petiole** 20–50 cm long, swollen and woolly at the base, with numerous stout spines; **leaflets** fourteen to one hundred and fifty on each leaf, 20–30 cm x 1.5–4 cm, linear to lanceolate, crowded, thin textured, the margins rolled back, apex drawn out and attenuate. **Male cones** 20–30 cm x 10–12 cm, cylindrical, brown; **sporophylls** with two tiny horns; **peduncle** about 10 cm long. **Female cones** 20–35 cm x 10–12 cm, cylindrical, grey; **sporophylls** with two prominent stout horns; **peduncle** about 12 cm long, woolly. **Seeds** about 2 cm x 1.8 cm, smooth.
**Distribution and Habitat:** Endemic to north-central Veracruz in Mexico. This species grows in a range of forested habitats including wet tropical rainforests and drier types from lowland regions to moderate elevations in the mountains.
**Notes:** *C. mexicana* is the type of the genus and was described in 1846 from material cultivated in Paris. It grows in a range of habitats and is quite variable in some morphological features including: trunks ranging from buried to fully emergent and globose to cylindrical; leaflets ranging from broad to narrow and from lax to stiffly

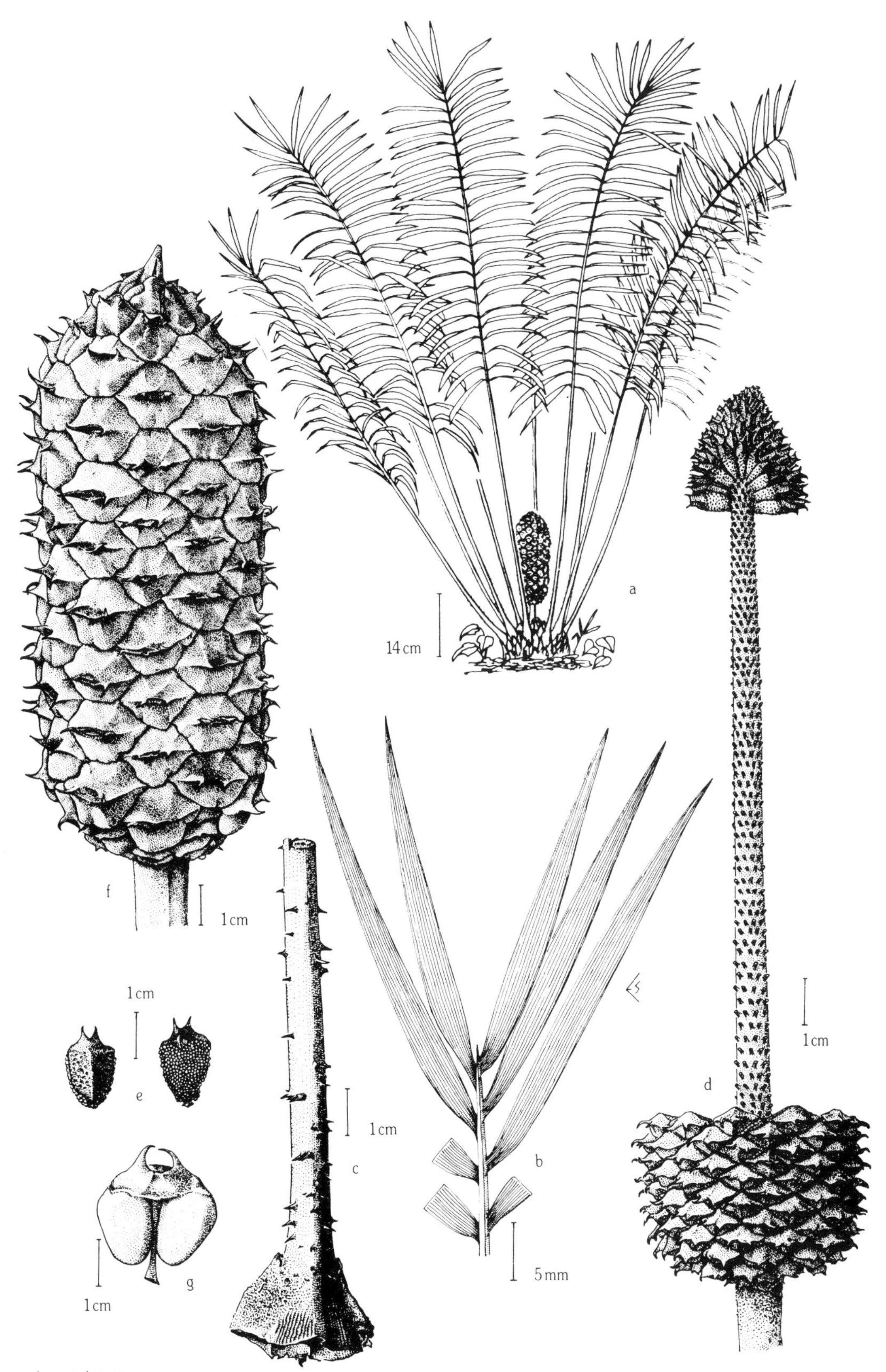


*Ceratozamia mexicana*
With permission from *Flora de Veracruz*, Fascicle 26, February 1983. Illustration by Edmundo Saavedra. **a.** adult female plant, **b.** detail of terminal leaflets, **c.** detail of base of petiole, **d.** male cone, **e.** male sporophylls, **f.** female cone, **g.** female sporophyll with ovules.

*Ceratozamia mexicana.*

erect; and prickles on the petioles and rhachises ranging from numerous to absent. This species is often linked with *C. robusta* but plants are less vigorous than that spe cies and have smaller fronds and cones and attenuate apices on the leaflets.
**Cultivation:** Suited to tropical, subtropical and warm temperate regions. Plants grow best in shade and should be given ample room to develop.
**Propagation:** From fresh seed.

Male cone of *Ceratozamia mexicana.*

## Ceratozamia microstrobila

Vovides & Rees
(with small cones)

**Description:** A small cycad with a subterranean ovoid to nearly cylindrical trunk to about 24 cm long and 10 cm across. **Young leaves** bright green. **Mature leaves** 2–4 on each plant, 50–70 cm long, erect to arching, dark green, shiny, glabrous; **petiole** 10–25 cm long, lacking prickles, with long hairs at the base; **leaflets** twenty to thirty on each leaf, 15–18 cm x 2.8–3.2 cm, lanceolate, leathery, dark green and shiny above, paler beneath, margins entire, partly recurved, apex acute. **Male cones** 15–17 cm x 2–2.3 cm, ovoid to barrel-shaped, brown, shortly hairy; **sporophylls** 0.5–0.7 cm x 0.4–0.5 cm, with two

Female cone of *Ceratozamia mexicana.*

horns 1–2 mm long; **peduncle** 4–5 cm long, hairy, lacking prickles. **Female cones** 5–6 cm x 4–4.4 cm, barrel-shaped, greenish brown, usually solitary; **sporophylls** 1.2–1.3 cm x 1.7–2.1 cm, peltate, shortly hairy on the edges, with two horns 1–2 mm long; **peduncle** 4–6 cm long, shortly hairy, lacking prickles. **Seeds** 1.8–1.9 cm x 1.2–1.4 cm, ovoid, smooth.

**Distribution and Habitat:** Endemic to San Luis Potosi, Mexico, where known only from the mountain Ejido las Abritas growing at about 850 m altitude. Plants are found in a transition zone between low deciduous forest and mixed oak woodland, and occur in humus-rich, shallow, reddish clay soil on limestone outcrops.

**Notes:** This species was described in 1983 from material collected in eastern Mexico. Some researchers believe that *C. microstrobila* falls within the range of variation of *C. latifolia* and is hence best regarded as a synonym of that species. Particularly interesting were the observations made on cultivated plants which suggested that those grown in a deep shade and high moisture regime develop like *C. latifolia* whereas those in bright light and drier conditions develop like *C. microstrobila*. It seems apparent that the relationships between these two species require further study.

**Cultivation:** Suited to cool tropical and subtropical regions. An attractive species which grows well in containers or in a sheltered position in the ground. Best in shade or filtered sun. Requires loamy soil and responds to mulches, light fertiliser applications and watering during dry periods.

**Propagation:** From fresh seed.

*Ceratozamia microstrobila*

*Ceratozamia miqueliana.*

D. W. STEVENSON

Leaf of *Ceratozamia norstogii.*

Male cone of *Ceratozamia norstogii.*

# *Ceratozamia miqueliana*

H. Wendl.
(after Friedrich Anton Wilhelm Miquel, nineteenth century Professor of Botany at Utrecht and Leiden)

**Description:** A small to medium-sized cycad with a cylindrical trunk to 1 m tall (swollen at the base) and 20 cm across. **Young leaves** hairy and covered with a powdery bloom. **Mature leaves** five to nine in an erect crown, 0.8–1.8 m long, dark green, flat in cross-section, straight in profile or recurved near the apex, of uniform width throughout except at the apex; **petiole** 30–50 cm long, swollen at the base, with numerous prickles; **leaflets** ten to eighteen on each leaf, widely spaced, inserted at about 90° to the rhachis; **median leaflets** 22–29 cm x 4–6.5 cm, asymmetrically obovate to broadly oblanceolate, thin textured and almost papery, dark green, apex acuminate and strong asymmetrical. **Male cones** 12–15 cm x 3–4 cm, cylindrical; **peduncle** hairy. **Female cones** 8–10 cm x 5–6 cm, ovoid-cylindrical; **peduncle** hairy. **Seed** details lacking.
**Distribution and Habitat:** Endemic to Veracruz, Mexico, where it grows between 60 m and 800 m altitude with a disjunct population on the coast of the Gulf of Mexico. This species grows in cool, humid, semi-deciduous rainforest.
**Notes:** Described in 1854 from Mexican plants of imprecise origin which were possibly cultivated in one or more botanic gardens in Europe. Since none of the original type material could be located a new type was designated in 1986 based on material collected at Veracruz, Mexico. In this species the leaflets are usually widely spaced; very rarely an individual leaf may have some clustered leaflets. *C. miqueliana* is closest to *C. euryphyllidea* but can be distinguished by its smaller, papery leaflets which lack the undulate margins.
**Cultivation:** Suited to cool tropical and subtropical regions. *C. miqueliana*, which can be grown in containers or the open ground, requires cool, protected conditions and well-drained soil. It responds favourably to mulching, light fertiliser applications and watering during dry periods.
**Propagation:** From fresh seed.

# *Ceratozamia norstogii*

D. Stevenson
(after Knut Norstog, contemporary American research worker specialising in cycads)

**Description:** A small cycad which in nature develops a slender trunk to about 50 cm long and 10 cm across. **Young leaves** densely covered with brown hairs. **Mature leaves** a few in an attractive crown, 1–2 m long, dark green, smooth, glabrous; **petiole** 20–30 cm long, swollen

at the base, armed with sharp prickles; **rhachis** spirally twisted; **leaflets** 100–160 on each leaf, 20–50 cm x 0.3–1 cm, linear, sessile, the margins inrolled, apex long tapered, acute to acuminate. **Male cones** 20–25 cm x 5–8 cm, usually solitary, cylindrical, tawny brown; **sporophylls** 1.2–1.5 cm x 0.5–0.8 cm, broadly wedge-shaped, with two horns 1–3 mm long; **peduncle** 2–5 cm long, woolly. **Female cones** 20–40 cm x 9–12 cm, usually solitary, cylindrical, olive green; **sporophylls** 2–4 cm x 1.5–3.5 cm, peltate with two stout horns. **Seeds** 2–3.5 cm x 1.2–1.5 cm, ovoid, smooth, white when ripe.
**Distribution and Habitat:** Endemic to Chiapas, Mexico, where it grows on mountain slopes in shady forest dominated by species of *Pinus* and *Quercus*. The prevailing climate has a distinct dry period and the leathery inrolled leaflets of *C. norstogii* are able to cope with the seasonal dry conditions.
**Notes:** This species, although only named in 1982, was first collected in 1925 and was recognised as being distinct by three researchers prior to its being described. The type locality is Sierra Madre del Sul in Chiapas. Plants can be confused with *C. zaragozae* which also has spirally twisted leaves but *C. norstogii* has longer fronds (to 2 m long), ovoid seeds and the apices of its leaflets contract more abruptly to a point. Although the spirally twisted rhachis is usually mentioned as distinctive in this species, it should be noted that some populations do not have this feature.
Natural populations of this species were decimated by overcollecting soon after its description. Following this poaching, many researchers adopted a policy of not including detailed localities when describing new species of cycads.
**Cultivation:** Suited to subtropical and warm temperate regions. *C. norstogii* is generally uncommon in cultivation and is a collector's item. Plants are most suited to the warm humid climates of the tropics and subtropics. They grow readily in conditions suitable for the genus. Young plants look attractive as container subjects.
**Propagation:** From fresh seed.

*Ceratozamia robusta.*

# *Ceratozamia robusta*

Miq.
(robust, vigorous)

**Description:** A medium to large cycad which in nature develops a trunk to 2 m tall and 30 cm across. **Young leaves** pale green. **Mature leaves** eight to thirty per plant, 1.5–3 m long, light green, smooth, glabrous; **petiole** 20–60 cm long, swollen and woolly at the base, with numerous, stout prickles; **leaflets** 100–200 on each leaf, 20–30 cm x 2–5 cm, linear-lanceolate to lanceolate, falcate, light green above, paler beneath, thin-textured to almost papery, apex acute, the margins rolled back. **Male cones** 30–50 cm x 10–14 cm, cylindrical, grey-brown; **sporophylls** with two tiny horns; **peduncle** about 10 cm long, woolly. **Female cones** 30–50 cm x 10–14 cm, cylindrical, grey; **sporophylls** with two prominent stout horns; peduncle about 15 cm long, woolly. **Seeds** about 2.5 cm x 2 cm, ovoid, smooth.
**Distribution and Habitat:** Endemic to Belize, Guatemala and Mexico (Oaxaca and Veracruz). This species grows in the understorey of wet, humid tropical rainforests.
**Notes:** This species was described in 1847 from plants cultivated in Europe and originating from somewhere in Mexico. Since none of the original type specimens could be located a new type was chosen in 1986 from material collected in Chiapas. *C. robusta* is the most vigorous and largest species in the genus. It has been included with *C. mexicana* by some authorities but is much more vigorous with a larger trunk, longer leaves and cones and acute tips on the leaflets. It has the widest distribution of any species of *Ceratozamia* and is also somewhat variable. Plants from the forests of Belize and Guatemala are the largest of all and often have crowns of relatively lax

*Ceratozamia robusta*

leaves. Those from Veracruz are smaller and less vigorous than other variants and those from Chiapas fall somewhere in between.
**Cultivation:** Suited to tropical and warm subtropical regions. Young plants are suited to containers but must be planted in the ground if they are to achieve their potential. Requires protection from excessive hot sun and an abundance of water and humidity.
**Propagation:** From fresh seed.

## *Ceratozamia zaragozae*

Medellin-Leal
(after General Ignacio Zaragoza)

**Description:** A small cycad which in nature develops an ovoid trunk to about 20 cm tall and 11 cm across. **Young leaves** with brown woolly hairs. **Mature leaves** two to five per plant, 0.2–0.95 m long, dark green and shiny, smooth, glabrous; **petiole** 15–25 cm long, swollen and densely woolly at the base, nearly smooth and with very few prickles; **rhachis** spirally twisted; **leaflets** sixteen to forty-three on each leaf, 5–25 cm x 0.3–1 cm, linear-lanceolate, straight or somewhat falcate, leathery, dark green and shiny above, paler beneath, the margins slightly inrolled, apex drawn out. **Male cones** 10–20 cm x 2–3 cm, subcylindrical to cylindrical, brown; **sporophylls** 8–11 mm x 3–6 mm with two horns 2–3 mm long; **peduncle** 9–14 cm long, woolly. **Female cones** 8–12 cm x 6–7 cm, ovoid to nearly cylindrical, usually solitary, brown, glabrous; **sporophylls** 2.2–3.7 cm x 2–2.5 cm, with two small horns separated by a roughened area. **Seeds** about 2 cm across, spherical, prominently ribbed.
**Distribution and Habitat:** Endemic to San Luis Potosi, eastern Mexico, growing at about 1800 m altitude. Of limited distribution this species grows on hillsides of rhyolite in shady forests dominated by species of *Pinus* and *Quercus*.
**Notes:** This species was described in 1963 from material collected at about 1800 m altitude in the Sierra de la Equiteria, San Luis Potosi. As in *C. norstogii*, the leaves of this species are spirally twisted but, *C. zaragozae* has shorter leaves (to 1 m long), almost spherical, ribbed seeds and drawn out, attenuate tips to the leaflets. It is endangered in its native state where it is threatened by illegal collecting.
**Cultivation:** Suited to subtropical and warm temperate regions. This species is uncommonly grown and is a collector's item. Attractive in the ground or as a container subject.
**Propagation:** From fresh seed.

# GENUS *Chigua*

The recently described genus *Chigua* consists of two species, both endemic to Colombia in South America. This is the only genus of cycads to be endemic to South America.

Although the first specimen of this genus was collected as early as 1918, it was thought to be aberrant and was tentatively placed in *Zamia*. Subsequent field trips re-collected this species and resulted in the discovery of another with similar characters.

## DISTRIBUTION OF *CHIGUA*

**Derivation:** The genus *Chigua* was described by Dennis Stevenson in 1990 and is derived from a Spanish transliteration of an Indian name which is commonly used for cycads in both Central America and South America.

**Generic Description:** Terrestrial cycads with a slender, subterranean unbranched trunk. Leaf bases falling free from the trunk after senescence. New leaves emerging singly, glabrous. Mature leaves pinnate, more or less oblong in outline, straight, flat in cross-section. Petiole swollen at the base, bearing spine-like prickles. Rhachis straight, bearing prickles. Leaflets articulate at the base, opposite to nearly opposite, flat, with prominent marginal teeth. Midrib and lateral veins prominent. Male cones one to a few, cylindrical, erect, shortly hairy, long pedunculate; sporophylls peltate, the outer face hexagonal. Female cones one to a few, broadly cylindrical to barrel-shaped, hairy, very long-pedunculate; sporophylls hexagonal, with a conspicuous bump at each angle. Seeds radiospermic, ovoid, the sarcotesta pink to red.

**Notable Generic Features:** Species of *Chigua* are very closely related to *Zamia* but can be readily distinguished by the leaflets which have a prominent midrib and forked lateral veins which extend at a steep angle from the midrib to the leaflet margins. Other useful generic features include:

- leaflets articulate with toothed margins;
- sporophylls hexagonal with raised areas on the sporophyll face of both male and female cones; and
- the longest known cycad peduncle.

**Recent Studies:** The only comprehensive paper on this genus was written by Dennis Stevenson (see '*Chigua*, a New Genus in the Zamiaceae, with comments on its Biogeographic Significance', *Memoirs of the New York Botanical Garden* 57 (1990): 169–172).

**Notes:** There is some contention about the placement of this genus with some researchers claiming it would be better placed within the genus *Zamia*.

**Habitat:** Species of *Chigua* are essentially found in the lowlands, growing in wet, humid conditions of the understorey of primary rainforest.

**Conservation:** Little is known about the conservation problems of species of *Chigua* however they are probably similar to those of *Zamia*. Because of the unique position of the genus, there is certain to be a strong demand by collectors which will place considerable pressure on natural populations. Because of these sorts of problems, precise localities of the types were omitted from the original publication.

**Cultivation:** Although very rare in cultivation the species of *Chigua* grow readily in loamy soil if given warm, humid conditions and bright filtered light. They are cold sensitive and are best suited to tropical and warm subtropical regions.

**Propagation:** From seed which is best sown fresh. The seeds have an after-ripening period of one to two months and take about six to twelve months to germinate.

## KEY TO SPECIES

Leaflets 15–25 cm x 3–5 cm, lanceolate .... ***C. restrepoi***
Leaflets 30–35 cm x 1–1.5 cm, linear to linear-lanceolate .................................................. ***C. bernalii***

# *Chigua bernalii*

D. Stevenson
(after Rodrigo Bernal, the discoverer of this species)

**Description:** A small cycad with an ellipsoid subterranean trunk to 30 cm long and 10 cm across. **Mature leaves** two or three per plant, 1.6 m long, erect; **petiole** 1–1.4 m long, armed with robust prickles; **leaflets** 60–110 per leaf, 3–3.5 cm x 1–1.5 cm, linear to linear-lanceolate, thin textured and papery, tapered to the base, the margins wavy and with prominent teeth. **Male** and **female cones** unknown.
**Distribution and Habitat:** Endemic to Colombia where found growing in lowland rainforest at about 150 m altitude.
**Notes:** This species, described in 1990, is apparently known only from the type collection which was made in 1986.
**Cultivation and Propagation:** As for the genus.

Leaf of *Chigua restrepoi*.

Male cone of *Chigua restrepoi*.

# *Chigua restrepoi*

D. Stevenson
(after Padre Sergio Restrepo, an amateur botanist who helped locate this species; he was subsequently assassinated)

**Description:** A small cycad with an ellipsoid subterranean trunk to 40 cm long and 15 cm across. **Mature leaves** two or three per plant, 1.2–1.8 m long, erect; **petiole** 0.6–0.8 m long, armed with robust prickles on the lower surface; **leaflets** 40–60 on each leaf, 3–5 cm x 1.5–2.5 cm, lanceolate, thin textured and papery, sessile, tapered to the base, the margins with prominent teeth. **Male cones** 4–5 cm x 2 cm, cylindrical, covered with reddish brown hairs; **sporophylls** about 5 cm x 2 cm; **peduncle** to 10 cm long, glabrous. **Female cones** 12–15 cm x 4–5 cm, cylindrical, covered with reddish brown hairs; **peduncle** to 30 cm long, glabrous. **Seeds** about 1.8 cm long, ovoid, red at maturity.
**Distribution and Habitat:** Endemic to Colombia where found growing in lowland rainforest at about 120 m altitude.
**Notes:** This species, the type of the genus, was first collected in 1918 but was not rediscovered until 1986 when it was collected twice and again in 1987, the latter being used as the type collection. Very few plants of this species have ever been seen in the wild.
**Cultivation and Propagation:** As for the genus.

D. W. STEVENSON

D. W. STEVENSON

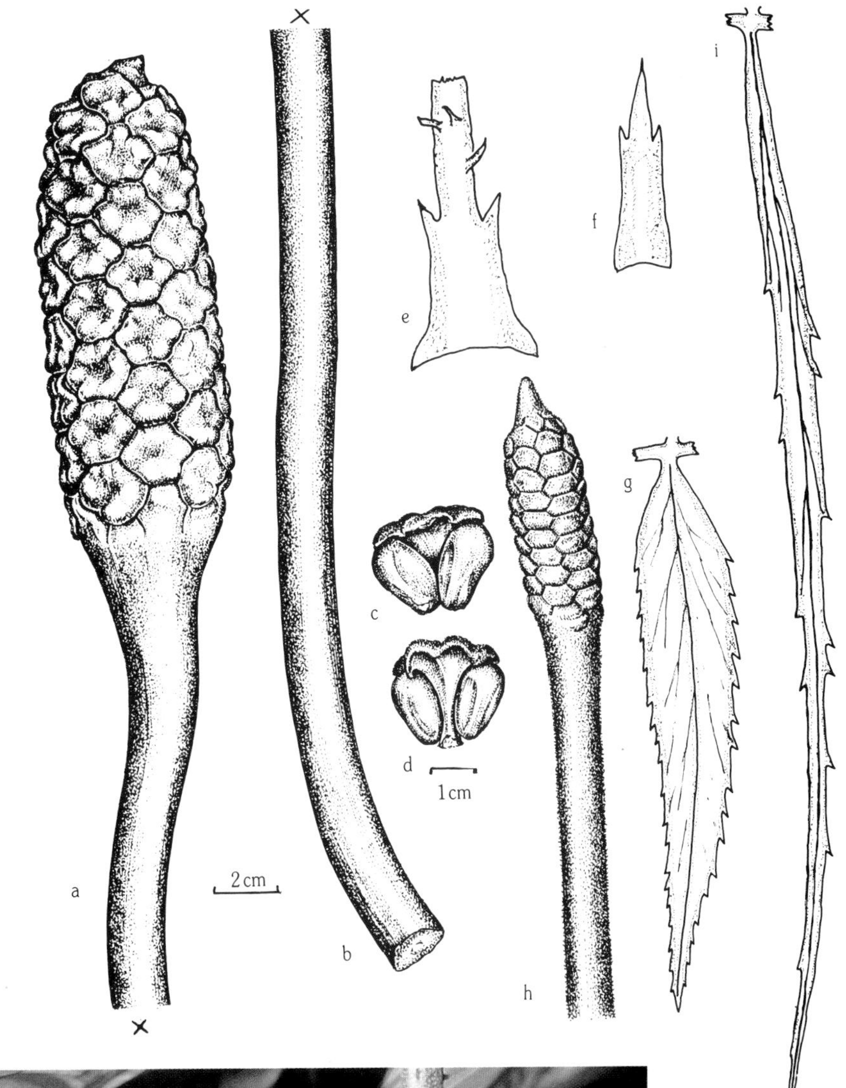


Reprinted by permission from Memoirs of the New York Botanical Garden 57:171, a,d,h, by Priscilla Fawett; e,g,i by Dennis Stevenson, copyright 1990 New York Botanical Garden. *Chigua restrepoi:* **a.** female cone, **b.** peduncle, **c.** female sporophyll with ovules, from beneath, **d.** female sporophyll with ovules, from above, **e.** leaf base with stipules, **f.** cataphyll, **g.** leaflet. *Chigua bernalii:* **h.** male cone, **i.** leaflet.

Female cone of *Chigua restrepoi.*

GENUS

# *Cycas*

The number of species in the genus *Cycas* is uncertain because of a high degree of confusion which exists arising from the absence of any comprehensive generic study. In fact, the situation has been compounded by the almost incomprehensible conclusions reached after a study by Julius Schuster in 1932 (if you are game see 'Cycadaceae' in A. Engler (ed.), *Das Pflanzenreich* 99(1932): 1–168). Numerous names have been applied within this genus but until a comprehensive modern study is undertaken, the status of many of these must be regarded as being questionable. Schuster opted for the use of numerous infraspecific taxa which frequently resulted in convoluted combinations and often drew together taxa which had very tenuous relationships with each other. Within the genus a number of obvious or accepted taxa can be identified and in this publication all of these have been treated at specific rank.

The genus *Cycas* probably consists of about forty species. Some of these are very widely distributed but the majority are relatively localised in their distribution. It should be noted that despite the present nomenclatural confusion of *Cycas*, there are still undescribed species known which require formal recognition. For notes on fossil species of *Cycas* see page 28.

**Derivation:** The genus *Cycas* was described by Carl von Linnaeus in 1753. The generic name is derived from the Greek word *koikas* apparently used by Theophrastus for a kind of palm. This was apparently transliterated at some stage to *kykas* and hence *Cycas*.

**Generic Description:** Terrestrial cycads with a trunk which is either ovoid and subterranean or slender to stout and emergent. Leaf bases retained on the trunk at senescence. New leaves erect with circinnate leaflets, emerging singly or in flushes, glabrous or with hairs which are shed with age. Mature leaves pinnate, oblong in outline, flat or vee-ed in cross-section, the older leaves spreading or deflexing after a flush of growth. Cataphylls prominent, arising in alternating flushes with the leaves, often rigid and pungent-tipped. Petioles swollen and hairy at the base, often with a lateral series of short, rigid, thorn-like processes borne more or less in opposite pairs, these being reduced lower leaflets. Rhachis lacking prickles, not twisted, straight or recurved in profile. Leaflets decurrent at the base, alternate to nearly opposite, with a prominent midrib, mostly evenly spaced except for the lower leaflets, straight or falcate, margins usually entire, sometimes serrulate in the distal third, lacking a callous base. Cones markedly dissimilar in shape and size. Male cones cylindrical; sporophylls arranged in a typical cone. Female cones loose and open; sporophylls arranged in a loose grouping surrounding the vegetative apex of the stem, with a linear stalk and an expanded apical lobe which may be entire, pinnatifid or deeply lobed on the margins. Ovules two to eight (rarely one) on each sporophyll, attached to the linear portion. Seeds platyspermic, ovoid to oblate, or rounded, the sarcotesta not usually highly coloured but bright red in *C. taiwaniana*.

**Notable Generic Features:** Species of *Cycas* can be readily recognised by their leaflets which have a prominent midrib and lack any obvious secondary veins. Other features useful in recognising members of this genus include:

- a subterranean or emergent trunk clothed with persistent leaf bases;
- young parts hairy;
- new leaves not coiled but the leaflets coiled like watchsprings;



*Cycas conferta* in habitat.
K. HILL

- leaflets not articulate at the base;
- lower leaflets abruptly reduced to a lateral series of paired, short, rigid thorn-like processes;
- male sporophylls arranged in a cone;
- female sporophylls arranged in a loose crown surrounding the vegetative apex of the stem;
- female sporophylls consisting of a linear stalk and an expanded apical lobe; and
- female sporophylls usually bearing more than two ovules.

**Infrageneric Classification:** Because of the confused taxonomic situation in the genus *Cycas*, the attempts at infrageneric classification have been of limited use. For example J. Schuster in 'Cycadaceae', *Das Pflanzenreich* 99(1932): 64–84, proposed three sections and two subsections but recent studies have shown that one of these sections and one subsection are illegitimate under the rules of botanical nomenclature. Schuster's two legitimate sections are:

- Section *Asiorientales*: Based on *C. revoluta*.
- Section *Indosinenses*: Based on *C. siamensis*.

After a study of Thailand cycads published in the *Natural History Bulletin of the Siam Society* 24(1971): 163–75, T. Smitinand proposed two sections within the genus *Cycas*.

- Section *Cycas* (most species): Members of this section have emergent trunks, simple, undivided leaflets and oblong to oblong-ovoid male cones with acuminate sporophylls.
- Section *Stangerioides* (*C. micholitzii*): Members of this section have subterranean or shortly emergent trunks, simple or dichotomously forked leaflets, and slightly oblong male cones with leathery, apiculate sporophylls.

The above classifications are useful but it is obvious that further research is still needed into relationships between species within this genus.

**Recent Studies:** There have been no recent studies of major consequence carried out on this genus, although a few new species have been described over the last decade. A revision of the Australian species of *Cycas* is presently being undertaken by Ken Hill of the Royal Botanic Gardens, Sydney, New South Wales. The first paper resulting from this study was published in *Telopea* in 1992.

**Habitat:** Species of *Cycas* grow in a wide range of habitats from coastal and near-coastal lowlands to hills and ranges which may be considerable distances from the coast. Many species grow in sparse forests and woodlands, a few in grassland and a substantial number on rocky slopes and escarpments where the vegetation is sparse. Fires are common in some areas where *Cycas* grow. A few species from arid areas may become completely deciduous each year for a short period prior to the production of new leaves.

**Cultivation:** As a general rule, species of *Cycas* are popular subjects for cultivation, probably because of their palm-like appearance and predictable dimensions. When young they are excellent in containers and the more attractive species can be used for indoor decoration and in glasshouses and conservatories. Most species are of tropical origin and very few are sufficiently cold-hardy to survive in temperate regions. As a general rule, mature plants transplant readily although plants of some species may take many months to produce new leaves. Seedlings of most species are vigorous and established readily, but those of Australian species which originate in a seasonally dry tropical climate may be slow and difficult, especially in cooler climates.

Plants of the majority of species achieve best appearance in conditions of bright light or full sun. This is particularly true for the blue-leaved species which only maintain good colouration in hot, dry climates. When grown in shady conditions or in cool or excessively humid climates, plants of these species quickly lose their attractive blue appearance. Air movement should be unimpeded as some species, especially those with a glaucous bloom, can suffer from the attacks of leaf fungi which thrive in stagnant conditions. Root-rotting fungi can cause the death of small or large *Cycas* plants unless soil drainage is excellent. Humus-rich, acid soils are satisfactory for most species, with relatively few showing a preference for neutral to slightly alkaline soils. Practices such as regular mulching, light fertiliser applications and watering during dry periods all produce healthy growth.

**Propagation:** Most species of *Cycas* are propagated solely from seed. This has a limited period of viability and for best results should be sown fresh. The seeds have an after-ripening period of six to twelve months and germination usually takes twelve to eighteen months (see also page 84). It should be noted by growers that three species of *Cycas* (*C. circinalis*, *C. rumphii* and *C. thouarsii*) have seeds which float when viable. A few species, such as *C. revoluta*, can be propagated by some vegetative techniques (see Chapter 9).

## KEY TO SPECIES

Because of the confused state of the taxonomy of this genus no attempt has been made to compile a key.

## DISTRIBUTION OF *CYCAS*

# *Cycas angulata*

R. Br.
(angular)

**Description:** A large cycad with a massive blackish trunk to 12 m tall and 40 cm across, swollen at the base, with occasional offsets produced on the trunk and sparse suckers arising from the base, growing in clumps of up to six stems. **Young leaves** blue, bearing soft, dark brown hairs. **Mature leaves** numerous in an arching to rounded crown, 1–1.7 m x 20–30 cm, blue-green to grey green, glossy, shallowly vee-ed in cross-section, arching or recurved in profile, often twisted near the apex; **petiole** 30–50 cm long, strongly swollen at the base, glabrous or with grey hairs; **leaflets** 180–320 on each leaf, crowded, evenly distributed throughout, inserted on the rhachis at 40–60° (the lower ones at about 90°); **median leaflets** 14–23 cm x 0.4–0.5 cm, linear, straight or slightly twisted, hairy at the base, glaucous and waxy beneath, margins slightly recurved and thickened, midrib yellow, apex abruptly acuminate with a green mucro about 1 mm long; **lower leaflets** reduced, abruptly replaced by a series of short, yellow thorns. **Male cones** 15–25 cm x 12–15 cm, narrowly-ovoid, yellowish brown, hairy; **sporophylls** 3–4 cm x 1.5–2 cm, ovate, with a short, erect point. **Female cones** loose and open, covered with waxy bloom; **sporophylls** 30–50 cm long, covered with a waxy bloom and grey to orange hairs; **apical lobe** 6–8 cm x 4.5–5.5 cm, ovate-lanceolate, margins with short yellow teeth, apex drawn out into a long-acuminate point; **ovules** four to twelve on each sporophyll. **Seeds** 4.5–6 cm x 4–5 cm, ovoid to globular, yellow to brown, not powdery.
**Distribution and Habitat:** Endemic to Australia where it is known from near Borroloola in the Northern Territory, growing in the lower reaches of the Wearyan, Foelsche and Robinson Rivers; also in Queensland on Mornington Island and the Wellesley group. In Queensland this species is not known from the mainland with unconfirmed reports of the species from the vicinity of Running Waters being referrable to *C. brunnea*. In the Northern Territory *C. angulata* forms colonies in sparse grassy woodland with the cycad dominating the vegetation and often being the tallest plant species in the community. The soils are sandy and the climate is tropical with hot, wet, humid summers and long, dry, mild winters. Summer temperatures have a day/night regime to 38°C and 25°C respectively. Winters have a day/night temperature regime to 30°C and 15°C respectively. Rainfall is 1200–1500 mm per annum with nearly all falling in summer.
**Notes:** Described in 1810 from material collected from Bountiful Island, near Mornington Island in the Gulf of Carpentaria, this is the largest species of *Cycas* in Australia. The seeds are an important item in the diet of local Aborigines. Similar to *C. brunnea* but with narrower,

K. HILL

K. HILL

*Cycas angulata* in habitat.

more crowded, greener leaflets having more strongly recurved margins and with larger, non-powdery seeds.
**Cultivation:** Suited to tropical regions with a seasonally dry climate. Best planted in full sun. Established specimens transplant readily. Seedlings are relatively fast growing and establish readily in tropical regions but elsewhere are slow growing and can be difficult to establish. Tolerance of frost is not high.
**Propagation:** From seed and by removal of suckers.

sh of new leaves
*Cycas angulata.*

Male cone of *Cycas angulata.*

K. HILL

## *Cycas armstrongii*

Miq.

(after John Armstrong, nineteenth century collector for Kew Gardens, resident in the Northern Territory)

**Description:** A medium-sized cycad with an erect, often branched slender trunk to 4 m tall and 15 cm across, with offsets produced frequently on the trunk and occasional basal suckers, growing in clumps of up to five stems. **Young leaves** bright green, covered with brown hairs. **Mature leaves** numerous in an obliquely erect to spreading crown, 0.6–1.2 m x 20–25 cm, bright grassy green, flat in cross-section, straight in profile; **petiole** 15–30 cm long, slender; **leaflets** 150–250 on each leaf, moderately spaced, evenly distributed throughout, middle ones inserted on the rhachis at about 70°, distal ones at about 30°, proximal ones at about 90°; **median leaflets** 15–20 cm x 0.6–0.7 cm, linear, straight, leathery, glabrous, midrib yellow, margins flat, apex long-acuminate, with a yellow mucro about 1 mm long; **lower leaflets** abruptly replaced by a series of short thorns. **Male cones** 12–20 cm x 10–12 cm, ovoid, rusty brown, hairy; **sporophylls** 3–4 cm x 1.8–2.3 cm, ovate, with a prominent erect apex about 1.5 cm long. **Female cones** loose and open, hairy; **sporophylls** 15–20 cm long, hairy; **apical lobe** 4–5 cm x 2–3 cm, triangular, margins with short, sharp teeth, apex acuminate; **ovules** four on each sporophyll. **Seeds** 2–4 cm x 2–4 cm, globose, orange to brown.

**Distribution and Habitat:** Endemic to Australia where occurring in northern parts of Western Australia and the Northern Territory. This species commonly grows in open grassy woodland forming sparse colonies. The climate is monsoonal with hot, wet, humid summers and long, dry, mild winters. Summer temperatures have a day/night regime to 38°C and 25°C respectively. Winters have a day/night temperature regime to 30°C and 15°C respectively. Rainfall is about 1200–1500 mm per annum with nearly all falling over summer.

**Notes:** This species was described in 1868 from material collected at Port Essington on the northern coast of the Northern Territory. It was for many years confused with *C. media* but can be distinguished by its tall, slender trunk, brighter green leaves and different male and female cones. Fires occur almost annually in the habitat where this species grows. Plants are normally deciduous in the early part of the dry season and produce new leaves after two to four months. The Aborigines use the seeds of this species as food, after suitable treatment. *C. lane-poolei* is similar and has been confused with *C. armstrongii* but is more robust with larger leaves, larger leaflets and larger seeds. The relationship between *C. papuana* from New Guinea and *C. armstrongii* is in need of investigation.

**Cultivation:** Suited to tropical regions which have a seasonally dry climate. Plants will grow in full sun or filtered sun. Mature specimens transplant readily. Seedlings are very slow growing and can be tricky to establish, especially in areas away from the tropics. Cultivated plants may shed their leaves annually and should then be kept on the dry side.

**Propagation:** From seed or by removal of basal suckers or trunk offsets.

Female sporophylls and developing seeds of *Cycas armstrongii*.

## *Cycas baguanheensis*

L. K. Fu & S.Z. Cheng

(from the area of Baguan Forest, Sichuan, China)

**Description:** A small to medium-sized cycad with an erect trunk to 2.5 m tall and 30 cm across. **Mature leaves** numerous in an obliquely erect to widely spreading crown, 0.6–1 m x 20–35 cm, dark bluish green, flat in cross-section, straight in profile; **petiole** 4–8 cm long; **leaflets** 100–200 on each leaf, moderately crowded, evenly spaced throughout, inserted on the rhachis at about 50°; **median leaflets** 11–17 cm x 0.4–0.5 cm, linear, straight, leathery, glabrous, margins flat, apex

*Cycas armstrongii.*

acuminate; **basal leaflets** abruptly replaced by a series of rigid thorns about 0.3 cm long. **Male cones** 20–40 cm x 8–10 cm, cylindrical to fusiform or ellipsoid, rusty brown; **sporophylls** 4–6 cm x 2–3 cm, nearly triangular, covered with brown hairs, apex pointed. **Female cones** loose and open, reddish brown, densely hairy; **sporophylls** 15–25 cm, long, densely hairy; **apical lobe** 3–5 cm x 1.5–2.5 cm, ovoid, margins deeply lobed; **ovules** two to six on each sporophyll. **Seeds** 2.5 cm x 2–2.5 cm, oblate, slightly ridged, golden yellow to reddish yellow.
**Distribution and Habitat:** Endemic to China where occurring in the Sichuan Province. This species grows on forested slopes and close to streams.
**Notes:** Described in 1981, this species is closely related to *C. panzhihuaensis* but has smaller leaflets that have flat, non-revolute margins and densely hairy sporophylls with the hairs being persistent. It is cultivated to a limited extent in China but is very rarely grown elsewhere.
**Cultivation:** Suited to temperate and cool subtropical regions. A cold-hardy species which tolerates heavy frosts. Will grow in partial shade to full sun.
**Propagation:** From seed.

## *Cycas balansae*

Warb.
(after B. Balansa, original collector)

**Description:** Very few details are recorded in the original description of this poorly known species. Details available are as follows: Trunk to 2 m tall; **leaflets** linear-lanceolate; male cones about 20 cm x 5–6 cm, shortly stalked; **sporophylls** 1.6–2.7 cm x 1.2 cm.
**Distribution and Habitat:** Known from Vietnam where it probably grows in sheltered forests. The climate is tropical with hot, humid summers and warm, moist to dry winters.

Colony of *Cycas basaltica* in habitat.

K. HILL

**Notes:** This species was described in 1900 from material collected near Hanoi. Its relationships with *C. inermis*, *C. undulata* and *C. siamensis* is in need of critical study.
**Cultivation:** Suited to tropical and warm subtropical regions. Probably requires a sheltered situation, humid conditions and regular watering. Tolerance of frost is unlikely to be high.
**Propagation:** From seed.

# *Cycas basaltica*

C. Gardner
(growing on soils of basalt origin)

**Description:** A medium-sized cycad with an erect trunk to 3 m tall and 45 cm across, swollen at the base. **Young leaves** silvery, covered with brown hairs. **Mature leaves** numerous in an obliquely erect crown, 0.5–1.5 m x 25–30 cm, dark green to grey green, flat in cross-section, straight or incurved in profile; **petiole** 10–20 cm long, concave at the base; **leaflets** numerous, moderately crowded, evenly distributed throughout, middle **leaflets** inserted on the rhachis at 80–90°, proximal **leaflets** reflexed; **median leaflets** 10–17 cm x 0.6–0.7 cm, linear, straight or slightly falcate, leathery, dark green and hairy above, paler and hairy beneath, midrib white, margins thickened and slightly recurved, base decurrent, apex often decurved, abruptly acuminate, with a yellow mucro about 1 mm long; **lower leaflets** deflexed, reduced in size, gradually reducing to a series of rigid, yellow thorns. **Male cones** 15–20 cm x 6–9 cm, narrowly ovoid to conical, brown, hairy; **sporophylls** 2–3 cm x 1–1.5 cm, deltoid, brown, hairy, with an erect apical point about 1 cm long. **Female cones** loose and open, covered with rusty brown hairs; **sporophylls** 10–15 cm long, brown, hairy; **apical lobe** 3–5 cm x 2–3 cm, narrowly ovate, the margins entire; **ovules** four on each sporophyll. **Seeds** 2–2.5 cm x 2–2.3 cm, nearly globular, somewhat flattened, brown.
**Distribution and Habitat:** Endemic to Australia where it is found in the Kimberley Region of northern Western Australia and on some adjacent offshore islands. This species grows in sparse woodland and among grass and boulders on low, rocky hills. The soils are red clays and loams of basaltic origin. The climate is tropical with hot, wet, humid summers having a day/night temperature regime to 36°C and 26°C respectively. The winters are long and dry with a day/night temperature regime to 29°C and 14°C respectively. The wettest areas have a rainfall of 1000 mm per annum with drier areas averaging 600 mm, mostly falling in summer.
**Notes:** This species was described in 1923 from material collected in basalt hills near the Lawley River of northern Western Australia. It is very closely related to *C. furfuracea* but can be distinguished from that species by its stout trunk swollen at the base, its broad leaflets being held flat and which are still hairy even when mature and its nearly globular, relatively small seeds.
**Cultivation:** Suited to tropical regions which have a seasonally dry climate. Best grown in full sun. Mature specimens transplant readily. Seedlings are very slow growing and can be tricky to establish, especially away from the tropics. Tolerates light frosts.
**Propagation:** From seed.

# *Cycas beddomei*

Dyer
(after Colonel R. H. Beddome, former Conservator of Forests, Madras, India)

**Description:** A small cycad with a trunk to 0.4 m tall and 35 cm across, growing in clumps of up to eight stems, with suckers produced from the base and offsets developing on the trunk. **Mature leaves** numerous in an obliquely erect to spreading crown, 1–1.2 m x 20–36 cm, pale green, flat in cross-section, straight in profile; **petiole** 10–15 cm long, four-angled in cross-section, with tufts of hairs at the base; **leaflets** 100–200 on each leaf, moderately crowded and evenly distributed throughout, inserted on the rhachis at about 45°; **median leaflets** 12–18 cm x 0.2–0.35 cm, linear, straight, leathery, glabrous, margins strongly revolute, apex acuminate or mucronate; **basal leaflets** reduced and abruptly replaced by a series of widely spaced tiny thorns about 1 mm long. **Male cones** 20–30 cm x 5–7.5 cm, ovoid, brownish, with pink hairs; **sporophylls** 2–3.5 cm x 1–2 cm, triangular-rhomboid, apex with an upcurved point about 2 cm long. **Female cones** loose and open, brown; **sporophylls** 10–20 cm long, covered with pink to red hairs; **apical lobe** 6–7.5 cm x 2–2.5 cm, ovate-lanceolate, the margins with numerous linear teeth about 2 cm long, the apex drawn out into an acuminate point of similar length; **ovules** two to four on each sporophyll. **Seeds** 3–4 x 3–4 cm, almost globose, yellow to brown.
**Distribution and Habitat:** Endemic to the Eastern Ghats mountain range in southern India. This species grows in skeletal soils on the forested slopes of hills at altitudes between 300 m and 900 m. The forests are mainly deciduous and the cycads are reported to prefer sunny situations. The climate is tropical with short hot, humid summers and long dry winters.
**Notes:** This species was described in 1883 after having been previously confused with *C. revoluta*. Like that species its leaflets have strongly revolute margins but it can be distinguished by its dwarf habit, light green leaves and differently shaped male and female sporophylls which bear pink to red hairs. *C. beddomei* is drought-tolerant and able to survive relatively long dry periods which occur regularly where it grows. This species is becoming

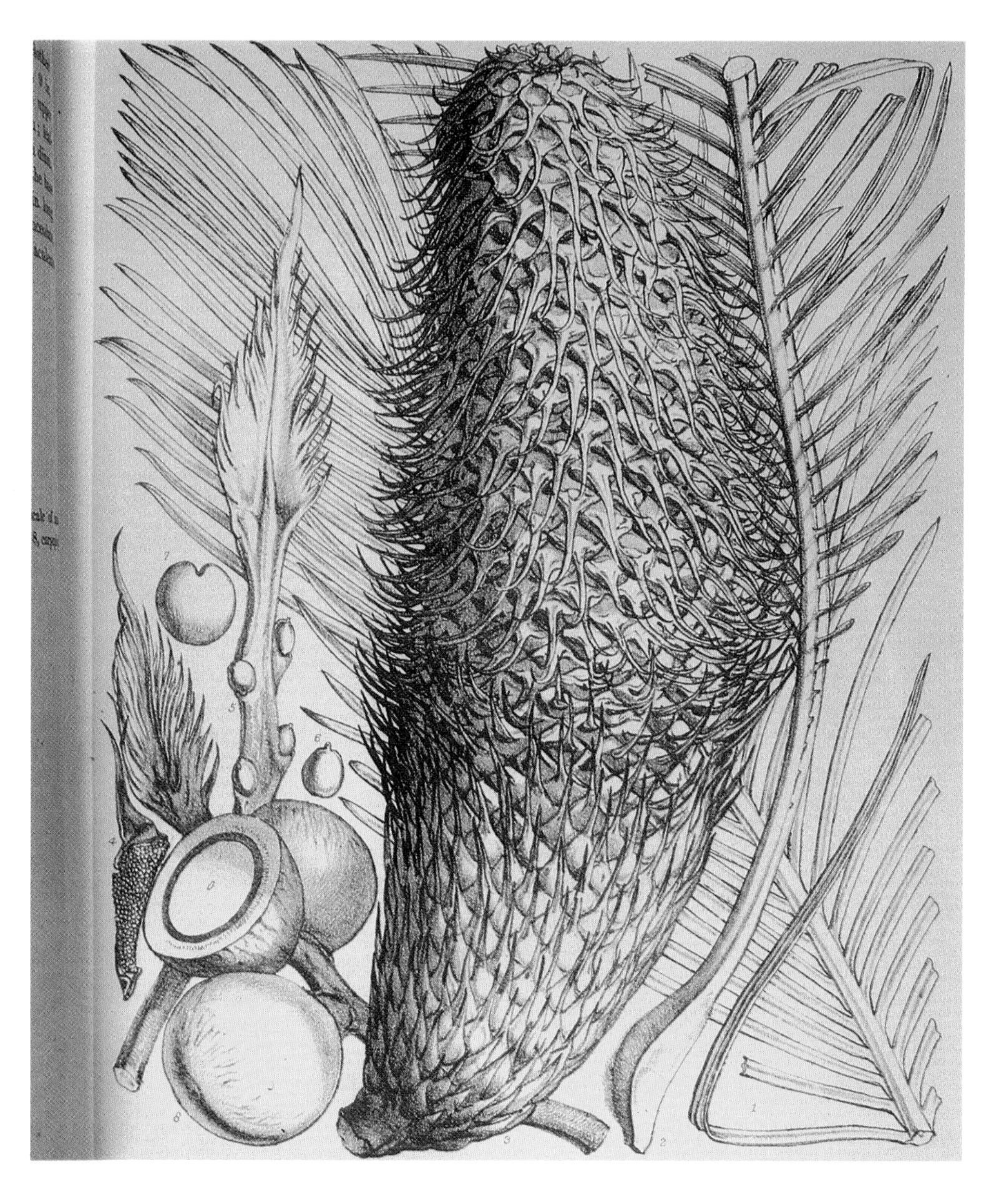

*Cycas beddomei*
From the *Transactions of the Linnaean Society*, London, 5: 85, plate 17, (1883)

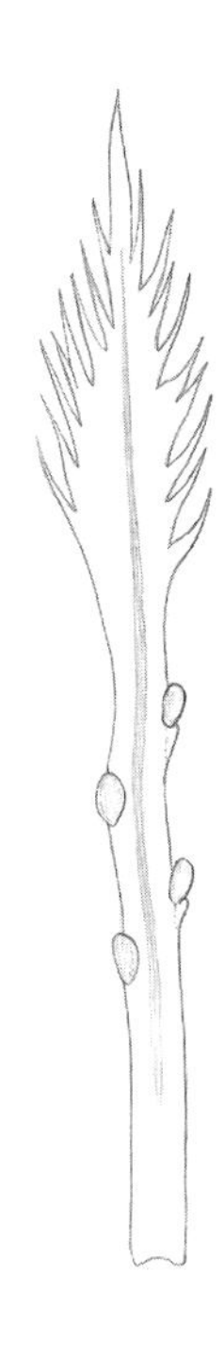

Sporophyll of *Cycas beddomei*

uncommon to rare due to clearing and the fact that the young male cones are heavily collected annually by local people and used for medicinal purposes.
**Cultivation:** Suited to tropical and subtropical regions. Will grow in full sun, partial sun or shade. Requires excellent drainage and apparently withstands light frosts.
**Propagation:** From seed and by removal of suckers and trunk offsets.

## *Cycas brunnea*

K. Hill
(brown, in reference to the brown hairs on the young growth)

**Description:** A medium to large cycad with a trunk about 2 m tall and 30 cm across (rarely to 5 m tall), with occasional offsets produced on the trunk and sparse suckers arising from the base. **Young leaves** bluish, densely covered with dark chocolate brown hairs. **Mature leaves** 1.2–1.8 m x 20–30 cm, numerous in an arching to rounded crown, blue to blue green, becoming grey-green, glossy, vee-ed in cross-section, arching in profile; **petiole** 30–60 cm long, swollen at the base, glabrous or with a few grey hairs; **leaflets** 160–240 on each leaf, crowded, evenly distributed throughout, inserted on the the rhachis at about 50°; **median leaflets** 17–27 cm x 0.6–0.8 cm, linear, obliquely erect, margins slightly recurved, apex acuminate; **lower leaflets** reduced, abruptly replaced by a series of short, yellow thorns. **Male cones** about 20 cm x 13 cm, narrowly ovoid to cylindrical, brown, hairy; **sporophylls** about 5.5 cm x 2 cm, ovate, with an erect point to 2.8 cm long. **Female cones** loose and open; **sporophylls** about 30 cm long, bearing orange hairs; **apical lobe** about 7 cm x 2.5 cm, triangular, margins with regular short teeth, apex long-acuminate; **ovules** four to six on each sporophyll. **Seeds** about 4 cm x 3 cm, ovoid to

globose, yellow to orange brown, powdery.
**Distribution and Habitat:** Endemic to Australia where it occurs in the vicinity of Lawn Hill to the south-west of Burketown, near the Gulf of Carpentaria in north-western Queensland. It also occurs on Wollogorang Station in adjacent parts of the Northern Territory. In Queensland it grows among sparse, stunted forest on exposed limestone ridges and also near streams in tall woodland and gallery forest. In the Northern Territory it grows on soils derived from sandstone. The climate is tropical with variable and erratic, wet summers having day temperatures to 35°C and night temperatures to 25°C. Winters are dry with day temperatures to 27°C and night temperatures to 13°C. Rainfall averages 700 mm per annum with occasional drier years.
**Notes:** This species was described in 1992 from material collected at Running Waters on Lawn Hill Creek in north-western Queensland. It is related to *C. angulata* but can be distinguished by its broader, flatter, more widely spaced, blue to grey-green, glossy leaflets which have slightly recurved margins and smaller, powdery seeds. The appearance of this species is markedly influenced by the habitat where it grows. Plants growing on exposed limestone ridges reach about 4 m tall and have blue leaves, whereas those in sheltered gallery forests grow to about 6 m tall and have green leaves in a larger crown.
**Notes:** Suited to tropical regions, especially those which have a seasonally dry climate. Best grown in sun or filtered sun. Requires excellent drainage and tolerates light frosts only. Seedlings are slow growing and can be difficult to establish.
**Propagation:** From seed and by removal of offsets from the trunk.

## *Cycas cairnsiana*

F. Muell.
(after Sir William Cairns, former governor of Queensland)

**Description:** A small to medium-sized cycad with an erect or leaning trunk to 3 m tall and 20 cm across, rarely to 5 m tall. Crown covered with yellowish brown wool. **Young leaves** whitish to pale blue, densely covered with loose, orange-brown hairs. **Mature leaves** about twenty in a stiffly spreading crown, 0.6–1 m x 4–7 cm, pale blue to bluish green, deeply vee-ed in cross-section, stiff and arching in profile, recurved in the distal third, often up-curved near the apex; **petiole** 18–27 cm long, pale bluish green, waxy, glabrous; rhachis of similar colouration; **leaflets** 180–280 on each leaf, moderately crowded, evenly distributed throughout although those towards the base more widely spaced, obliquely erect and forming a deep vee, inserted on the rhachis at about 40°; **median leaflets** 18–20 cm x 0.2–0.35 cm, linear, straight, stiff, rigid, covered with a waxy bloom, margins strongly recurved, apex acuminate with a yellow mucro about 1 mm long; **lower leaflets** reduced, abruptly replaced by a series of very short thorns. **Male cones** 16–20 cm x 7–10 cm, cylindrical to narrowly ovoid; **sporophylls** 3–5 cm x 0.7–1 cm, narrowly ovate, with an erect apical spine about 1 cm long. **Female cones** loose with open, densely covered with a white powdery bloom; **sporophylls** 14–21 cm long, covered with powdery bloom and loose brown hairs; **apical lobe** 4–7 cm x 1.5–2.5 cm, ovate, margins with a few small teeth, apex acuminate; **ovules** four to six on each sporophyll. **Seeds** 3–4 cm x 3–3.5 cm, ovoid to ellipsoid, bluish white due to a dense covering of powdery bloom.

K. HILL

K.

Male cone of *Cycas brunnea*.

*Cycas brunnea* in habitat.

K. HILL

Plants of *Cycas calcicola* reshooting after fire.

**Distribution and Habitat:** Endemic to Australia where it occurs in the Newcastle Range to the north-east of Einasleigh in north-eastern Queensland. This species grows on low hills among large granite boulders. The cycad plants occur in exposed situations among grass and sparse low shrubs. The climate is tropical with wet summers having a day/night temperature regime of 37°C to 21°C respectively. Winters are dry with a day/night temperature regime to 25°C to 8°C respectively, although frosts are common in some parts. Rainfall is about 600 mm per annum, falling mainly in summer.
**Notes:** This species was described in 1876 from material collected in the Newcastle Range by G. E. Armit. One of the most distinctive of all Australian *Cycas*, it can be recognised by the markedly blue, stiff, recurved leaves, with erect, narrow leaflets having strongly recurved margins. The powdery blue to white seeds are also notable. This species has been treated in horticultural literature as *Cycas* species 'Mt. Surprise' or *Cycas* species 'Champion's Blue Surprise'. Because of the specific epithet (*cairnsiana*) it has been commonly (and erroneously) believed that this species originated in the Cairns hinterland and consequently the species has been confused with *Cycas platyphylla*.
**Cultivation:** Suited to tropical regions with a seasonally dry climate. This species grows best and attains its most attractive appearance in warm to hot, dry conditions. While plants will grow in coastal districts, they struggle in the humid conditions and lose their distinctive glaucous appearance. Requires excellent drainage and tolerates moderate frosts. Seedlings are slow growing and can be difficult to establish under unsuitable conditions.
**Propagation:** From seed.

## *Cycas calcicola*

Maconochie
(growing on limestone)

**Description:** A small to medium-sized cycad with an erect trunk to 3 m tall and 30 cm across. **Young leaves** erect silvery or silvery blue. **Mature leaves** numerous in an obliquely erect crown, 0.6–1.2 m x 10–27 cm, dark green, flat in cross-section, straight in profile, with rusty, woolly hairs; **leaflets** 200–400 on each leaf, moderately spaced and evenly distributed throughout except those towards the base, inserted on the rhachis at about 40°; **median leaflets** 8–12 cm x 0.2–0.4 cm, linear, straight or slightly falcate, leathery, glabrous or hairy above, shortly hairy beneath, margins revolute, base narrowed and slightly angled but not decurrent, apex sharply mucronate with a spinose mucro about 1 mm long; **basal leaflets** reduced in size and more widely spaced, abruptly re-

K. HILL

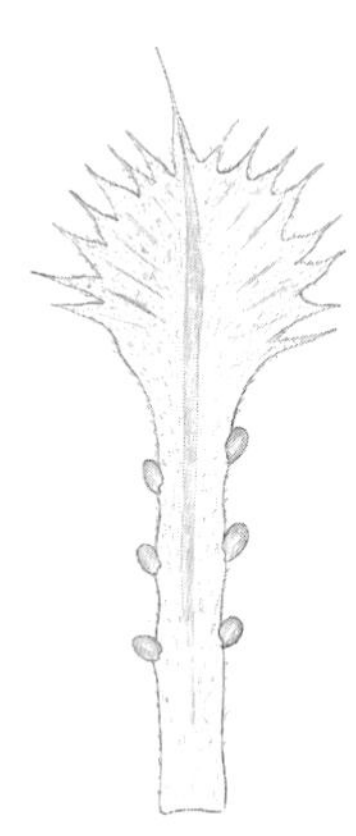

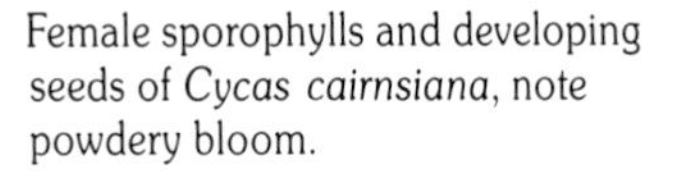

Female sporophylls and developing seeds of *Cycas cairnsiana*, note powdery bloom.

Sporophyll of *Cycas chamberlainii*

placed by a series of rigid thorns about 0.3 cm long. **Male cones** 17–26 cm x 5–6 cm, narrowly ovoid, grey, hairy; **sporophylls** 1.5–2.5 cm x 1.2 cm, deltoid, with a hooked tip. **Female cones** loose and open, hairy; **sporophylls** 10–15 cm long, rusty brown, hairy; **apical lobe** 2–2.5 cm x 0.8–1.1 cm, ovate, entire or with about eight papery marginal teeth; **ovules** two to six on each sporophyll. **Seeds** 3–3.5 cm x 2.5–2.7 cm, oblate to ovoid, brown, slightly glaucous.
**Distribution and Habitat:** Endemic to Australia where it occurs in a few disjunct localities in northern parts of the Northern Territory. This species grows in sparse, stunted woodland either on or close to limestone outcrops or on slopes and sandy flats derived from sandstone, usually where the water table is close to the surface. The climate is tropical with wet, humid summers having a day/night temperature regime to 38°C and 25°C respectively. Winters are dry with a day/night temperature regime to 30°C and 15°C respectively. Rainfall is about 750 mm per annum, mainly during summer.
**Notes:** Described in 1978 from material collected from a colony to the north of Katherine in the Northern Territory. *C. calcicola* is one of the most ornamental of the Australian cycads and is renowned for its spectacular flushes of silvery to bluish grey leaves which are produced after the almost annual fires which sweep the area. The species is not readily confused with any other Australian *Cycas*, it being distinguished by the silvery to bluish grey young leaves, the mature leaves being more or less straight in profile and flat in cross-section, and the narrow leaflets with revolute margins. This species is regarded as being threatened in its natural habitat.
**Cultivation:** Suited to tropical regions which have a seasonally dry climate; much less successful in equatorial tropical and subtropical regions. Large specimens transplant readily. Seedlings are slow and may be difficult to establish. Plants are best grown in full sun with unimpeded air movement and excellent drainage. Will grow in acid or slightly alkaline soils.
**Propagation:** From seed.

# *Cycas chamberlainii*

W. H. Brown & R. Kienholz
(after Charles Joseph Chamberlain, former Professor of Botany specialising in cycads, University of Chicago)

**Description:** A large cycad with a slender trunk to 8 m tall and 20 cm across, with occasional offsets produced on the trunk. **Young leaves** light green, bearing brown hairs. **Mature leaves** numerous in an obliquely erect to spreading crown, 1.2–1.6 m x 40–55 cm, dark green, glossy, flat in cross-section, straight in profile; **petiole** 15–30 cm long, hairy at the base; **leaflets** 160–200 on each leaf, moderately crowded, evenly distributed throughout, inserted on the rhachis at about 60°; **median leaflets** 20–30 cm x 0.9–1.1 cm, linear-lanceolate, straight or falcate, leathery, hairy towards the base, margins flat, base decurrent on the rhachis, apex acuminate; **lower leaflets** reduced, abruptly replaced by a series of short thorns about 0.3 cm long. **Male cones** 10–15 cm x 6–8 cm, ovoid, brown, hairy; **sporophylls** 3–3.5 cm x 1.5–2 cm, wedge-shaped, with an erect apical point about 1 cm long. **Female cones** loose and open, brown, densely hairy; **sporophylls** 15–22 cm long, densely hairy; **apical lobe** 5–6.5 cm x 4.5–6 cm, ovate to broadly triangular, the margins with four to ten spine-like lobes to 1.8 cm long, the apex acuminate; **ovules** four to six on each sporophyll. **Seeds** 3.5–4 cm x 2.5–3 cm, ovoid to obovoid, flattened, greenish brown.
**Distribution and Habitat:** Endemic to the Philippines where it occurs in mountainous regions of the island of Luzon. It grows on steep, rocky ridges at about 800 m altitude. The climate is tropical with hot, humid, wet summers and mild to cool, relatively dry winters.
**Notes:** This species was described in 1925 from material collected on Mt Arayat on the island of Luzon. It has been confused with *C. rumphii* but can be distinguished by its very slender trunk, with smaller leaflets and much smaller seeds.
**Cultivation:** A very handsome species which has proved to be fast growing. Suited to tropical and subtropical regions. Will grow in full sun or filtered sun. Responds to watering during dry periods. Needs protection from excessively cold weather.
**Propagation:** From seed.

# *Cycas chevalieri*

J. Leandri
(after M. Chevalier, the original collector of the species)

**Description:** Details of trunk not recorded. **Mature leaves** about 0.5 m x 30–50 cm; **leaflets** 17–27 cm x 1.8 cm, linear-lanceolate, straight or slightly falcate, dark green and glabrous, midrib prominent, margins flat, apex acute. **Male cones** not recorded. **Female cones** brown, hairy; **sporophylls** 7–12 cm long, covered with short

Male cone of *Cycas circinalis.*

brown hairs; **apical lobe** about 4 cm x 2 cm, ovate, the margins with numerous pointed lobes about 3 cm long; **ovules** two to four on each sporophyll. **Seeds** about 2 cm across, globose.
**Distribution and Habitat:** Occurring in northern parts of Vietnam and in adjacent areas of China. Details of habitat have not been recorded.
**Notes:** This poorly known species was described in 1931 from incomplete material collected by M. Chevalier in at least two localities.
**Cultivation:** Suited to tropical and warm subtropical regions.
**Propagation:** From seed.

## *Cycas circinalis*
L.
(coiled like a watchspring or a young fern frond)

**Description:** A medium-sized to large cycad with an erect, rarely branched trunk to 5 m tall and 45 cm across, usually suckering from the base and commonly growing in large clumps. **Young leaves** bright green, shiny. **Mature leaves** numerous in an obliquely erect to widely spreading crown, 1.5–3 m x 40–50 cm, dark green, flat in cross-section, straight in profile; **petiole** 35–50 cm long;

Clump of *Cycas circinalis.*

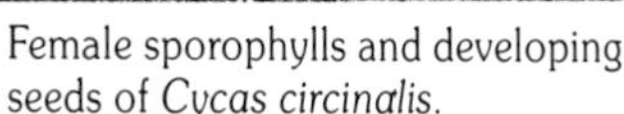

Female sporophylls and developing seeds of *Cycas circinalis*.

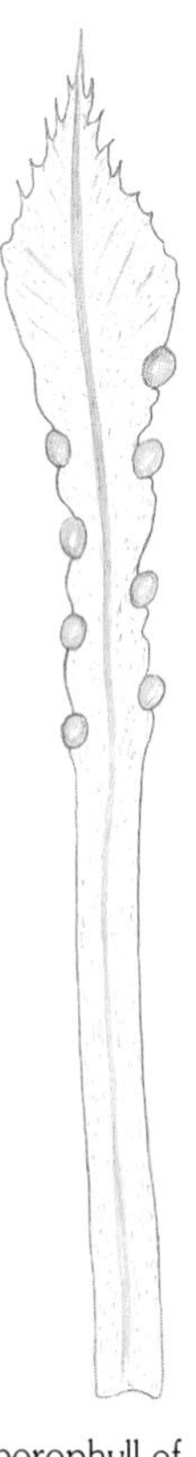

Sporophyll of *Cycas circinalis*

Early development of female sporophylls and ovules of *Cycas circinalis*.

**leaflets** 160–240 on each leaf, moderately spaced and not appearing crowded, evenly distributed throughout, inserted on the rhachis at about 80°; **median leaflets** 20–35 cm x 0.9–1.2 cm, linear-lanceolate, straight or slightly falcate, leathery, glabrous, upper surface shiny, margins flat, midrib raised on upper surface, apex acuminate; **lower leaflets** reduced and abruptly replaced by a series of rigid thorns about 0.4 cm long, the lower ones deflexed. **Male cones** 30–80 cm x 10–14 cm, cylindrical to narrowly ovoid, brown, hairy; **sporophylls** 3.5–5 cm x 1–2 cm, ovate, internally glabrous, externally hairy, with an upcurved point about 2.5 cm long. **Female cones** loose and open, hairy; **sporophylls** 20–30 cm long, rusty brown, hairy; **apical lobe** 7–10 cm x 2–3 cm, rhomboid, margins with numerous sharp, narrow teeth-like lobes; **ovules** six to twelve on each sporophyll. **Seeds** 2–5 cm x 3.5–4 cm, oblate to ovoid, pale reddish yellow, with a spongy flotation layer.

**Distribution and Habitat:** Occurring in India, Burma, Thailand, Sri Lanka, Malaysia, Indonesia and various islands of the South Pacific region. This species commonly forms dense stands in coastal and near-coastal districts but also occurs in inland areas and in India is found in mountainous regions up to 1000 m altitude. Plants grow in situations ranging from full sun to nearly dense shade. *C. circinalis* is essentially a species from the tropics growing in areas with hot, wet humid summers and mild dry winters.

**Notes:** *C. circinalis* was described in 1658 from material collected on the Malabar coast of south-western India. It is one of the first cycads named, and is the type of the genus *Cycas*. Because its seeds float (see page 60) and retain their viability for long periods, they are widely distributed by rivers and ocean currents. Thus, where suitable conditions exist, the species has become established in the coastal communities of many countries. These populations regenerate, are often very well established and the plants adapt over generations in various small ways to their own particular environmental circumstances and local climate. Thus, plants of this species from various countries may differ to varying degrees even though the basic characteristics of the species are still present. Nothing is static in nature and the situation is further complicated by new arrivals being regularly carried on the ocean currents. On reaching maturity these recent plants can hybridise with established plants and the resultant complex range of variation may defy taxonomic separation. Thus, *C. circinalis* must be regarded as an extremely variable species and one which has had a very confused botanical history. Its closest congener is

*C. rumphii* the seeds of which also float. The latter species, however, has much larger seeds than those of *C. circinalis*.
**Cultivation:** Suited to tropical and warm subtropical regions. A popular species which is widely planted in public and private gardens of the tropics. Grows best in a sunny situation and is adaptable to a range of well-drained soils. Plants will tolerate some cold but need protection from heavy frosts.
**Propagation:** From seed or by the removal of trunk offsets.

## *Cycas conferta*

S. Chirgwin
(crowded, in reference to the leaflets)

**Description:** A medium-sized to large cycad with a blackish, emergent trunk to 5 m tall and 40 cm across with occasional offsets produced on the trunk and sparse suckers arising from the base. **Young leaves** whitish from a dense covering of hairs. **Mature leaves** numerous in an obliquely erect, bushy crown, 0.8–1.3 m x 20–30 cm, bluish green to green, flat in cross-section, straight in profile, often twisted near the apex; **petiole** 15–30 cm long, swollen and woolly at the base; **leaflets** 150–230 per leaf, crowded, those towards the base more widely spaced, inserted on the rhachis at about 70°; **median leaflets** 10–16 cm x 0.4–0.6 cm, linear, straight, margins flat or slightly recurved, midrib yellowish, apex acuminate; **lower leaflets** abruptly replaced by a series of short, yellow thorns. **Male cones** about 20 cm x 12 cm, cylindrical, brown, hairy; **sporophylls** about 4 cm x 2 cm, ovate, with a short, erect point. **Female cones** loose and open, hairy; **sporophylls** 30–35 cm long, ovate, hairy; **apical lobe** 6–8 cm x 4.5–5.5 cm, ovate, margins shortly toothed, apex acuminate; **ovules** four to six on each sporophyll. **Seeds** about 5 cm x 5 cm, brownish.
**Distribution and Habitat:** Endemic to Australia where it occurs in coastal areas to the west of Darwin. This species grows in moist soils close to swampy areas and

Male cone of *Cycas conferta*.

K. HILL

*Cycas circinalis* From *Rumphia* 4, plate 176, (1849)

K. HILL

*Cycas couttsiana* in habitat.

low-lying sites where ground water is not far from the surface, often in coastal dunes. It grows in extensive colonies and may extend to the high water mark in some parts. The vegetation is sparse woodland. The climate is tropical with wet summers having a day/night temperature regime to 33°C and 25°C respectively. Winters are long and dry with a day/night temperature regime to 30°C and 19°C respectively. Rainfall is about 1500 mm per annum, falling mainly in summer.
**Notes:** This species, which has affinities with *C. armstrongii*, can be distinguished by its narrower, much thinner textured leaflets and the dense hairs on the young leaves. Fires occur almost annually where this species grows. Plants are normally deciduous for a short period during the dry season. Plants of this cycad have been cultivated as *Cycas* species 'Cox Peninsula'.
**Cultivation:** Suited to tropical regions which have a seasonally dry climate; should grow well in coastal situations. Requires full sun or filtered sun, unimpeded drainage and regular watering during dry periods. Seedlings are slow growing and may be difficult to establish, especially in areas away from the tropics. Cultivated plants may shed their leaves and should then be kept on the dry side.
**Propagation:** From seed or by removal of basal suckers or trunk offsets.

## *Cycas couttsiana*

K. Hill
(after Pat and David Coutts, cycad enthusiasts who discovered the species)

**Description:** A medium-sized to large cycad with an erect trunk about 3 m tall and 25 cm across (some records are to 10 m tall). **Young leaves** blue to blue green, powdery, densely covered with greyish white and orange-brown hairs. **Mature leaves** numerous in an obliquely erect to spreading crown, 1–1.5 m x 25–30 cm, blue to blue green, shallowly to deeply vee-ed in cross-section, straight in profile; **petiole** 17–21 cm long, bluish, powdery, usually with grey hairs; rhachis of similar colouration; **leaflets** 180–270 on each leaf, dull blue green, moderately crowded, those towards the base more widely spaced, obliquely erect and forming a broad vee, inserted on the rhachis at about 40–80°; **median leaflets** 15–20 cm x 0.4–0.6 cm, linear, straight, stiff, rigid, covered with a waxy bloom, margins slightly recurved to nearly flat, apex acuminate; **lower leaflets** reduced, abruptly replaced by a series of short thorns. **Male cones** 15–20 cm x 7–9 cm, cylindrical; **sporophylls** about 3 cm x 1.2 cm, ovate, with an erect spical spine about 1 cm long. **Female cones** loose and open, densely covered with grey hairs; **sporophylls** 15–25 cm long, covered with powdery bloom; **apical lobe** 4–6 cm x 2–3 cm, narrowly triangular, margins with a few small, regular teeth, apex acuminate; **ovules** four to six on each sporophyll. **Seeds** 3.5–4.5 cm x 3–3.8 cm, ovoid, greenish to yellowish brown with a covering of powdery bloom.
**Distribution and Habitat:** Endemic to Australia where it occurs in the southern parts of the Gregory Range to the north of Hughenden in northern Queensland. This species grows on low hills and slopes in open grassy woodland. Soils are shallow to deep red sandy loams. The climate is tropical with wet summers having a day/night temperature regime to 37°C and 21°C respectively. Winters are dry with a day/night temperature regime to 25°C and 8°C respectively. Frosts may occur in some areas in winter. The rainfall is about 600 mm per annum falling mainly in summer.
**Notes:** This species was described in 1992 from material collected on the upper reaches of the Stawell River in the Gregory Range in northern Queensland. It has been cultivated as *Cycas* species 'Glen Idle Blue' and has affinities with *C. cairnsiana* but is taller growing and lacks the revolute leaf margins of that species. It is basically distinguished from other Australian species by the relatively broad, dull blue leaflets with slightly recurved to almost flat margins and the prominent powdery covering on the seeds.
**Cultivation:** Suited to tropical and subtropical regions with a warm to hot, seasonally dry climate. Plants attain their best appearance in warm, dry conditions and while they will grow in coastal districts they struggle in the

humid conditions and lose their distinctive glaucous appearance. Requires excellent drainage, unimpeded air movement and tolerates moderate frosts. Seedlings are slow growing and can be difficult to establish under unsuitable conditions.
**Propagation:** From seed.

## *Cycas dilatata*

Griffith
(broadened, expanded, widened — in reference to the apical lobes of the female sporophylls)

**Description:** A small cycad with a trunk about 0.5 m tall and 15 cm across. **Mature leaves** numerous in an obliquely erect to spreading crown, 1–1.3 m x 15–25 cm, dark green, flat in cross-section,; **petiole** 30–40 cm long, glabrous, channelled; **leaflets** numerous, dark green, leathery, straight; **median leaflets** 15–20 cm x 0.6–0.7 cm, linear, margins flat, apex blunt or notched; **basal leaflets** reduced and abruptly replaced by a series of thorns. **Male cones** not recorded. **Female cones** loose and open, hairy; **sporophylls** about 15 cm long, with rusty hairs; **apical lobe** 4–5 cm x 3–5 cm, nearly heart-shaped, the margins with numerous long teeth; **ovules** four to eight on each sporophyll. **Seeds** about 2.5 cm x 2 cm, ovoid, yellow, smooth.
**Notes:** This poorly known species was described in 1854 but details of its original locality were not included with the type description. It was probably collected from India and its affinities seem to be with *C. pectinata*.
**Cultivation:** Suited to tropical and warm subtropical regions. Probably requires a sheltered position, warmth, humidity and regular watering. Frost tolerance is unlikely to be high.
**Propagation:** From seed.

## *Cycas furfuracea*

W. V. Fitzg.
(covered with loose, woolly scales)

**Description:** A small to medium-sized cycad with an erect trunk to 2.5 m tall and 40 cm across, with suckers produced from the base. **Young leaves** bluish, covered with brown woolly hair. **Mature leaves** numerous in an obliquely erect to spreading crown, 0.6–0.9 m x 35–45 cm, stiff and rigid, bluish green, shallowly vee-ed in cross-section, straight in profile; **petiole** 10–15 cm long; **leaflets** numerous, crowded, evenly distributed throughout, obliquely erect and forming a shallow to deep vee, inserted on the rhachis at about 50°; **median leaflets** 15–20 cm x 0.6–0.8 cm, linear, straight or slightly falcate, leathery, bluish green and glabrous above, greyish and hairy beneath, midrib yellow and strongly keeled beneath, margins thickened, flat, apex abruptly acuminate with a yellow mucro about 1 mm long; **lower leaflets** reduced, gradually reducing to a series of yellow thorns. **Male cones** 30–45 cm x 6–12 cm, narrowly ovoid, covered with woolly brown hairs; **sporophylls** 3–5 cm x 1–2 cm, triangular to ovate, with an erect point about 2 cm long. **Female cones** loose and open, covered with brown, woolly hairs; **sporophylls** 20–35 cm long, hairy; **apical lobe** 3–5 cm x 2–3 cm, ovate, the margins with short, spiny teeth; **ovules** four on each sporophyll. **Seeds** 2–4 cm x 2–3 cm, globose, yellow with grey hairs.
**Distribution and Habitat:** Endemic to Australia where it is restricted to the King Leopold Ranges and islands offshore from the Prince Regent River area of northern Western Australia. This species grows near the summits of sandstone bluffs and escarpments and on protected rocky slopes and gullies. The climate is tropical with wet summers having a day/night temperature regime to 36°C and 29°C respectively. Winters are dry having a day/night temperature regime to 26°C and 14°C respectively. Rain-

*Cycas furfuracea* in habitat.

K. HILL

fall is about 900 mm per annum, falling mainly in summer.
**Notes:** This species was described in 1918 from material collected in the King Leopold Ranges of northern Western Australia. It is closely related to *C. basaltica* but can be readily distinguished by the stout trunk and stiff leaves with the broad, leaflets (hairy beneath) which are held obliquely erect to form a shallow to deep vee.
**Cultivation:** Suited to tropical regions which have a seasonally dry climate. Requires a sunny position and excellent drainage. Mature specimens transplant readily. Seedlings are very slow growing and can be difficult to establish, especially away from the tropics. Tolerates light frosts.
**Propagation:** From seed and by removal of suckers.

## *Cycas gracilis*

Miq.
(graceful, slender)

This species was described in 1863 from material collected near Cape Upstart in central-eastern Queensland, Australia, by Baron Ferdinand von Mueller. The species occurs in the natural range of *Cycas media* and examination of a photo of the type and field collections by Ken Hill shows that the two are conspecific, with *C. media* being the accepted name.

## *Cycas guizhouensis*

Lan & R. F. Zou
(from the Province of Guizhou, China, the type area)

**Description:** A small cycad with a trunk to 0.6 m tall and 30 cm across. **Mature leaves** numerous in a spreading crown, 1–1.6 m x 20–36 cm, flat in cross-section, straight in profile; **petiole** 40–50 cm long; **leaflets** numerous, moderately crowded, evenly distributed throughout, inserted on the rhachis at about 45°; **median leaflets** 8–20 cm x 0.8–1.2 cm, linear-lanceolate, straight or falcate, dark green and shiny above, pale green beneath, midrib raised, glabrous, margins slightly revolute, base unequal, apex acuminate; **lower leaflets** abruptly replaced by a series of short thorns about 1.5 mm long. **Male cones** not recorded. **Female cones** loose and open, yellowish brown to rusty brown, hairy; **sporophylls** 14–20 cm long, densely hairy; **apical lobe** 6–7 cm x 7–8 cm, nearly round, the margins with numerous tapered lobes to 4.5 cm long, apex ending in a tapered lobe to 4.5 cm long; **ovules** two to eight on each sporophyll. **Seeds** about 3 cm x 3 cm, nearly globose, yellow with reddish brown mucro.
**Distribution and Habitat:** Endemic to China where it occurs in the south-western parts of the Province of Guizhou, growing mainly in the valleys of the Nanpan River and Quingshui River. It grows on slopes, often in colonies at altitudes of 400 m to 800 m. The soils are gravels or those derived from limestone. The climate is mild with cool, moist summers and moist, frost-free winters.
**Notes:** This species was described in 1983 from material cultivated in the hospital of Xingyi, Guizhou, China.
**Cultivation:** Suited to temperate and subtropical regions. This species, which is apparently fast growing, withstands cold periods but needs protection from frosts.
**Propagation:** From seed.

## *Cycas inermis*

Lour.
(unarmed, in reference to the petioles)

**Description:** A small cycad with a subterranean or emergent trunk to 1 m tall and 20 cm across. **Mature leaves** numerous in a widely spreading crown, 0.5–1 m x 20–30 cm, dark green, flat in cross-section; **petiole** 20–30 cm long, lacking thorns; **leaflets** 100–180 on each leaf, crowded; **median leaflets** 10–20 cm x 0.5–0.7 cm, linear-lanceolate, straight, leathery, glabrous, dark green above, paler beneath, margins flat, apex mucronate; **lower leaflets** ending abruptly, not replaced by a series of thorns. **Male cones** about 30 cm x 8 cm, covered with brown hairs; **sporophylls** 2.5–3.5 cm x 1.6–2 cm, deltoid, with an apical point. **Female cones** loose and open; details of **sporophylls** and seeds lacking.
**Distribution and Habitat:** Known from Vietnam where it probably grows in sheltered forests. The climate is tropical with hot, humid summers and warm, moist to dry winters.
**Notes:** This poorly known species was described in 1793 from material collected near Saigon by the French botanist Joao de Loureiro. It is apparently related to *C. siamensis* but the petioles lack any thorns.
**Cultivation:** Suited to tropical and warm subtropical regions. Probably requires a sheltered position, warmth, humidity and regular watering during dry periods. Frost tolerance is unlikely to be high.
**Propagation:** From seed.

## *Cycas jenkinsiana*

Griffith
(after Major Jenkins, collector of cycads in northern India)

**Description:** A small cycad with an erect trunk (often

branched) to 1 m tall and 15 cm across. **Mature leaves** numerous in an obliquely erect to spreading crown, 1–1.3 m x 20–30 cm, dark green, flat in cross-section, straight in profile; **petiole** 30–40 cm long, channelled, with brown wool; **leaflets** numerous, dark green, leathery, strongly falcate, apex blunt or notched; **median leaflets** 15–20 cm x 0.5–0.7 cm, linear, margins flat; **basal leaflets** reduced and abruptly replaced by a series of thorns. **Male cones** not recorded. **Female cones** loose and open, hairy; **sporophylls** about 14 cm long, with rusty, woolly hairs; **apical lobe** 4–6 cm x 2–2.5 cm, ovate-lanceolate, the margins with numerous long teeth, the apex drawn out into an acuminate point; **ovules** two to four on each sporophyll. **Seeds** 3–3.6 cm x 2.4–3 cm, ovoid, yellowish brown, smooth.
**Distribution and Habitat:** Known from the State of Assam in northern India. Details of habitat are not recorded.
**Notes:** This species was described in 1854 from material collected about Gowahatty in Lower Assam. It is a poorly known species the relationships of which need elucidating, but the strongly falcate leaflets would seem to be significant.
**Cultivation:** Suited to tropical and warm subtropical regions. Probably requires a sheltered situation, humid conditions and regular watering. Tolerance of frost is unlikely to be high.
**Propagation:** From seed.

# Cycas hainanensis

C. J. Chen
(from the island of Hainan, China)

**Description:** A medium-sized cycad with an erect trunk to 3.5 m tall and 35 cm across. **Mature leaves** numerous in an obliquely erect to spreading crown, 0.5–1.5 m x 20–40 cm, dark green, flat in cross-section, straight in profile; **petiole** short; **leaflets** moderately spaced, evenly distributed; **median leaflets** 15–25 cm x 1–1.4 cm, lanceolate, leathery, glabrous, margins flat, apex acuminate; **lower leaflets** reduced and abruptly replaced by numerous, crowded, rigid thorns. **Male cones** not recorded. **Female cones** loose and open; **sporophylls** hairy; **apical lobe** 10–12 cm x 4–5 cm, broadly ovate to rhomboid, the margins pinnately lobed. **Seeds** not recorded.
**Distribution and Habitat:** Apparently endemic to the Chinese island of Hainan (Hainandao) in the Gulf of Tonkin, South China Sea.
**Notes:** This poorly known species, apparently related to *C. rumphii* and *C. taiwaniana*, was described in 1975. It grows in rainforests and sheltered areas of other moist forests in good soils. The climate is tropical with hot, humid summers and mild winters. Rainfall is above 2000 mm per annum distributed throughout the year.
**Cultivation:** Suited to temperate and subtropical areas. Very frost-sensitive and also needs protection from direct hot sun.
**Propagation:** From seed.

# Cycas immersa

W. G. Craib
(covered up, embedded, immersed)

This species was described in 1912 from material collected in northern Thailand by A. F. G. Kerr. Later studies show that this species is identical with *C. siamensis*, the latter name having priority according to botanical rules.

# Cycas kennedyana

F. Muell.
(after Sir Arthur Kennedy, former governor of Queensland)

This species was described in 1882 from material collected in the Normanby Ranges to the south-west of Bowen in Queensland by Eugene Fitzalan. The species occurs within the natural range of *C. media* and examination of the type and field collections by Ken Hill show that the two are conspecific, with *C. media* being the accepted name.

# Cycas lane-poolei

C. Gardner
(after Charles Edward Lane-Poole, former Conservator of Forests in Western Australia)

**Description:** A medium-sized cycad with an unbranched, erect trunk to 5 m tall and 25 cm across. **Young leaves** covered with thick, orange, woolly hairs. **Mature leaves** numerous in a rounded, spreading crown, 0.5–1 m x 25–35 cm, dark green and shiny, flat in cross-section, straight in profile; **petiole** 20–30 cm long, slender, the base with thick, orange, woolly hairs; **leaflets** 150–200 on each leaf, dark green and shiny above, paler beneath, evenly distributed throughout, middle ones inserted on the rhachis at about 45°; **median leaflets** 10–15 cm x 0.6–1.1 cm, linear, straight, leathery, glabrous, midrib yellow, margins flat, apex acuminate, with a yellowish brown mucro; **lower leaflets** abruptly replaced by a series of short stout thorns. **Male cones** 15–25 cm x 12–14 cm, ovoid, brown, hairy; **sporophylls** 4–6 cm x 2–2.5 cm, ovate, with a prominent, erect apex about 1.5 cm long. **Female cones** loose and open, hairy; **sporophylls** 15–25 cm long, hairy; **apical lobe** 4–5 cm x 4–4.5 cm, ovate to triangular, margins with sharp teeth about 0.4 cm long, apex acuminate; **ovules** four on each sporophyll.

*Cycas lane-poolei* after fire.

Male cone of *Cycas lane-poolei*.

**Seeds** 5–6 cm x 4–5 cm, ovoid to globular, yellowish green, powdery when young.
**Distribution and Habitat:** Endemic to Australia where it occurs in the Kimberley Ranges of northern Western Australia, growing on the Mitchell Plateau and on Mount Elizabeth Station. The species forms sparse colonies in open forest and woodland, growing in sandy soil. The climate is tropical with hot, wet summers having a day/night temperature regime to 36°C and 27°C respectively. Winters are long and dry with a day/night temperature regime to 29°C and 19°C respectively. The rainfall averages 1000 mm per annum, with most falling in summer.
**Notes:** This species was described in 1923 from material collected near the source of the Moran river to the north-east of Mt Hann in the Kimberley district of northern Western Australia. Plants shed their leaves for a period during the dry season. Some researchers believe that this species is conspecific with *C. armstrongii*, but the relationship between the two is in need of further detailed study. *C. lane-poolei* appears to be much more robust than *C. armstrongii* with larger leaves, longer leaflets and larger seeds.
**Cultivation:** Suited to tropical regions which have a seasonally dry climate. Best in full sun, semi-shade or filtered sun. Seedlings are slow growing and are difficult to establish in areas away from the tropics. Requires excellent drainage.
**Propagation:** From seed.

## *Cycas macrocarpa*

Griffith
(with large fruit or seeds)

**Description:** A medium-sized cycad with an erect trunk to 3.5 m tall and 20 cm across, often branched. **Young leaves** covered with brown hairs. **Mature leaves** numerous in an obliquely erect to spreading crown, 1.5–2.5 m x 20–30 cm, dark green, flat in cross-section; **petiole** 20–40 cm long, glabrous, four-angled; **leaflets** numerous, dark green above, yellowish beneath, leathery, apex drawn out, attenuate; **median leaflets** 20–40 cm x 1–1.2 cm, broadly linear, margins recurved; **basal leaflets** reduced and abruptly replaced by a series of thorns. **Male cones** 30–40 cm X 12–14 cm, cylindrical, rusty hairy; **sporophylls** about 2 cm X 1.2 cm, wedge-shaped with a point about 0.8 cm long. **Female cones** loose and open, covered with rusty brown hairs; **sporophylls** 20–25 cm long, hairy; **apical lobe** 4–6 cm x 3–3.5 cm, triangular to wedge-shaped, hairy, margins with long spiny teeth, the apex drawn into a long point; **ovules** four to twelve on each sporophyll. **Seeds** 5–7 cm x 3.5–4.5, ellipsoid, green to yellow.
**Distribution and Habitat:** Known from Malacca. Details of habitat are not recorded.

**Notes:** This species was described in 1854 from material collected by a Mr Westerhout in September 1842 between Ayer Punnus and Tabong. It was apparently found at only one locality (near a Musulman tomb) and a note accompanying the original description suggested that it may have been introduced. Its status needs elucidating but the hairy fronds, recurved leaflet margins and large seeds would seem to be significant.
**Cultivation:** Suited to tropical and warm subtropical regions. Probably requires warm, humid conditions and regular watering. Tolerance of frost is unlikely to be high.
**Propagation:** From seed.

## *Cycas media*

R. Br.
(medium, in the middle)

**Description:** A medium-sized cycad with an erect trunk to 3.5 m tall and 20 cm across (rarely to 6 m tall), with occasional offsets produced on the trunk. **Young leaves** pale green, with loose pale brown hairs. **Mature leaves** numerous in an obliquely erect to rounded crown, 1–1.8 m x 20–30 cm, bright green to yellowish green, glossy, flat in cross-section, straight in profile; **petiole** 30–40 cm long, with grey hairs; **leaflets** 160–300 on each leaf, moderately spaced, evenly distributed throughout or the lower ones more widely spaced, inserted on the rhachis at about 80°; **median leaflets** 15–26 cm x 0.6–1 cm, linear, straight or slightly falcate, glossy, leathery, glabrous or hairy, midrib often yellowish, margins flat, apex abruptly acute with a short mucro; **lower leaflets** reduced, abruptly replaced by a series of short thorns. **Male cones** 15–25 cm x 10–15 cm, ovoid, brown, hairy; **sporophylls** 2.6–3.5 cm x 1–1.2 cm, ovate, with a short, erect point to 1 cm long. **Female cones** about 40 cm across, loose and open, hairy; **sporophylls** 25–35 cm long, hairy; **apical lobe** 3.5–9 cm x 2–3 cm, ovate to ovate-lanceolate, the margins with numerous short teeth, apex long acuminate; **ovules** four to ten on each sporophyll. **Seeds** 3–4 cm x 2.5–3.5 cm, ovoid to almost globular, orange-yellow to brownish, not powdery.
**Distribution and Habitat:** Occurs in Australia and south-eastern Papua New Guinea. In Australia it is widely distributed in north-eastern and central-eastern Queensland, from near the top of Cape York Peninsula, to near Rockhampton; also on some islands of the Torres Strait such as Moa Island (see notes below under variation). At various times this species has been erroneously recorded

Colony of *Cycas media* in habitat.

Old male cone of *Cycas media*.

Early development of female sporophylls and ovules of *Cycas media*.

Female sporophylls and developing seeds of *Cycas media*.

from the Northern Territory and Western Australia. The species is common in coastal and near-coastal localities, although on Cape York Peninsula plants also occur some distance from the coast. The favoured habitat is rocky hillsides and slopes with the plants growing under sparse sclerophyll forest. Occasional specimens are found on rainforest margins and within small patches of rainforest. The climate is tropical with hot, wet, humid summers having a day/night temperature regime to 30°C and 24°C respectively. The drier winters have a day/night temperature regime to 27°C and 13°C respectively. The rainfall is about 1500 mm per annum, falling mainly in summer.

**Variation:** Plants provisionally included in this taxon from northern parts of Cape York Peninsula, some Torres Strait Islands and south-eastern parts of Papua New Guinea have leaflets which are strongly vee-ed in cross-section. Their status is currently under investigation. Plants from the vicinity of Cooktown have broader, sickle-shaped leaflets and may represent possible hybrids between *C. media* and *C. silvestris*. Zones of hybridism occur where populations of *C. media* overlap with closely related taxa. In these areas plants may exhibit considerable variation in morphological features as a result of complex hybridisation. Such zones of overlap occur with *C. platyphylla* near Irvinebank in north-eastern Queensland and with *C. ophiolitica* in areas between Mackay and St Lawrence in central-eastern Queensland.

**Notes:** This species was described in 1810 from material collected on the Cumberland Islands offshore from Mackay in Queensland. It is the commonest species of *Cycas* in eastern Australia and usually grows in sparse to dense colonies. It can be distinguished by its glabrous, glossy green, flat leaflets and relatively small seeds (see also *C. megacarpa* and *C. silvestris*). Fires are common where this species grows and after such an occurrence the plants produce a spectacular flush of young, pale fronds which mature to a fresh, bright green. The young leaflets of this species are attacked by a species of chry-

somelid beetle (*Lilioceris nigripes*) and on some plants, all or most of the leaves may be severely disfigured. The larvae of a small blue butterfly (*Theclinesthes miskini*) also feed on the leaflets. The seeds, although poisonous, were an important source of food for the Aborigines who ate them after suitable treatment. The trunks of this species were at one time harvested commercially for starch in a limited, local industry. *Cycas kennedyana* F. Muell. and *C. normanbyana* F. Muell. are synonymous (see separate entries).
**Cultivation:** Suited to tropical, subtropical and warm temperate regions. A hardy species which will grow in sun or shade in a variety of well-drained soils. Large specimens transplant readily but may take some time to produce new leaves. Seedlings may be slow and difficult to establish. Tolerates light to moderate frosts.
**Propagation:** From seed.

*Cycas megacarpa* in habitat.

## *Cycas megacarpa*

K. Hill
(with large fruit or seeds)

**Description:** A small to medium-sized cycad with an erect trunk to 3 m tall and 15 cm across (rarely to 6 m tall). **Young leaves** light green, densely covered with brown hairs. **Mature leaves** numerous in an erect to rounded crown, 0.5–1.1 m x 18–25 cm, mid-green to dark green, glossy, flat or slightly keeled in cross-section, straight to arching in profile; **petiole** 20–35 cm long, glabrous or with grey hairs; **leaflets** 120–170 on each leaf, crowded, the lower ones more widely spaced, inserted on the rhachis at about 50°; **median leaflets** 12–20 cm x 0.5–0.75 cm, linear, straight, leathery, midrib yellowish, margins flat, apex acute to acuminate; **lower leaflets** reduced, abruptly replaced by a series of short thorns. **Male cones** about 18 cm x 17 cm, cylindrical, brown, hairy; **sporophylls** about 3.5 cm x 1.1 cm, ovate, with a short erect point to 0.8 cm long. **Female cones** loose and open, hairy; **sporophylls** about 25 cm long, hairy; **apical lobe** 4–7 cm x 2–3 cm, ovate, the margins with short regular teeth, apex acuminate; **ovules** two to four on each sporophyll. **Seeds** 4–6 cm x 3.5–4.5 cm, ovoid, greenish to light brown, not powdery.
**Distribution and Habitat:** Endemic to Australia where it occurs in central Queensland between Mount Morgan south of Rockhampton and Goomeri near Gladstone. The species grows on hills and slopes in tall open forest which has a grassy understorey and also in rainforest. Soils are shallow clay loams which are often stony. The climate is subtropical to tropical with hot, humid summers having a day/night temperature regime to 30°C and 22°C respectively. Winters are mild and dry with a day/night temperature regime to 22°C and 12°C respectively. Rainfall is about 1200 mm per annum and is spread throughout the year with the heaviest falls in summer.
**Notes:** This species was described in 1992 from material collected near Miriam Vale in central-eastern Queensland. It is closely related to *C. media* but can be distinguished by a much more slender trunk, smaller leaves and larger seeds. A hybrid interzone occurs in the northern part of its range (around Mount Morgan) where *C. megacarpa* overlaps with *C. ophiolitica*. Plants from this area may

Female sporophylls and developing seeds of *Cycas megacarpa*.

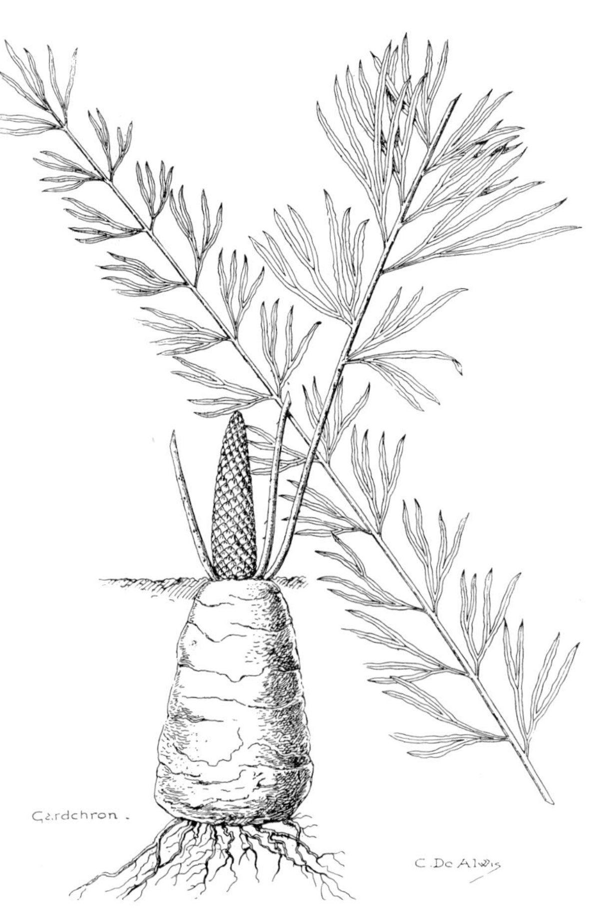


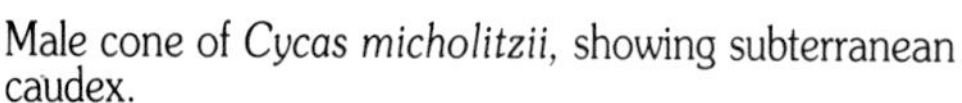

Male cone of *Cycas micholitzii*, showing subterranean caudex.
Both illustrations from *Gardeners Chronicle* 38:142–144, (1905)

*Cycas micholitzii* (a) male cone, (b) lower surface of male cone, showing anthers, (c) upper surface of scale of male cone, showing anthers, (d) carpophyll with immature fruits, (e) pinnules

have even larger seeds than from populations further south. Plants from this region of hybridism may be in cultivation as *Cycas* species 'Mount Morgan'. Fires are a common, often annual, occurrence where this species grows.

**Cultivation:** Suited to subtropical and warm temperate regions. An adaptable species which grows well in filtered sun or semi-shade in a range of well-drained soils. Tolerates light to moderate frosts. An excellent plant for a large container.

**Propagation:** From seed.

## *Cycas micholitzii*

Dyer var. ***micholitzii***

(after W. Micholitz, original collector of the species who worked for the English nursery, Sanders)

**Description:** A small cycad with an unbranched, subterranean trunk which is swollen at the base. **Mature leaves** two to five in an erect to spreading sparse crown, 2–3 m long, bright green, flat in cross-section; **petiole** 20–35 cm long; **leaflets** 15–20 cm long, each forked two or three times into narrow, linear, somewhat irregular segments, widely spaced and not crowded, inserted at about 50° to the rhachis, margins entire, apex acute to acuminate; **lower leaflets** reduced to small thorns. **Male cones** 12–18 cm x 4–6 cm, linear-ovoid, erect, pale brown; **sporophylls** with a very short acuminate apex. **Female cones** hairy; **sporophylls** 8–12 cm long, the appendage deeply divided into about fifteen linear-tapered, pointed lobes. **Seeds** about 2 cm x 1.8 cm, oblate.

**Distribution and Habitat:** Known with certainty from south-western China (Guanxi Province) near the border with Vietnam and possibly also occurring in adjacent areas of Laos and Vietnam. It grows in shady situations in rainforest in tropical areas which have a hot, humid climate. The rainfall is above 1500 mm per annum distributed throughout the year.

**Notes:** *C. micholitzii* var. *micholitzii* is remarkable for its relatively small stature within the genus and also its dichotomously divided leaflets which impart to the leaves a most un-*Cycas*-like appearance. Described in 1905 from material collected by Micholitz from the area then known as Annam, it remained virtually unknown until its rediscovery a few years ago. It is still a rare and poorly known species. It is recorded that plants occur widely dispersed in their natural habitat and do not grow in colonies.

**Cultivation:** Suited to tropical and warm subtropical regions. This species, which is very rare in cultivation, requires warm, humid conditions with a buoyant atmosphere. Plants apparently appreciate some protection from direct, hot sun and are very frost sensitive.
**Propagation:** From seed.

## *Cycas micholitzii* var. simplicipinna

Smitinand
(with simple, undivided leaflets)

**Description:** A small cycad with a subterranean or emergent trunk to 0.3 m long and 30 cm across. **Mature leaves** about six in an erect to spreading sparse crown, 1–1.5 m x 30–40 cm, bright green, flat in cross-section, straight in profile; **petiole** 15–24 cm long; **leaflets** 30–40 on each leaf, moderately spaced and not crowded, evenly distributed throughout, inserted on the rhachis at about 45°; **median leaflets** 18–22 cm x 1–2.5 cm, linear-lanceolate, papery textured, straight or slightly falcate, glabrous, margins wavy, apex acuminate, somewhat prickly; **lower leaflets** hardly reduced, abruptly replaced by a few short thorns. **Male cones** 15–21 cm x 2–4 cm, narrowly cylindrical, reddish brown, hairy; **sporophylls** about 1.4 cm x 1 cm, spathulate, relatively soft with a short apical point, hairy inside; **peduncle** 2–3.5 cm long, rusty hairy. **Female cones** loose and open, rusty brown, hairy; **sporophylls** 8–12 cm long, rusty hairy; **apical lobe** 4–4.5 cm x 3–3.5 cm, rhomboid, hairy, margins with long tapered, pointed lobes; **ovules** two on each sporophyll. **Seeds** 2.5–2.7 cm x 2 cm, ellipsoid, dark green, usually only one maturing on each sporophyll.
**Distribution and Habitat:** Occurring in northern Thailand and western China; possibly also found in Burma and Laos. This species grows in evergreen forests of mountainous regions at altitudes of 600–1100 m. The climate is tropical being hot, wet and humid in summer and mild in winter.
**Notes:** Although named as a variety of *C. micholitzii* in 1971, this taxon is obviously distinct and worthy of specific rank.
**Cultivation:** Suited to tropical and subtropical regions. Best grown in a protected, shady or partially shady position. Requires free drainage and responds to mulches and watering during dry periods. Tolerance of cold is very low.
**Propagation:** From seed.

## *Cycas normanbyana*

F. Muell.
(after the Marquis of Normanby, former Governor of Queensland)

This species was described in 1874 from material collected by Eugene Fitzalan in hills near the estuary of the Burdekin River in central-eastern Queensland, Australia. Because of the specific epithet, its habitat has been erroneously linked to the Normanby Ranges which are further south (see entry for *C. kennedyana*, page 143). The species occurs within the natural range of *C. media* and

K. HILL

Colony of *Cycas ophiolitica* in habitat.

L

Ripening seeds of *Cycas ophiolitica*.

examination of the type and field collections by Ken Hill show that the two are conspecific, with *C. media* being the accepted name.

## *Cycas ophiolitica*

K. Hill
(growing on soils of serpentine origin)

**Description:** A small to medium-sized cycad with an erect trunk to 2 m tall and 20 cm across (rarely to 4 m tall), often swollen at the base, with occasional offsets produced on the trunk. **Young leaves** powdery blue to blue, densely covered with greyish and some brown hairs. **Mature leaves** numerous in a dense, spreading to rounded crown, 1–1.4 m x 20–30 cm, blue green to dark green, glossy, deeply vee-ed in cross-section, arching in profile; **petiole** 18–35 cm long, hairy; **leaflets** 170–220 on each leaf, crowded, the lower ones more widely spaced, inserted on the rhachis at about 60°; **median leaflets** 15–24 cm x 0.6–0.75 cm, linear, straight, leathery, margins flat or slightly recurved, glabrous or hairy, apex acute; **lower leaflets** reduced, abruptly replaced by a series of short thorns. **Male cones** about 17 cm x 8 cm, cylindrical, brown, hairy; **sporophylls** about 3.5 cm x 1.4 cm, ovate, with a short, erect point to 1.1 cm long. **Female cones** loose and open, hairy; **sporophylls** about 30 cm long, hairy; **apical lobe** 4–7 cm x 2–3 cm, ovate, the margins with short, regular teeth, apex acuminate; **ovules** two to six on each sporophyll. **Seeds** about 3 cm x 3 cm, ovoid to globose, light brown, often powdery.
**Distribution and Habitat:** Endemic to Australia where it occurs between Marlborough and Rockhampton in central-eastern Queensland. The species grows on hills and slopes in sparse, grassy open forest. Although this species reaches its best development on red clay soils near Marlborough it is more frequently found on shallow, stony, infertile soils which are developed on sandstone and serpentine rocks. The climate is tropical with hot, humid summers having a day/night temperature regime to 32°C and 22°C respectively. Winters are mild and dry with a day/night temperature regime to 24°C and 10°C respectively. Rainfall is about 1500 mm per annum falling mainly in summer and autumn.
**Notes:** This species was described in 1992 from material collected near Marlborough in central-eastern Queensland. Its distribution lies between that of *C. media* to the north and *C. megacarpa* in the south. The species can be distinguished from *C. media* by its narrower, more crowded, more or less glaucous leaflets and hairy petioles, and from *C. megacarpa* by its narrower, more crowded leaflets and much smaller seeds. Hybrid intergrades between *C. ophiolitica* and the two adjacent taxa occur where their ranges overlap. Plants cultivated as *Cycas* species 'Mount Morgan' are in fact hybrids between *C. ophiolitica* and *C. megacarpa*.
The most distinctive form of *C. ophiolitica* occurs in the vicinity of Marlborough and this has been cultivated and detailed in literature as *Cycas* species 'Marlborough Blue'. Plants growing in sunny situations on red clay soils derived from serpentine rock have the bluest leaves and these populations have suffered badly from poaching.
**Cultivation:** Suited to tropical, subtropical and warm temperate regions. Although easy to grow, plants of this species only attain good colouration when grown in warm to hot dry climates. In coastal districts they struggle in the humid conditions and lose their distinctive glaucous appearance. Requires a sunny position, excellent drainage and unimpeded air movement. Tolerates light frosts only. Seedlings are generally slow growing.
**Propagation:** From seed and by removal of offsets from the trunk.

## *Cycas panzhihuaensis*

L. Zhou & S. Y. Yang
(from the mining town of Panzhihua, China)

**Description:** A small to medium-sized cycad with an erect trunk 2.5–3 m tall and 15–25 cm across. **Mature leaves** sparse to numerous in an obliquely erect to spreading crown, 0.7–1.2 m x 20–40 cm, older leaves deflexed, bluish to grey-green, flat in cross-section, straight in profile; **petiole** 4–8 cm long; **leaflets** 140–210 on each leaf, moderately crowded, evenly spaced throughout, inserted on the rhachis at about 40°; **median leaflets** 12–23 cm x 0.6–0.7 cm, linear to linear-lanceolate, straight or slightly falcate, leathery, glabrous, margins slightly revolute, apex acuminate; **lower leaflets** abruptly replaced by a series of rigid thorns about 0.2 cm long. **Male cones** 25–45 cm x 8–11 cm, cylindrical to fusiform, slightly curved, brownish yellow; **sporophylls** 4.5–6 cm x 2–2.8 cm, nearly triangular, becoming deflexed, covered with yellowish brown hairs, apex with a rigid beak. **Female cones** loose and open, yellowish brown to rusty brown; **sporophylls** 15–24 cm long, rusty brown, initially hairy, becoming glabrous; **apical lobe** 3.5–5.5 cm x 2–3 cm, ovoid to rhomboid, deeply lobed and appearing pinnate with 30–40 segments each to 4 cm long; **ovules** two to six on each sporophyll. **Seeds** 2.5–3 cm x 2–2.5 cm, oblate, reddish yellow.
**Distribution and Habitat:** Endemic to China where it occurs in a number of areas in the south of the Sichuan Province and in the Yunnan Province. It grows in open situations on the sides of steep rocky hills at altitudes between 1100 m and 2000 m. The climate is subtropical with hot, humid summers and, because of the high altitudes, the winters are cold and dry with severe frosts being common. Rainfall is 760 mm per annum with most

falling between June and October.
**Notes:** This species was described in 1981. One substantial colony grows on the site of a large mine, and extensive efforts have been made to ensure its survival by transplanting mature plants and propagating seedlings. The species is also common in a gazetted reserve nearby from which the seeds are collected annually and propagated in a nearby botanic garden. These are planted out locally in gardens and may also be used for revegetation in wild areas.
This cycad is very tolerant of periods of dryness and it also withstands fires which occur frequently where it grows.
**Cultivation:** Suited to temperate and cool temperate regions. A very cold hardy species which will tolerate heavy frosts and snow. Also withstands dryness. Best grown in an open situation and requires excellent drainage.
**Propagation:** From seed.

## *Cycas papuana*

F. Muell.
(from Papua New Guinea)

**Description:** A medium-sized cycad with an erect, often branched, slender trunk to 4 m tall and 25 cm across. **Young leaves** bright green, with loose brown hairs. **Mature leaves** numerous in an obliquely erect to spreading crown, 0.5–2m x 15–25 cm, bright green, flat in cross-section, straight in profile; **petiole** 10–15 cm long, slender, lacking thorns; **leaflets** 100–200 on each leaf, moderately spaced, evenly distributed throughout; **median leaflets** 10–20 cm x 5–17 mm, linear-lanceolate, papery, shiny green above, paler beneath, midrib yellow, margins flat, apex acuminate; **lower leaflets** ending abruptly, not replaced by a series of short thorns. **Male cones** 10–20 cm x 8–12 cm, cylindrical-ovoid, yellowish brown, with brown hairs; **sporophylls** 3–3.5 cm x 1.3–2 cm, wedge-shaped, with a prominent erect apex about 1.5 cm long. **Female cones** with reddish brown hairs; **sporophylls** 10–25 cm long, hairy; **apical lobe** 3.5–5 cm x 1–1.8 cm, rhomboid, distal margins with sharp teeth 1–4 mm long, apex acuminate; **ovules** two to four on each sporophyll. **Seeds** 3.5–5.5 cm x 2.5–2.8 cm, ovoid to globose, orange to brownish, not powdery.
**Distribution and Habitat:** Apparently endemic to the western Province of southern parts of Papua New Guinea particularly in the region of the Fly River delta. It forms sparse colonies in open woodland and savannah, frequently growing with kunai grass. The climate is tropical with hot, wet, humid summers and long, dry, mild winters.
**Notes:** This species was described in 1876 from material collected in the vicinity of the Fly River. It is very similar to *C. armstrongii* and studies may show the two to be conspecific. Fires are a common occurrence where this species grows and plants may be deciduous for part of the dry season.
**Cultivation:** Suited to tropical regions which have a seasonally dry climate. Best in full sun or filtered sun. Requires excellent drainage.
**Propagation:** From seed or by removal of trunk offsets.

## *Cycas pectinata*

Griffith
(spreading like the teeth of a comb, in reference to the lobes on the apex of the female sporophylls)

**Description:** A medium-sized to large cycad with an erect trunk to 4 m tall and 30 cm across, often branched in its upper parts. **Mature leaves** numerous in an erect to widely spreading crown, 1.5–2 m x 30–50 cm, dark green, flat in cross-section, recurved in profile; **petiole** 20–35 cm long; **leaflets** 80–180 on each leaf, moderately spaced, evenly distributed throughout, inserted on the rhachis at about 40°; **median leaflets** 14–25 cm x 0.4–1.2 cm, linear-lanceolate, straight or slightly falcate, leathery, glabrous, margins flat, apex mucronate; **lower leaflets** reduced and abruptly replaced by a series of widely spaced, rigid thorns about 0.2 cm long. **Male cones** 30–40 cm x 12–15 cm, cylindrical to narrowly ovoid, brown, hairy; **sporophylls** 3.5–5 cm x 1–2.4 cm, triangular, apex thickened with an upcurved point about 4 cm long. **Female cones** loose and open, brown; **sporophylls** 15–20 cm long, covered with yellowish brown, silky hairs; **apical lobe** 6–7.5 cm x 6–7.5 cm, broadly triangular to nearly orbicular, the margins with numerous long, almost spiny, lobes or teeth (to 2 cm long), the apex drawn out into an acuminate point about 2.5 cm long; **ovules** four to six on each sporophyll. **Seeds** 4–6 cm x 3–4.5 cm, ovoid, orange to yellow.
**Distribution and Habitat:** Occurring in northern India, Bangladesh, Burma, Vietnam, Laos, Cambodia, Thailand and Malaysia. This species grows in mountainous areas to about 600 m altitude. It is usually found growing in exposed sunny positions in poor soil and may also grow on limestone outcrops. It is also sometimes found in evergreen forests. The climate is tropical with wet, humid summers and mild, dry winters. The rainfall is above 1000 mm per annum with most falling over summer.
**Notes:** This species, described in 1854, is in need of a new specific epithet since the combination *C. pectinata* was first validly published in 1876. Although distinctive, this species is included with *C. siamensis* by some authors. *C. pectinata* can be distinguished by its trunk not being abruptly swollen at the base, by the fusiform to ovoid male cones and the broad apex of the female sporophylls with the margins having long teeth. It is re-

K. HILL

Cycas platyphylla in habitat.

**Cultivation:** Suited to tropical and warm subtropical regions. Grows readily in full sun to partial sun or filtered sun in well-drained soil. Frost-tolerance is not high.
**Propagation:** From seed.

# *Cycas platyphylla*

K. Hill
(with broad leaflets — in reference to the apical lobe of the female sporophyll)

**Description:** A small to medium-sized cycad with an erect, unbranched trunk to 2 m tall and 30 cm across (rarely to 4 m tall), often swollen at the base. **Young leaves** glaucous to strongly bluish-green, densely covered with orange brown, woolly hairs. **Mature leaves** numerous in a stiffly spreading crown, 0.5–1.1 m x 10–20 cm, glaucous green to green or yellow green, somewhat shiny, broadly and deeply vee-ed in cross-section, straight in profile or gently recurved near the apex; **petiole** 12–24 cm long, glabrous; **leaflets** 120–260 on each leaf, 10–17 cm x 0.4–0.6 cm, linear, straight or slightly falcate, leathery, hairy, margins recurved, apex abruptly acute; **lower leaflets** abruptly replaced by a series of short thorns. **Male cones** 15–20 cm x 8–11 cm, ovoid, brown, hairy; **sporophylls** 2–3 cm x 1–1.3 cm, ovate, with an erect point to 0.9 cm long. **Female cones** about 40 cm across, loose and open, hairy; **sporophylls** 20–30 cm long, loosely hairy; **apical lobe** 5–8 cm x 1.5–3.5 cm, ovate to nearly orbicular, hairy, the margins with numerous, short, crenate lobes, apex long-acuminate; **ovules** four to six on each sporophyll. **Seeds** 3–4 cm x 2.5–4 cm, ovoid to globular, green to yellowish, densely powdery.
**Distribution and Habitat:** Endemic to Australia where it is restricted to the Petford district on the western side of the Atherton Tableland in north-eastern Queensland, and a disjunct population about 250 km to the south on Wandovale Station. Plants grow in scattered groups among grass in sparse open forest, often on gentle rocky slopes. The climate is tropical with hot, wet, humid summers having a day/night temperature regime to 31°C and 21°C respectively. Winters are cool to cold and dry with a day/night temperature regime to 25°C e and 9°C respectively. Rainfall is about 800 mm per annum falling mainly in summer.
**Notes:** This species was described in 1992 from material collected in the vicinity of Petford in north-eastern Queensland. It has long been confused with *C. cairnsiana* from which it can be distinguished by the initially blue leaves which age to yellowish green and the broader leaflets with recurved margins. Plants can be readily separated from *C. media* by the recurved leaflet margins. An extensive hybrid interzone exists between this species and adjacent populations of *C. media*. Plants from these areas exhibit extreme morphological variation. A significant and attractive feature of plants of *C. platyphylla* is a spectacular flush of hairy new leaves.
**Cultivation:** Suited to highland regions of the tropics, subtropical and temperate regions. Plants attain their most attractive appearance in warm to hot, dry conditions. They

K. HILL

Male cone of *Cycas platyphylla*.

corded as being a very drought resistant species.
One variety from Vietnam, *C. pectinata* var. *elongata* J. Leandri, has marginal lobes of the female sporophylls longer than in the typical variety.

are best grown in a sunny situation and require excellent drainage. In shady or humid conditions the plants lose their attractive appearance and may suffer from debilitating attacks by leaf fungi. Tolerates moderately heavy frosts.
**Propagation:** From seed.

## *Cycas pruinosa*

Maconochie
(covered with a powdery bloom)

Old male cone of *Cycas pruinosa*.

**Description:** A small cycad with an erect trunk to 2 m tall and 40 cm across. **Young leaves** hairy. **Mature leaves** numerous in a widely spreading to arching or recurved crown, 0.8–1 m x 15–30 cm, grey-green to bluish grey, deeply vee-ed or broadly U-shaped in cross-section, especially in the proximal two-thirds of the leaf, gently recurved in profile; **petiole** 4–6 cm long, tetragonal in cross-section; **leaflets** 120–240 on each leaf, crowded and evenly spaced throughout except those towards the base, inserted on the rhachis at about 50°, obliquely erect or upcurved, those towards the apex curved forwards and the tips crowded; **median leaflets** 11–20 cm x 0.2–0.4 cm, linear, straight or slightly falcate, leathery, glabrous, margins strongly revolute, base narrowed but not decurrent, apex pungent, mucro about 2 mm long, yellow; **lower leaflets** reduced in size and more widely spaced, abruptly replaced by a series of rigid thorns about 0.3 cm long. **Male cones** 38–50 cm x 8–9 cm, narrowly cylindrical, bluish green maturing to light brown, shortly hairy; **sporophylls** 1.5–2 cm x 1.5 cm, deltoid, with a hairy appendage about 1.4 cm long. **Female cones** loose and open, rusty brown, hairy, the sporophylls incurved; **sporophylls** 27–30 cm long, rusty brown, hairy; **apical lobe** 14–15 cm x 1–1.5 cm, ovate, with twenty to thirty teeth to 2.5 cm long; **ovules** four on each sporophyll. **Seeds** 3–4 cm x 3–3.5 cm, oblate, covered with grey or white powder.
**Distribution and Habitat:** Endemic to Australia where it occurs in the Kimberley region of northern Western Australia, particularly around Argyle, Kununurra and the Ord River. This species grows among grass and low shrubs on rocky slopes and ridges. The climate is tropical with hot, wet, humid summers having a day/night temperature regime to 36°C and 27°C respectively. Winters are mild to cool and dry with a day/night temperature regime to 29°C and 19°C respectively. Rainfall averages 1000 mm per annum, mostly falling in summer.
**Notes:** Described in 1978 from material collected at Ternonis Gorge in the Durack Range of Western Australia. This species consists of numerous, small, relict populations and is widely distributed in the Kimberley region. It grows on soils arising from a diverse range of rock types including sandstone (Deception Range and Durack Range), granite (Bedford Downs), metamorphics (Ord River) and limestone (Napier Range). Populations may consist predominantly of all-blue specimens, such as occur around the Ord Dam, blue-green to glaucous specimens intermixed with bright green plants, such as occur in the Napier Range, or all of the plants may be green, as in the Deception Range population. *C. pruinosa* can be distinguished from other Australian species by the widely spreading leaves, the erect leaflets with strongly revolute margins, the powdery bloom on the seeds and the apical lobes of the female sporophylls which have a very long,

Female sporophylls and developing seeds of *Cycas pruinosa*.

drawn out apex and numerous marginal teeth.
**Cultivation:** Suited to tropical regions which have a seasonally dry climate. Established specimens transplant readily. Seedlings are slow and may be difficult to establish. Plants are best grown in full sun, with free air movement and excellent drainage. Best in acid soils.
**Propagation:** From seed.

Female sporophylls of *Cycas revoluta*.

## *Cycas revoluta*

Thunb.
(with recurved margins, in reference to those of the leaflets)

**Description:** A medium-sized cycad with a trunk to 3 m tall and 35 cm across, with numerous suckers arising from the base and occasional offsets produced on the trunk. **Young leaves** bright green, bearing brown hairs. **Mature leaves** numerous in a widely spreading, flat crown, 0.6–1.5 cm x 20–25 cm, dark green, flat in cross-section, straight in profile; **petiole** about 10 cm long, four-sided in cross-section; **leaflets** numerous, crowded, evenly distributed throughout, inserted on the rhachis at about 45°; **median leaflets** 10–20 cm x 0.5–0.6 cm, linear, straight or slightly falcate, leathery, glabrous, upper surface shiny, margins strongly recurved, apex acuminate, prickly; **lower leaflets** abruptly replaced by a few short thorns. **Male cones** 10–40 cm x 4–6 cm, narrowly cylindrical to ovoid, brown, hairy, somewhat lax; **sporophylls** 2–3.9 cm x 1–1.7 cm, narrowly wedge-shaped, with a short, upcurved point about 0.5 cm long. **Female cones** loose and open, brown, hairy; **sporophylls** 10–20 cm long, densely covered with brown hairs; **apical lobe** 4–6 cm x 2 cm, ovate, margins deeply laciniate with twelve to eighteen tapered lobes about 2 cm long; **ovules** four to six on each sporophyll, densely hairy. **Seeds** 2–3.5 cm x 1.5–2.5 cm, oblate, loosely hairy, bright orange to red, occasionally yellowish.
**Distribution and Habitat:** Endemic to the Ryukyu Islands and the islands of Mitsuhama Ito and Satsuma to the south of the island of Kyushu, Japan. This species grows in extensive colonies on hillsides, often as the dominant component of the vegetation. The climate is mild in summer and cold in winter.
**Notes:** Although of restricted distribution, this species, which was described in 1782, has become the most commonly cultivated cycad in the world. The species can be distinguished immediately from most other *Cycas* by the strongly revolute margins of its leaflets. It is also the only species of *Cycas* to have hairy seeds. The seeds are used as a food source locally where it grows. This species is commonly known as 'Sago Palm' in the nursery trade.
**Cultivation:** Suited to temperate and subtropical regions but also commonly planted in the tropics. This species is prized for its ornamental appearance, hardiness and adaptability. As a potted cycad it is unexcelled with plants

Young leaves of *Cycas revoluta*.

Male cone of *Cycas revoluta*.

Large clump of *Cycas revoluta*.

Mature seeds and female sporophylls of *Cycas revoluta*.

growing successfully in the limited confines of a container for many years providing their requirements for nutrients and water are met. Often grown in glasshouses and conservatories but also an excellent indoor plant which will tolerate poor light and neglect. Prized as a bonsai subject especially in Japan where highly valued specimens many hundreds of years old exist in collections. Tolerates moderate to heavy frosts and specimens planted in the open ground are best placed in full sun. Watering during dry periods, mulches and regular light fertiliser applications are beneficial practices.

**Propagation:** From seed and by removal of basal suckers and trunk offsets.

Male cone of *Cycas rumphii.*

## *Cycas riuminiana*

Porte ex Regel
(after M. Riumin)

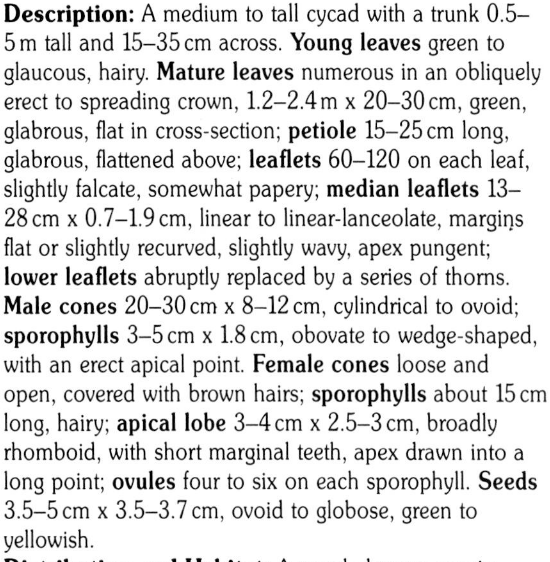

**Description:** A medium to tall cycad with a trunk 0.5–5 m tall and 15–35 cm across. **Young leaves** green to glaucous, hairy. **Mature leaves** numerous in an obliquely erect to spreading crown, 1.2–2.4 m x 20–30 cm, green, glabrous, flat in cross-section; **petiole** 15–25 cm long, glabrous, flattened above; **leaflets** 60–120 on each leaf, slightly falcate, somewhat papery; **median leaflets** 13–28 cm x 0.7–1.9 cm, linear to linear-lanceolate, margins flat or slightly recurved, slightly wavy, apex pungent; **lower leaflets** abruptly replaced by a series of thorns. **Male cones** 20–30 cm x 8–12 cm, cylindrical to ovoid; **sporophylls** 3–5 cm x 1.8 cm, obovate to wedge-shaped, with an erect apical point. **Female cones** loose and open, covered with brown hairs; **sporophylls** about 15 cm long, hairy; **apical lobe** 3–4 cm x 2.5–3 cm, broadly rhomboid, with short marginal teeth, apex drawn into a long point; **ovules** four to six on each sporophyll. **Seeds** 3.5–5 cm x 3.5–3.7 cm, ovoid to globose, green to yellowish.
**Distribution and Habitat:** A poorly known species apparently endemic to the Philippines. Details of habitat are not recorded.
**Notes:** This species was described in 1863 from plants cultivated in Moscow and Leningrad. Its affinities need elucidating.
**Cultivation:** Suited to tropical and warm subtropical regions. Probably requires a sheltered situation, humid conditions and regular watering. Tolerance of frost is unlikely to be high.
**Propagation:** From seed.

## *Cycas rumphii*

Miq.
(after G. F. Rumphius)

**Description:** A medium-sized to large cycad with an erect, commonly branched trunk to 7 m tall and 45 cm across. **Young leaves** light green. **Mature leaves** numerous in an obliquely erect to widely spreading crown, 1–2 m x 40–70 cm, dark green, flat in cross-section, straight in profile; **petiole** 30–45 cm long; **leaflets** 50–100 on each leaf, moderately spaced and not crowded, evenly distributed throughout, inserted on the rhachis at about 60°; **median leaflets** 20–38 cm x 1.2–1.9 cm, linear-lanceolate, straight or slightly falcate, leathery, glabrous, upper surface shiny, margins flat, midrib flat, apex acuminate; **lower leaflets** reduced and abruptly replaced by a series of rigid thorns about 0.4 cm long, the lower ones deflexed. **Male cones** 30–60 cm x 10–12 cm, cylindrical to narrowly ovoid, reddish brown, hairy; **sporophylls** 3.5–5 cm x 1.5–2 cm, ovate, hairy, with an upcurved point about 1 cm long. **Female cones** loose and open, reddish brown, hairy; **sporophylls** 20–25 cm long, reddish brown, hairy; **apical lobe** 6–9 cm x 2–2.5 cm, rhomboid, margins with small, often obscure teeth, apex acuminate; **ovules** six to ten on each sporophyll. **Seeds** 5–7.5 cm x 4–4.5 cm, oblong to narrowly ovoid, brownish, with a spongy flotation layer.
**Distribution and Habitat:** Occurring in coastal districts of India and the offshore islands, Sri Lanka, Malaysia, Moluccas, Guam, Fiji, New Caledonia, Solomon Islands, Indonesia and New Guinea. Also reported to occur in the southern provinces of China. This species mainly grows in coastal communities often on stabilised dunes. Plants grow in situations ranging from full sun to nearly dense shade. The climate is mostly tropical with hot, wet, humid summers and mild, moist to dry winters. Rainfall is often above 1500 mm per annum distributed throughout the year.
**Notes:** *C. rumphii* was described in 1859 from material collected in the Moluccas. It is essentially a species of coastal and near-coastal districts because the seeds, which float and retain their viability, are distributed widely on ocean currents. Distribution is compounded by the activities of humans since the species is used as a food source in many countries. The comments on distribution, adaptation and variation noted under *C. circinalis* are equally applicable to this species. *C. rumphii* and *C. circinalis* are very closely related and some botanists prefer to treat them as a single, variable species. While *C. rumphii* has much larger seeds than *C. circinalis*, other features, including, shorter leaves, fewer, less crowded leaflets, taller, more frequently branched trunks, shorter

Clump of *Cycas rumphii*.

points on the male sporophylls and much shorter teeth on the margins of the female sporophylls, have been noted by various authors as other differentiating criteria. The seeds of this species are rich in starch and, after suitable preparation, are eaten by people of various cultures.
**Cultivation:** Suited to tropical and warm subtropical regions. Very commonly grown in tropical parts of the world and widely planted as an ornamental in public and private gardens. Grows well in sun or partial shade and adapts to most well-drained soil types. Plants are tolerant of some cold but should be protected from frosts.
**Propagation:** From seed and by removal of trunk offsets.

## *Cycas scratchleyana*

F. Muell.
(after General Scratchley, first Goveror of British New Guinea)
**Description:** A small cycad with an erect, slender trunk to 1.5 m tall and 20 cm across. **Young leaves** bright green. **Mature leaves** numerous in an obliquely erect to spreading crown, 0.5–1 m x 20–25 cm, dark green and shiny, flat in cross-section, straight in profile; **petiole** 15–25 cm long, slender; **leaflets** 150–200 on each leaf; **median leaflets** 20–28 cm x 1–1.3 cm, linear, somewhat papery but rigid, dark green and shiny on both surfaces, paler beneath, margins flat, apex long-acuminate; **lower leaflets** abruptly replaced by a series of short thorns. **Male cones** about 28 cm x 13 cm, cylindrical; **sporophylls** 1.5–3 cm x 0.5–2.5 cm, wedge-shaped, lacking a pungent apical point. **Female cones** loose and open, covered with brown hairs; **sporophylls** about 12 cm long, hairy; **apical lobe** 4–5 cm x 2.5–3.2 cm, rhomboid, with a few marginal teeth; **ovules** two to four on each sporophyll. **Seeds** 4–5 cm x 2–3 cm, ovoid, green to brown.
**Distribution and Habitat:** Endemic to Papua New Guinea where known from inland regions at moderate elevations. It apparently grows on sheltered slopes in or close to rainforest.
**Notes:** This poorly known species was described in 1885 from material collected by W. Armit on Mt Bedford in the vicinity of the Jala River in Dedouri Country. Its relationship with other species of *Cycas* needs elucidating.
**Cultivation:** Suited to tropical and warm subtropical regions. Probably requires a sheltered situation, humid conditions and regular watering. Tolerance of frost is unlikely to be high.
**Propagation:** From seed.

## *Cycas siamensis*

Miq.
(from Siam)

**Description:** A small cycad with a subterranean or emergent trunk to 2 m long and 30 cm across, abruptly swollen at the base. **Young leaves** bearing red-brown hairs. **Mature leaves** numerous in a widely spreading crown, 0.6–1.2 m x 10–40 cm, dark green, flat in cross-section, recurved in profile; **petiole** 20–35 cm long; **leaflets** 70–180 on each leaf, fairly crowded, evenly distributed throughout, inserted on the rhachis at about 40°; **median leaflets** 10–20 cm x 0.5–0.7 cm, linear to linear-lanceolate, straight or slightly falcate, leathery, glabrous, margins flat, apex mucronate; **lower leaflets** reduced and abruptly replaced by a series of widely spaced, rigid

thorns about 0.2 cm long. **Male cones** 20–30 cm x 6–8 cm, narrowly oblong, brown, hairy; **sporophylls** 1.5–2 cm x 1.5–1.8 cm, linear-ovoid, with a slender upcurved point about 2 cm long. **Female cones** loose and open, brown; **sporophylls** 12–16 cm long, covered with tawny hairs; **apical lobe** 4–6 cm x 3–5 cm, more or less broadly rhomboid, the margins with numerous narrow teeth to 1 cm long, the apex drawn out into an acuminate point about 5 cm long; **ovules** two to four on each sporophyll. **Seeds** 3.5–4 cm x 3–3.5 cm, ovoid to oblong, brownish.
**Distribution and Habitat:** Occurring in northern India, Burma, Thailand, Laos, Vietnam and the Yunnan Province of China. This species grows in drier types of forests which contain a high proportion of deciduous species, although in China it is reported from broad-leaved evergreen forests. The soils are often stony and may be lateritic. The climate is tropical with hot, humid summers and mild, dry winters. The rainfall is up to more than 2000 mm per annum, falling mainly in summer.
**Notes:** This species was described in 1863 from material collected in Siam. It is very similar to *C. pectinata* (see that species for differences). *C. immersa* W. G. Graib, which was described in 1912 from material collected in Laos, is synonymous.
**Cultivation:** Suited to tropical and warm subtropical regions. Requires excellent drainage and can be grown in filtered sun, but may require protection from direct hot sun. Intolerant of dryness and responds to mulches and regular watering. Plants are not particularly tolerant of cold, especially frosts.
**Propagation:** From seed.

# *Cycas silvestris*

K. Hill
(growing in forests)

**Description:** A medium-sized cycad with a trunk to 4 m tall and 15 cm across, lacking offsets or suckers. **Young leaves** light green, densely covered with greyish and some brown hairs. **Mature leaves** 1–2 m x 20–30 cm, numerous in an obliquely erect to rounded crown, bright green to dark green, shiny, flat in cross-section, straight in profile; **petiole** 40–50 cm long, glabrous; **leaflets** 90–150 on each leaf, moderately spaced, evenly distributed throughout, inserted on the rhachis at about 80°; **median leaflets** 15–30 cm x 0.9–1.5 cm, linear, straight or slightly falcate, leathery, shiny, glabrous, midrib yellowish, margins flat, apex acute; **lower leaflets** reduced, abruptly replaced by a series of short thorns. **Male cones** not seen; **sporophylls** about 5 cm x 3 cm, ovate, with a short, erect point. **Female cones** loose and open, hairy; **sporophylls** about 30 cm long, with brown hairs; **apical lobe** about 4 cm x 2 cm, narrowly triangular, the margins with numerous short, crenate lobes, apex long-acuminate; **ovules** four to ten on each sporophyll. **Seeds** about 3.5 cm x 3 cm, almost globular, brownish yellow, not powdery.
**Distribution and Habitat:** Endemic to Australia where it occurs on Cape York Peninsula in north-eastern Queensland, mainly in the vicinity of the Olive River estuary and with scattered occurrences south to near the Pascoe River. This species is strictly coastal and grows on stabilised dunes in Melaleuca scrub and in littoral rain-forest which often has emergent Hoop Pines. Soils are white to grey, siliceous sands. The climate is tropical with hot, wet, humid summers having a day/night temperature regime to 30°C and 24°C respectively. Winters are mild and generally dry, although sporadic coastal showers may occur. Winter days range from 27°C to 13°C at night. Rainfall is about 1500 mm per annum, falling mainly over summer.
**Notes:** This species was described in 1992 from material collected near Bolt Head close to the Olive River estuary on north-eastern Cape York Peninsula. It is a species of restricted distribution which has a similar general appearance to plants of *C. circinalis* and *C. rumphii*, but lacks the large, floating, water-dispersed seeds of those species. Seedlings of *C. silvestris* also have very spiny petioles and its seeds are quite small. Its closest relative is probably *C. media*, from which it can be distinguished by the broader, shiny, relatively thin-textured leaflets.
**Cultivation:** Suited to tropical and subtropical regions, especially coastal localities. Requires well-drained soil and probably responds to mulches and watering during dry periods. Best in full sun or filtered sun. Frost-tolerance is unlikely to be high.
**Propagation:** From fresh seed.

# *Cycas szechuanensis*

W. C. Cheng & L. K. Fu
(from the area of Sichuan, China)

**Description:** A small to medium-sized cycad with a trunk to 3 m long and 40 cm across. **Mature leaves** numerous in a widely spreading crown, 0.5–1 m x 40–60 cm, dark green, flat in cross-section, straight or recurved in profile; **petiole** 15–25 cm long; **leaflets** fairly crowded, evenly distributed throughout, inserted on the rhachis at about 50°; **median leaflets** 20–30 cm x 1.2–1.3 cm, linear-lanceolate to lanceolate, straight or slightly falcate, leathery, glabrous, margins flat, apex mucronate; **lower leaflets** abruptly replaced by a few small thorns. **Male cones** about 25 cm x 6 cm, brown, hairy. **Female cones** loose and open, brown; **sporophylls** 19–23 cm long; **apical lobe** rounded, the margins deeply lobed; **ovules** six to eight on each sporophyll. **Seeds** not recorded.
**Distribution and Habitat:** Endemic to central China in

the Sichuan Province and Yunnan Province. It grows in sheltered forests in fertile soils. The climate is mild with significant periods of drizzly weather and fogs.
**Notes:** This poorly known species was described in 1975. Its closest relative is apparently *C. siamensis*.
**Cultivation:** Suited to temperate and cool subtropical regions. Plants are apparently not tolerant of dryness and respond to mulches and regular watering during dry periods.
**Propagation:** From seed.

## *Cycas taiwaniana*

Carruthers
(from Taiwan)

**Description:** A medium-sized cycad with a trunk to 3.5 m tall and 35 cm across, with occasional offsets produced on the trunk. **Mature leaves** few to numerous in an obliquely erect to widely spreading crown, 1–1.8 m x 40–50 cm, dark green, flat in cross-section, straight in profile; **petiole** 10–20 cm long, angular in cross-section; **leaflets** numerous, moderately crowded, evenly distributed throughout, inserted on the rhachis at about 50°; **median leaflets** 15–22 cm x 0.4–0.8 cm, linear, straight or slightly falcate, leathery, glabrous, upper surface shiny, margins flat and not revolute, apex acuminate with a yellow mucro about 2 mm long; **lower leaflets** reduced and more widely spaced, abruptly replaced by a series of short thorns. **Male cones** 30–50 cm x 9–10 cm, narrowly cylindrical to narrowly ovoid, stiffly erect, brown, hairy; **sporophylls** 3–4 cm x 1.5–1.8 cm, linear-spathulate, with a very short point. **Female cones** loose and open, brown, hairy; **sporophylls** 10–15 cm long, covered with brown hairs; **apical lobe** 4–7 cm x 4.5–5.5 cm, ovate, margins with numerous tapered lobes about 2 cm long; **ovules** two to four on each sporophyll. **Seeds** 4–4.5 cm x 2.5–3 cm, ellipsoid, bright red becoming black with age.
**Distribution and Habitat:** Occurring on the islands of Taiwan and Hainan (Hainandao) and south-eastern China around Canton (Guangzhou) and Fukien (Fujian). This species grows at relatively low altitudes (400–900 m) on steep, exposed hillsides in shallow, stony, skeletal soils. The climate is warm to hot and humid for much of the year.
**Notes:** This species was first collected on Taiwan (then Formosa) in 1867 by a Mr Swinhoe and described as new in 1893. It is recorded that the local people use the leaf to flavour tea. The species is now reduced to rarity in Taiwan as the result of clearing and overcollecting and is protected in a 300 ha nature reserve north-west of Taitung City.
**Cultivation:** Suited to tropical, subtropical and warm temperate regions. An attractive, fast-growing species which is best when planted in full sun. Requires unimpeded drainage and responds to watering and fertilisers. Tolerates light frosts only.
**Propagation:** From seed.

Cataphylls and emerging leaves of *Cycas taiwaniana*.

## *Cycas thouarsii*

R. Br. ex Gaudich.
(after Louis Marie Aubert du Petit-Thouars, eighteenth and nineteenth century French botanist)

**Distribution and Habitat:** A large cycad which develops an erect trunk to 10 m tall and 45 cm across, with offsets produced along the trunk. **Young leaves** light green, arising in whorls. **Mature leaves** bright green or bluish green, shiny, numerous (up to forty) in a rounded crown, 1.5–3 m x 30–60 cm, flat in cross-section, straight

Mature seeds and female sporophylls of *Cycas taiwaniana*.

D. W. STEVENSON

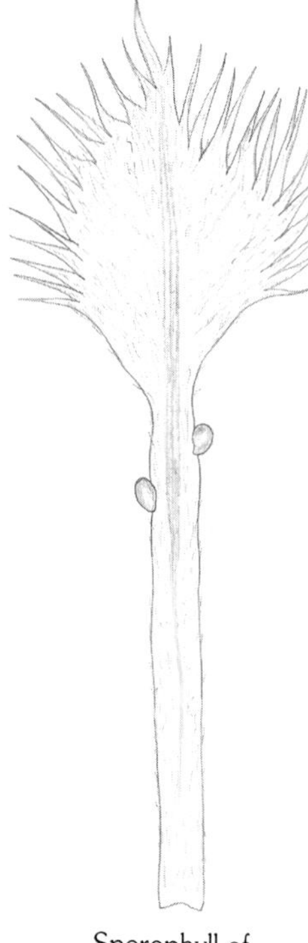

Sporophyll of
*Cycas wadei*

*Cycas taiwaniana*.

in profile; **leaflets** linear to linear-lanceolate, moderately spaced, leathery, falcate, apex acute, margins thickened, basal margin decurrent on the rhachis, inserted on the rhachis at 60–90°; **median leaflets** 20–38 cm x 0.7–2 cm; **basal leaflets** reduced in size, abruptly replaced by a series of rigid thorns about 0.5 cm long. **Male cones** 30–60 cm x 11–20 cm, cylindrical, yellowish orange, with a pungent odour; **sporophylls** spreading horizontally, the outer face triangular with the tip sharply upturned, velvety; **peduncle** 3–5 cm long, velvety. **Female cones** loose and open, velvety; **sporophylls** 30–35 cm long, velvety; **apical lobe** 7–10 cm x 2–3 cm, ovate-lanceolate, irregularly lobed; **ovules** eight to ten on each sporophyll. **Seeds** 5–6 cm x 4–5 cm, oblate, wrinkled, brick red, with a spongy flotation layer.

**Distribution and Habitat:** Occurring on the east coast of Madagascar, islands of the Comores and Seychelles Groups and on the east coast of Africa at the delta of the Zambesi River in Mozambique and further north in the vicinity of Dar-es-Salaam, Tanzania and in Kenya. This species grows in open forest and woodland usually in coastal and near-coastal districts. The climate is hot and humid in summer and mild and frost free in winter. Rainfall ranges from 1000 mm to 2000 mm per annum with a summer maximum.

**Notes:** The natural occurrence of this species on the continent of Africa is subject to conjecture, with some authors maintaining that because of its attractive habit and value as a food plant, the species was spread by Arab traders. These origins are probably now lost in time; however, at present the species is commonly grown along the east coast of Africa and appears to be naturalised (if not natural) on the coastal plain north of Dar-es-Salaam. Although this species is extremely similar to *C. rumphii* and *C. circinalis* it can be distinguished from both by the ovate-lanceolate apex on the female sporophylls and the larger red seeds.

The epithet of this species has an interesting history. Although it is named after the French botanist Louis du Petit-Thouars, he actually misidentified it as *C. circinalis* when he described and figured the species in 1804. The species was later recognised as distinct by Robert Brown and described by Charles Beaupre Gaudichaud in 1829. It is named in honour of Thouars.

Studies of the chromosomes of *C. thouarsii* suggest that it is the oldest living species of *Cycas* (and perhaps the oldest of all living Cycadales) and may have been around for about 140 million years.

**Cultivation:** Suited to tropical and subtropical regions. Although adaptable, best appearance is achieved in light shade in hot, humid conditions. Excess sun and a dry atmosphere will cause stunted leaves and a yellowish or bleached colouration. Very fast growing and in suitable conditions three or four flushes of leaves may occur in a year. Very frost-tender. Responds well to mulches and watering during dry periods.

**Propagation:** From seed and also by the removal of offsets which may take about twelve months to produce new leaves.

## Cycas undulata

Hort. ex Guadich.
(with wavy margins)

**Description:** A small to medium cycad with a trunk about 0.5 m long and 20 cm across. **Mature leaves** about thirty in an obliquely erect to spreading crown, 0.5–1 m x 20–30 cm, dark green, shiny, flat in cross-section, straight or slightly curved in profile; **petiole** 20–35 cm long, slender; **leaflets** 80–150 on each leaf, moderately spaced, evenly distributed throughout, inserted on the rhachis at about 80°; **median leaflets** 10–25 cm x 1–1.3 cm, linear-lanceolate, straight or slightly falcate, leathery, dark green and shiny, margins distinctly wavy, apex acuminate; **lower leaflets** reduced and abruptly replaced by a series of rigid thorns; details of cones and seeds lacking.

**Distribution and Habitat:** Apparently known from Vietnam where it probably grows in sheltered forests. The climate is tropical with hot, humid summers and warm, moist to dry winters.

**Notes:** This poorly known species was described in 1817–20 from material cultivated in Paris. It was collected on the nineteenth-century voyage of Henri Louis de Freycinet. Its relationships with *C. inermis*, *C. siamensis* and *C. balansae* are in need of critical study.

**Cultivation and Propagation:** As for *C. inermis*.

## Cycas wadei

Merr.
(after Dr H. Windsor Wade)

**Description:** A medium-sized cycad with a trunk to 1 m tall and 40 cm across, swollen at the base and often branching, with prominent rings on the trunk. **Young leaves** covered with cinnamon brown hairs. **Mature leaves** numerous in an obliquely erect to spreading crown, 0.5–0.8 m x 25–30 cm, dark green with prominent paler midribs to the **leaflets**, flat in cross-section, straight in profile; **petiole** 10–20 cm long, upper surface angular; **leaflets** about 200 on each leaf, moderately crowded, evenly distributed throughout, inserted on the rhachis at 70–80°; **median leaflets** 15–20 cm x 0.4–0.5 cm, linear to linear-lanceolate, straight or slightly falcate, rigid, leathery, margins flat, narrowed to the base, apex acuminate, pungent; lower leaflets abruptly replaced by a series of short thorns about 1.5 mm long. **Male cones** 40–70 cm x 9–10 cm, cylindrical, brown, hairy; **sporophylls** 2–3 cm x 1.5–2 cm, rhomboid, with a short point. **Female cones**

loose and open, rusty brown, hairy; **sporophylls** 18–22 cm long, densely covered with rusty brown hairs; **apical lobe** 8–10 cm x 6–8 cm, ovate, the margins with numerous, rigid, pointed lobes to 3.5 cm long, the apex long-acuminate; **ovules** two to four on each sporophyll. **Seeds** 3.5–4 cm x 2.5–3 cm, ovoid to ellipsoid, brown, shiny.

**Distribution and Habitat:** Endemic to the Philippines where it is restricted to the island of Palawan, growing in the vicinity of Culion. This species grows in open grasslands of lowland regions. The climate is tropical with hot, wet, humid summers and mild moist winters.

**Notes:** Discovered in 1902 but not described until 1936 after collection of suitable fertile material in 1931. Much of the area where this species occurred has been used as a leper colony and there is some doubt as to whether the species is still extant in the wild. Specimens were transplanted to the grounds of the Bureau of Science in Manila. The species can be distinguished by its tall branched trunks swollen at the base and its narrow, rigid leaflets with prominent pale midribs.

**Cultivation:** Suited to tropical and warm subtropical regions. Requires an open situation and watering during dry periods. Cold-sensitive and requires protection from frosts.

**Propagation:** From seeds and perhaps by offsets from the trunk.

## *Cycas* sp. 'Pine Creek'

**Description:** A medium-sized cycad with an erect, unbranched trunk to 4 m tall and 15 cm across. **Young leaves** bright green, lightly covered with loose, brown hairs. **Mature leaves** numerous in an obliquely erect to stiffly spreading crown, 0.5–1 m x 12–20 cm, light green to bright green, flat in cross-section, straight in profile; **petiole** 20–50 cm long, slender; **leaflets** 80–140 on each leaf, crowded and often overlapping, evenly distributed throughout or those towards the apex more crowded, middle ones inserted on the rhachis at about 60°, proximal ones nearly at right angles, all with a decurrent base; **median leaflets** 8–14 cm x 0.6–0.8 cm, linear, curved towards the apex of the frond, thin-textured, glabrous, midrib thick, white, margins flat, yellowish, apex acuminate, with a prominent yellow mucro; **lower leaflets** abruptly replaced by a series of short thorns. **Male cones** about 20 cm x 12 cm, ovoid, brown hairy; **sporophylls** about 3.5 cm x 2 cm, ovate, with an erect apex about 2 cm long. **Female cones** loose and open, sparsely hairy; **sporophylls** about 20 cm long; **apical lobe** about 5 cm x 3 cm, ovate-triangular, with short, sharp teeth, apex acuminate; **ovules** four on each sporophyll. **Seeds** about 6 cm x 4 cm, pear-shaped, brownish purple.

**Distribution and Habitat:** Endemic to Australia where it occurs in the vicinity of Pine Creek to the north of Katherine in the Northern Territory. It grows among large granite boulders on slopes and near streams in stunted woodland. Soils are grey gravelly loams. The climate is tropical and strongly seasonal with hot, wet, humid summers having a day/night temperature regime to 39°C and 25°C respectively. Winters are long dry and mild with a day/night temperature regime to 30°C and 18°C respectively. The rainfall is about 900 mm per annum, almost totally falling over summer.

**Notes:** An undescribed species which can be readily distinguished by the long slender petioles and the crowded, often overlapping bright green leaflets. Plants may be deciduous for a short period during the dry season.

**Cultivation:** A very attractive species which is best suited to tropical regions which have a seasonally dry climate. Will grow in full sun, filtered sun or semi-shade. Seedlings can be difficult to establish especially in areas away from the tropics.

**Propagation:** From seed.

GENUS

# *Dioon*

The genus *Dioon* consists of ten living species with all but one being restricted to Mexico, the odd species being *D. mejiae* which is found in Honduras. Those species which occur in Mexico are mostly found in mountainous areas. A couple of widespread species have roughly a north–south distribution whereas most of the others are very localised. For fossil species of *Dioon* see page 28.

**Derivation:** The genus *Dioon* was described by John Lindley in 1843. In this publication the spelling of *Dion* was used and Miquel, in 1848, corrected the spelling to *Dioon*. Recently, this has created considerable debate among cycad scholars as to the correct spelling of the genus. Briefly the arguments have been as to whether Lindley made a mistake in the original spelling (thus an orthographic error) or whether the spelling was intentional. An exact transliteration of the Greek words used to coin the genus would result in *Dioon* not *Dion*; however, Lindley compounded the problem by using the latter spelling in subsequent publications. After presentation of arguments both for and against the use of *Dion* and *Dioon*, the nomenclatural committee voted in favour of the latter, thus Dioon is the correct and mandatory spelling to be used for the genus. The name refers to the paired ovules, unfortunately a feature of the majority of cycad genera (from the Greek, *dis* 'two'; *oon* 'an egg').

**Generic Description:** Terrestrial cycads with a cylindrical or rarely ovoid trunk which is solitary (rarely branched), and partly subterranean or wholly emergent. Leaf bases either falling free or remaining attached after senescence. New leaves emerging in flushes, green or coloured, glabrous or hairy. Mature leaves pinnate, lanceolate in outline, straight or recurved, flat or vee-ed in cross-section. Petiole swollen at the base, glabrous or hairy, lacking any prickles. Rhachis straight or recurved near the apex, lacking any prickles. Leaflets non-articulate, decurrent at the base, opposite to nearly opposite, evenly or unevenly arranged along the rhachis, the lower ones reduced in size and pungent-tipped, entire or with marginal spines. Midrib absent, lateral veins immersed and obscure. Male cones one to a few, long-cylindrical, erect, hairy, pedunculate; sporophylls peltate. Female cones one to a few, ovoid, hairy, pedunculate; sporophylls much elongated. Seeds radiospermic, ovoid or subglobose, smooth or wrinkled, the sarcotesta whitish or cream.

**Notable Generic Features:** Species of *Dioon* can be recognised by their leaves which have non-articulate leaflets broadly decurrent on the rhachis and lack prickles on the petiole and rhachis. Other features include:

- sporophylls with more or less triangular outer facings lacking spines, horns or any other outgrowth;
- female sporophylls leaf-like; and
- ovules borne on a short stalk.

**Recent Studies:** The genus *Dioon* has been well studied in recent times and no major changes in taxonomy are anticipated. For a recent detailed treatment see Sergio Sabato and Paolo De Luca, 'Evolutionary Trends in *Dion* (Zamiaceae)', *American Journal of Botany* 72, no. 9 (1985): 1353–63.

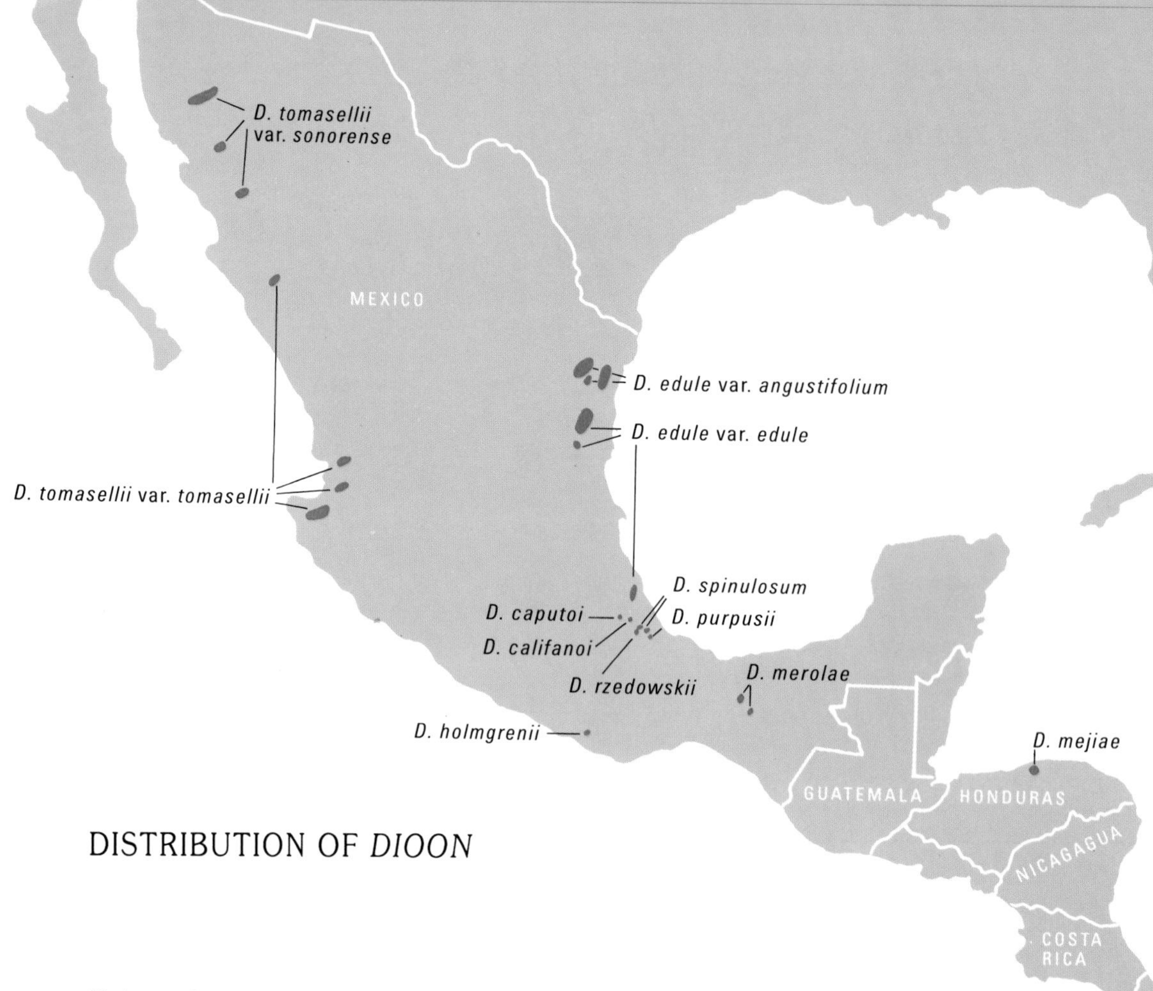


DISTRIBUTION OF *DIOON*

**Habitat:** Species of *Dioon* usually grow on steep rocky hillsides where the soils are skeletal and poor. Altitudinally they range from just above sea level to about 2500 m altitude. *D. spinulosum* and *D. rzedowskii* grow in moist habitats where temperature variations are minimal, whereas many of the other species grow in much drier habitats which experience wide fluctuations of temperature.

**Cultivation:** Species of *Dioon* have proved to be adaptable and can be grown either in containers or the ground. Most species will tolerate considerable exposure to bright sunlight even when small and fluctuations in temperature and humidity present few problems. *D. spinulosum* and *D. rzedowskii* are the least tolerant of direct hot sun, low humidities and extreme temperature fluctuations and may require extra protection. Those species originating from high altitudes or semi-arid environments will tolerate some exposure to frosts (especially if the plants are not in active growth) but some degree of frost protection is advisable for all species.

Well-drained, humus-rich, acid soils are satisfactory for most species with a couple (*D. spinulosum*, *D. rzedowskii*) showing preference for neutral to slightly alkaline soils. Mulches, watering during dry periods and regular light fertiliser applications are appreciated by all species.

**Propagation:** All species of *Dioon* are primarily propagated from seed. Seeds have a relatively short period of viability and for best results should be sown soon after they mature. This time can be judged by the condition of the female cone and the colouration of the seeds. If seeds are not to be sown quickly they should be stored in moist peatmoss to prevent dehydration. The seeds of *Dioon* have an after-ripening period of one to two months and seed germination for most species takes about twelve to eighteen months.

## KEY TO SPECIES

Adapted from Sabato and De Luca, 'Evolutionary Trends in *Dion* (Zamiaceae)', *American Journal of Botany* 72, No. 9 (1985): 1353–63.

**1** Leaflets falcate (at least the proximal ones) .................... ***D. tomasellii***
Leaflets straight ........ **2**

**2** Leaflets less than 15 mm wide ........ **3**
Leaflets more than 15 mm wide ........ **8**

**3** Leaflets with entire margins ........ ***D. edule***
Leaflet margins bearing spines ........ **4**

**4** Leaflets at right angles or nearly so to the rhachis ...... **5**
Leaflets at 45° or less to the rhachis ........ **6**

**5** Leaflets 10–12 mm wide, overlapping ........ ***D. merolae***
Leaflets 7–9 mm wide, not overlapping ........ ***D. holmgrenii***

**6** Leaflets less than 5 mm wide ........ ***D. caputoi***
Leaflets more than 5 mm wide ........ **7**

**7** Leaflets 7–8 mm wide, obliquely erect (vee-ed in cross-section) ........ ***D. califanoi***
Leaflets 8–10 mm wide, planar (flat in cross-section) ........ ***D. purpusii***

**8** Petiole lacking spines ........ ***D. rzedowskii***
Petiole spiny ........ **9**

**9** Leaflet margins entire ........ ***D. mejiae***
Leaflet margins spiny ........ ***D. spinulosum***

# *Dioon califanoi*

De Luca & Sabato
(after Professor Luigi Califano)

**Description:** A medium-sized cycad which in nature develops an erect or reclining trunk to 3 m tall and 30 cm across. **Young leaves** light green and woolly. **Mature leaves** stiff, rigid, numerous in an obliquely erect crown, 0.7–0.85 m long, dark green, broadest near the middle and tapered to each end, the rhachis strongly keeled, straight in profile, the leaflets held in a shallow vee when viewed in cross-section; **petiole** 10–12 cm long, swollen at the base, lacking prickles; **leaflets** 160–200 on each leaf, crowded and evenly distributed throughout, inserted at an acute angle to the rhachis and held obliquely erect to form a vee, rigid and pungently pointed, lacking marginal spines; **median leaflets** 6–7 cm x 0.7–0.8 cm, linear-lanceolate, lacking spines. **Male cones** 30–40 cm x 8–10 cm, long-cylindrical; **sporophylls** about 2.5 cm long. **Female cones** 40–50 cm x 20–25 cm, ovoid, densely woolly; **sporophylls** to 10 cm x 5 cm. **Seeds** 3–4 cm x 2–2.5 cm, ovoid, smooth, white or cream.
**Distribution and Habitat:** Endemic to Oaxaca, Mexico, where it grows between 1800 m and 2350 m altitude.

*Dioon califanoi.*

Plants grow on steep shady slopes in tropical, deciduous forests.
**Notes:** This distinctive species, described in 1979, can be recognised by the keeled rhachis of the fronds and the crowded, rigid leaflets held obliquely erect in a shallow vee. The trunks of old specimens have a reclining habit.
**Cultivation:** Suited to subtropical and warm temperate regions. A very attractive species which is uncommonly grown.
**Propagation:** From fresh seed.

## *Dioon caputoi*

De Luca, Sabato & Vazquez-Torres
(after Professor Guiseppe Caputo, Professor of Botany, University of Naples)

**Description:** A medium-sized cycad which in nature develops an erect to prostrate cylindrical trunk to 1 m tall and 25 cm across. **Young leaves** pale green, shortly hairy, often glaucous. **Mature leaves** numerous in an obliquely erect crown, 0.7–0.9 m long, stiff, rigid, dark green, flat in cross-section, straight in profile, broadest near the middle and tapered to each end; **petiole** 12–15 cm long, swollen at the base, lacking prickles; **leaflets** 100–140 on each leaf, moderately to widely spaced and evenly distributed throughout, inserted at 30–40° to the rhachis, margins rolled back entire or the upper margin with a single spine 2–3 mm long; **median leaflets** 8–10 cm x 0.4–0.5 cm, linear-lanceolate, rigid and pungently pointed. **Male cones** 30–40 cm x 9–10 cm, long-cylindrical; **sporophylls** about 4 cm long, shallowly bilobed. **Female cones** 30–40 cm x 20–25 cm, ovoid, densely hairy; **sporophylls** to 14 cm x 5 cm. **Seeds** 3–4 cm across, nearly globose, smooth, white.
**Distribution and Habitat:** Endemic to a small area of about one square kilometre in Puebla, Mexico, where it grows at about 2000 m altitude. Plants are found on steep slopes in stunted forest and grow in poor, shallow calcareous soil.
**Notes:** *D. caputoi*, which was described in 1980, is a very rare species that is known in the wild from only about fifty individuals. Seedlings are apparently rare and the species appears to be not reproducing to any substantial degree. This species grows in the driest habitat of any *Dioon*. Distinguishing features include the narrow leaflets spaced evenly and inserted at an acute angle on the rhachis but held in a flat plane. Its closest relative is *D. purpusii* but this species can be distinguished by its unevenly spaced leaflets which become crowded in the distal part of the leaf.
**Cultivation:** Suited to subtropical and warm temperate regions. Experience with this species is limited but results suggest that it prefers neutral to alkaline soils.
**Propagation:** From fresh seed.

## *Dioon edule* Lindley var. *edule*

(edible, in reference to the local use of the seeds)

**Description:** A medium-sized cycad with an erect or reclining trunk to 3 m tall and 30 cm across. **Young leaves** pale green, shortly hairy. **Mature leaves** numerous in a stiff, obliquely erect crown, 0.7–1.4 m long, stiff, rigid, light green or glaucous, flat in cross-section, straight in profile, broadest near the middle and tapered to each end; **petiole** 10–15 cm long, swollen at the base, lacking prickles; **leaflets** 120–160 on each leaf, moderately spaced but those towards the apex often crowded and overlapping, inserted at 60–90° to the rhachis, lacking marginal spines; **median leaflets** 6–12 cm x 0.6–0.9 cm, linear to linear-lanceolate, leathery, tapered and long-acuminate. **Male cones** 20–40 cm x 6–10 cm cylindrical; **sporophylls** 2–3 cm long, with upcurved pointed tips. **Female cones** 20–35 cm x 12–20 cm, ovoid, woolly; **sporophylls** 5–8.5 cm long; **peduncle** 8–12 cm long. **Seeds** 2.5–3.5 cm x 2–2.5 cm, ovoid.
**Distribution and Habitat:** Endemic to Mexico where widely distributed in the Sierra Madre Oriental mountain range between sea level and 1500 m altitude. The species is most commonly found growing in tropical deciduous forest and oak forests.
**Notes:** This species, described in 1843, is the type of the genus and was named from a specimen of imprecise Mexican origin which was cultivated in Britain. It is closest to *D. caputoi* and *D. tomasellii* from which it can be distinguished by the entire leaflets on mature leaves and smaller seeds. Young plants have distinctly spinulose leaflets. A number of variants of this species have been decribed but recent field studies have shown that *D. edule* is a variable species which encompasses all named variants with the exception of the variety treated below. Two notable variants which may be in cultivation are:

1. from Hidalgo and Queretaro where the plants have persistently hairy leaflets with strongly rolled back margins; and
2. from San Luis Potosi and southern Tamaulipas in which the young leaves are prominently glaucous.

Studies of leaflet widths in relation to distribution have shown that northern populations of this species have narrower leaflets and smaller seeds. These northern populations are treated below as a distinct variety (var. *angustifolium*). *D. edule* var. *edule* is found in central Veracruz, Queretaro, northern Hidalgo, San Luis Potosi and southern Tamaulipas. In Mexico the young seeds of this species are ground and cooked as tortillas and the fronds are used as decoration, often for religious ceremonies.
**Cultivation:** Suited to subtropical and warm temperate regions. Both varieties of *D. edule* have similar cultural requirements although *D. edule* var. *angustifolium* is comparatively rare in cultivation. *D. edule* var. *edule* is

*Dioon edule* From the *Botanical Magazine* 101, tab.6184, (1875)

very widely grown and has proved to be adaptable to a range of climates and situations. Plants grow best in full sun and need excellent drainage. Watering during summer is beneficial and unimpeded air movement is important to avoid foliage damage resulting from being excessively wet. Tolerant of light to moderate frosts.

**Propagation:** From fresh seed; occasionally plants can be propagated by basal suckers.

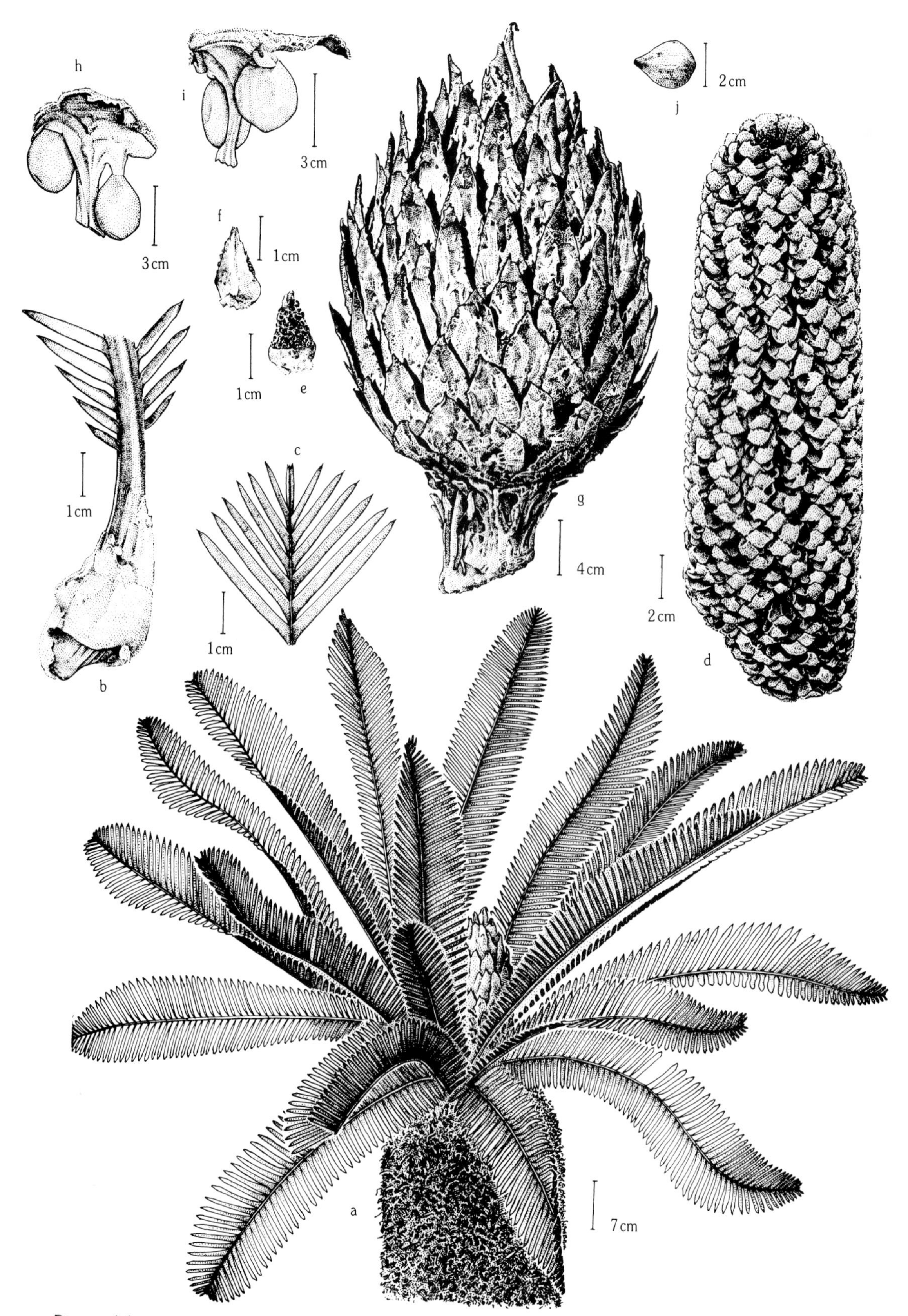


*Dioon edule*
With permission from *Flora de Veracruz*, Fascicle 26, February 1983. Illustration by Edmundo Saavedra. **a.** adult plant, **b.** detail of base of petiole, **c.** detail of terminal leaflets, **d.** male cone, **e. & f.** male sporophylls, **g.** female cone, **h. & i.** female sporophylls with ovules, **j.** seed.

Clump of *Dioon edule*.

Leaf of *Dioon edule*.

*Dioon edule*.

Female cone of *Dioon edule*.

*Dioon edule* var. *angustifolia* in habitat. D. W. STEVEN

## *Dioon edule* var. *angustifolium*

(Miq.) Miq.
(with narrow leaves)

**Description:** Differs from *D. edule* var. *edule* by its narrower **leaflets** (6–11 cm x 0.4–0.6 cm), which are generally glaucous when young and inserted at an acuter angle to the rhachis, and its smaller seeds (about 2 cm x 2 cm).
**Distribution and Habitat:** Endemic to northern parts of the Sierra Madre Oriental mountain range (Tamaulipas and Nuevo Leon) at between 200 m and 1500 m altitude. It grows in similar habitats to *D. edule* var. *edule*.
**Notes:** This variety was described as a species in 1848 (*D. angustifolium* Miq.) and later reduced to a variety of *D. edule* by the same author. Plants originating from Tamaulipas have leaflets which are longer than normal (to 16 cm), and are inserted at an acute angle to the rhachis.
**Cultivation and Propagation:** As for *D. edule* var. *edule*.

## *Dioon holmgrenii*

De Luca, Sabato & Vazquez-Torres
(after Dr Noel Holmgren, botanist, New York Botanic Gardens)

**Description:** A medium-sized to tall cycad which in nature develops an erect trunk to 6 m tall and 40 cm across. **Young leaves** light green, densely covered with short hairs. **Mature leaves** numerous in an obliquely erect crown, 1.3–1.5 m long, stiff, rigid, dark green, flat in cross-section, straight in profile, broadest near the middle and tapered to each end; **petiole** 13–15 cm long, swollen at the base, lacking prickles; **leaflets** 230–260 on each leaf, moderately spaced and evenly distributed throughout, inserted at about 60° to the rhachis, with one or two marginal spines on the lower side, one to four on the upper side; **median leaflets** 10–12 cm x 0.7–0.9 cm, linear-lanceolate, rigid, apex pungent, spinulose. **Male cones** 30–40 cm x 6–7.5 cm, long-cylindrical; **sporophylls** about 2 cm long, shallowly bilobed. **Female cones** 30–50 cm x 20–30 cm, ovoid, densely hairy; **sporophylls**

to 12 cm x 5 cm. **Seeds** 2.5–3 cm across, nearly globose, smooth, white or cream.
**Distribution and Habitat:** Endemic to Oaxaca, Mexico, where it grows between 650 m and 850 m elevation. Vegetation types are mainly humid forests dominated by species of *Quercus* and *Pinus*. Plants often grow near ravines in humus-rich soil.
**Notes:** This species, described in 1981, can be distinguished from others by its flat fronds with narrow, spinulose leaflets which are evenly spaced along the rhachis and are inserted at an oblique angle.
**Cultivation:** Suited to cool tropical and subtropical regions. This attractive species is uncommon in cultivation and is mainly a collector's item.
**Propagation:** From fresh seed.

# *Dioon mejiae*

Standley & L. O. Williams
(after Dr Isidoro Mejia, original collector)

**Description:** A small to medium-sized cycad which in nature develops a trunk to 1 m tall and 25 cm across. **Young leaves** light green and with very long hairs. **Mature leaves** numerous in an obliquely erect crown, 1–2 m long, stiff, rigid, dark green, flat in cross-section, straight in profile, broadest near the middle and tapered to each end; **petiole** 2–15 cm long, swollen at the base, prickly; **leaflets** about 200 on each leaf, moderately spaced and evenly distributed throughout, inserted at 90° to the rhachis, the margins with a few spines 1–2 mm long; **basal leaflets** with prominent apical spines; **median leaflets** 12–22 cm x 1.2–1.7 cm, linear-lanceolate, dark green, rigid and pungently pointed. **Male cones** 30–80 cm x 8–10 cm, long-cylindrical; **sporophylls** about 2.5 cm long. **Female cones** 30–50 cm x 20–30 cm, ovoid, pale brown, densely hairy; **sporophylls** to 18 cm x 6 cm. **Seeds** 3–3.5 cm x 2 cm, ovoid, roughened.
**Distribution and Habitat:** Occurs in Honduras growing in a dry rocky canyon at an elevation of about 750 m. It has also recently been collected in north-central Nicaragua.
**Notes:** This species was first collected in 1910 and seedlings raised from this collection and grown in a garden in Honduras formed the basis for its description in 1950. In Honduras it is known from a single locality which is protected by the government. The seeds are eaten as tortillas by the local inhabitants after boiling and grinding and the leaves are used for decoration. Seedlings of this species are similar to those of *D. spinulosum* but have a prickly petiole. The seeds have an unusual tail-like appendix on the chalazal end.
**Cultivation:** Suited to tropical and subtropical regions. *D. mejiae* is one of the more commonly grown members of this genus. A very hardy species which will withstand considerable exposure to sun as well as some dryness of the roots.
**Propagation:** From fresh seed or from suckers which are freely produced and transplant readily.

# *Dioon merolae*

De Luca, Sabato & Vazquez-Torres
(after Aldo Merola, Professor of Botany, University of Naples, Italy)

**Description:** A medium-sized to large cycad which in nature develops an erect or reclining trunk to 6 m tall and 40 cm across. **Young leaves** light green, covered with woolly hairs. **Mature leaves** numerous in an obliquely erect crown, 0.8–1 m long, stiff, rigid, dark green, flat in cross-section, straight or slightly recurved in profile, broadest near the middle and tapered to each end; **petiole** 7–10 cm long, swollen at the base, lacking prickles; **leaflets** about 240 on each leaf, crowded and overlapping with the tips often curved towards the apex of the

*Dioon holmgrenii* in habitat.

A. MORETTI

*Dioon purpusii.*

*Dioon merolae.*

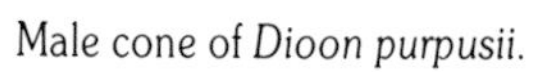

Male cone of *Dioon purpusii.*

Leaf of *Dioon purpusii.*

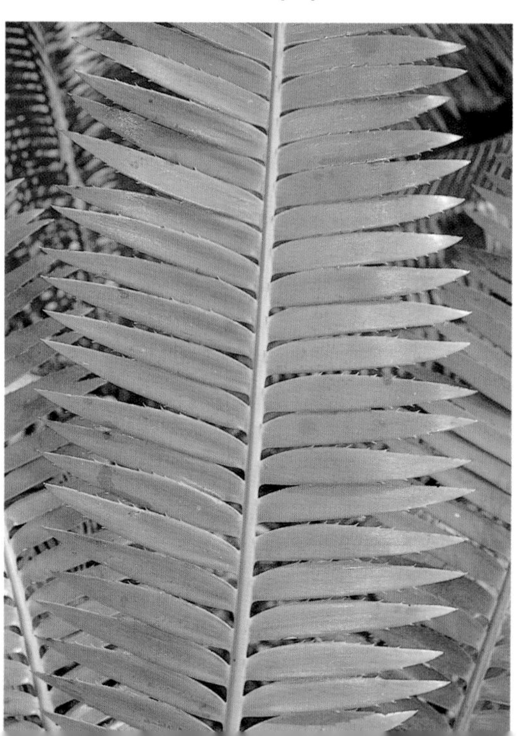

frond, inserted at about 50° to the rhachis, the **lower leaflets** entire, those in the upper part with one or two marginal spines on the lower side, one to four on the upper side; **median leaflets** 7–9 cm x 1–1.2 cm, linear-lanceolate, rigid and pungently pointed. **Male cones** 30–40 cm x 8–10 cm, long-cylindrical, brown; **sporophylls** about 2.5 cm long, bilobed. **Female cones** 40–45 cm x 20–25 cm, ovoid, densely woolly; **sporophylls** to 12 cm x 5 cm. **Seeds** about 3 cm long, ovoid, smooth, white or cream.
**Distribution and Habitat:** Endemic to Chiapas, Mexico, where it grows in the mountains between 900 m and 1200 m altitude. It occurs in forests dominated by species of *Pinus* and *Quercus* and also in tropical semi-deciduous forests.
**Notes:** This species was described in 1981 from specimens collected in 1979. The first collections of the species, however, were made in 1909. These specimens were never formally described although the name '*D. dohenyi*' became attached to them in the literature. *D. merolae* is a distinctive species which can be recognised by its fronds which are flat in plane and have crowded, overlapping, rigid leaflets which arise from the rhachis at an acute angle. Those of *D. edule*, with which it is often confused, are attached to the rhachis nearly at right angles. It is reported that old plants of this species often have reclining trunks and young plants always have woolly fronds with strongly spinulose leaflets.
**Cultivation:** Suited to subtropical and warm temperate regions. This attractive species is rarely grown and is mainly a collector's item.
**Propagation:** From fresh seed.

## *Dioon purpusii*

Rose
(after G. A. Purpus, original collector)

**Description:** A medium-sized to large cycad which in nature develops a trunk to 5 m tall and 40 cm across. Apex of trunk woolly. **Mature leaves** numerous in an obliquely erect crown, 0.8–1.6 m long, stiff, rigid, dark green, flat in cross-section on mature plants, obliquely erect and vee-ed in young plants, straight in profile, broadest near the middle and tapered to each end; **petiole** 5–20 cm long, somewhat four–angled, swollen at the base, lacking prickles; **leaflets** 150–260 on each leaf, moderately spaced in the lower half but crowded in the upper part of the leaf, especially near the apex, inserted at about 45° to the rhachis, entire or with two or three spines on the upper margin and one or two on the lower margins, rigid and pungently pointed; **median leaflets** 7–11.5 cm x 0.8–1 cm, linear-lanceolate, leathery, long-acuminate. **Male cones** 20–30 cm x 0.7–0.8 cm, long-cylindrical; **sporophylls** with the apex pointing sharply upwards. **Female cones** 38–44 cm x 16–20 cm, broadly ovoid, woolly; **sporophylls** 10–15 cm x 5–7 cm. **Seeds** 3–4 cm long, ovoid, smooth, white.
**Distribution and Habitat:** Endemic to Oaxaca, Mexico, where it is found at about 1300 m elevation. The species grows in tropical deciduous forests on the sides of steep canyons.
**Notes:** This species was described in 1909 and an updated description provided in 1979 following an expedition to the type locality. *D. purpusii* has spinulose leaflets like *D. spinulosum* but is distinguished by its broader leaflets which are not regularly spaced on the frond. The species is reported to be common in the type locality and is reproducing freely. Old plants often have reclining and branching trunks.
**Cultivation:** Suited to subtropical and warm temperate regions. An attractive species which is moderately well-known in cultivation. Grows well and has proved to be adaptable.
**Propagation:** From fresh seed and to a lesser degree by division of suckers.

## *Dioon rzedowskii*

De Luca, A. Moretti, Sabato & Vazquez-Torres
(after Dr Jerzy Rzedowski, contemporary Mexican botanist)

**Description:** A medium-sized to large cycad which in nature develops an erect or reclining trunk to 5 m tall and 40 cm across. Apex of trunk densely woolly. **Young leaves** pale green and hairy. **Mature leaves** numerous in an obliquely erect crown, 1.4–1.8 m long, stiff, rigid, dark green, leathery, glabrous, flat in cross-section, straight in profile, broadest near the middle and tapered to each end; **petiole** 10–15 cm long, swollen at the base, lacking prickles; **leaflets** about 160 on each leaf, moderately spaced and evenly distributed throughout, inserted at right angles to the rhachis, stiff, rigid, lacking marginal spines although those towards the frond apex sometimes have one or two; **median leaflets** 14–19 cm x 1.8–2.1 cm, asymmetrically linear-lanceolate, the margins spineless, apex asymmetrically long-acuminate. **Male cones** 30–50 cm x 8 cm, cylindrical; **sporophylls** about 3 cm long, deltoid. **Female cones** 60–80 cm x 20–25 cm, ovoid, densely woolly; **sporophylls** about 10 cm x 5 cm. **Seeds** 5–6 cm x 3 cm, linear-ovoid, smooth, white or cream.
**Distribution and Habitat:** Endemic to Oaxaca, Mexico, where it grows between 650 m and 850 m altitude. The species occurs in small groups with individual plants growing in the crevices of limestone cliffs where humus collects. The surrounding vegetation is short and the cycads are exposed to full sun for most of the day.
**Notes:** This species was described in 1980. It is reported that young plants always have spines on their leaflet margins whereas those on mature plants lack spines. Young

A. MORETTI

*Dioon rzedowskii* in habitat.

to mature plants have erect trunks but old plants have a reclining habit. The female cones of this species are large and become pendulous with age, hanging below the crown of leaves. Unique features of this species include the fronds being flat in profile and the leaflets being asymmetrically lanceolate, tapered to a long apical point and lacking marginal spines.
**Cultivation:** Suited to subtropical and warm temperate regions. Plants of this species require humid conditions and partial shade or filtered sun. They are adaptable to a range of situations and soil types although favouring those of neutral to alkaline pH. Plants are tolerant of considerable exposure to sun.
**Propagation:** From fresh seed.

## *Dioon spinulosum*

Dyer
(with small spines, in reference to the leaflet margins)

**Description:** A medium-sized to very large cycad which in nature develops an erect or reclining trunk to 16 m tall and 40 cm across, although trunks are most commonly 5–10 m tall. **Young leaves** slightly hairy, translucent blue green, often covered with a yellowish bloom. **Mature leaves** numerous in a graceful, arching to rounded crown, 1.5–2 m long, curved, light green, flat in cross-section, shallowly recurved in profile, broadest near the middle and tapered to each end; **petiole** 2–15 cm long, swollen at the base, lacking prickles; **leaflets** 140–240 on each leaf, moderately spaced and evenly distributed throughout, inserted at 80–90° to the rhachis, the upper margins with three to ten spines, the lower with two to six; **median leaflets** 15–20 cm x 1.4–2 cm, lanceolate, leathery, the apex drawn out, acuminate, sometimes curved upwards. **Male cones** 40–55 cm x 7–10 cm, cylindrical; **sporophylls** 3–4 cm long, triangular; **peduncle** 15–20 cm long, hairy. **Female cones** 30–90 cm x 22–35 cm, ovoid, woolly; **sporophylls** 7–8 cm. long; **peduncle** 30–45 cm long. **Seeds** 4–5 cm x 3–3.5 cm, ovoid, smooth, white to yellowish.
**Distribution and Habitat:** Endemic to Mexico where once widely distributed but now localised in the lowlands of the Sierra Madre Oriental mountain range at between 20 m and 300 m altitude. Plants grow in shady ravines and canyons developed on limestone formations and covered with tropical evergreen forest.
**Notes:** This species was described in 1884 from material collected at Tuxtla, Yucatan, Mexico. It may be locally common and in some parts is the dominant understorey plant, growing as it does in dense patches. It is one of the tallest of all cycads with trunks reaching in excess of 15 m and its graceful crown is reminiscent of some species of small palms. It also has among the largest cones of any cycad with female cones weighing in excess of 25 kg being recorded. Being so heavy these cones commonly cause the peduncle to bend and the cones slip between the leaves and hang below the crown. At maturity, these cones burst with a loud noise and scatter the seeds. In Mexico the seeds are gathered and used by the local inhabitants. Young seeds are ground up and cooked as tortillas whereas old seeds may be used as toys or ornaments. Cut fronds are used for decoration in various religious ceremonies.
*D. spinulosum* is similar to both *D. merolae* and *D. meijiae*. *D. spinulosum* is the tallest of the three, has strongly spinulose leaflets (sparsely spinulose in *D. merolae* and *D. meijiae*) and graceful, arching leaves in a rounded crown (stiffly rigid in the other two). Seedlings of *D. spinulosum* resemble those of *D. mejiae* but lack prickles on the petioles.
**Cultivation:** Suited to warm tropical and subtropical regions. A popular species prized for its attractive crown. Plants require warm humid conditions and grow well in partial shade or filtered sun. A wide range of soils are suitable with those of a neutral to slightly alkaline pH being preferred. Frost-tolerance is very low.
**Propagation:** From fresh seed.

*Dioon spinulosum.*

Male cone of *Dioon spinulosum.*

Leaf of *Dioon spinulosum.*

Female cone of *Dioon spinulosum.*

## *Dioon tomasellii* var. *tomasellii*

De Luca, Sabato & Vazquez-Torres
(after Dr Ruggero Tomaselli, former Professor of Botany, University of Pavia, Italy)

**Description:** A small to medium-sized cycad which in nature develops a cylindrical trunk to 1 m long (mostly below the ground), covered with persistent leaf bases, rarely branched at the base. **Young leaves** hairy. **Mature leaves** numerous in an obliquely erect crown, 1–2 m long, stiff, rigid, light green or glaucous, flat in cross-section, straight in profile, broadest near the middle and tapered to each end; **petiole** 6–35 cm long, swollen at the base, lacking prickles; **leaflets** 140–180 on each leaf, moderately spaced, inserted at 90° to the rhachis, falcate, marginal spines lacking or present, the apices often slightly decurved; **median leaflets** 10–18 cm x 0.7–1.2 cm, lanceolate, tapered and long-acuminate. **Male cones** 25–50 cm x 6–10 cm, cylindrical; **sporophylls** about 3.5 cm long. Female cone 20–30 cm x 15–20 cm, ovoid, woolly; **sporophylls** about 4.5 cm long, the outer tip reflexed. **Seeds** 2–2.5 cm x 2.5–3.5 cm, ovoid, cream to red.
**Distribution and Habitat:** Endemic to western Mexico where it is found in mountainous districts between 950 m and 1100 m altitude. It grows in forests dominated by species of *Pinus* and *Quercus*.
**Notes:** This species, originally confused with *D. purpusii*, was described in 1984. The new species is readily recognised by the falcate leaflets, a feature obvious in seedlings as well as adult plants. Another useful feature is the tip of the sporophylls of the female cone being reflexed. Studies of leaflet widths in relation to distribution have shown that northern populations of this species have narrower leaflets and different seedling characteristics. These northern populations are treated as a distinct variety (var. *sonorense*). *D. tomasellii* var. *tomasellii* is found in Guerrero, Michoacan, Jalisco, Nayarit and Durango.
**Cultivation:** Both varieties of *D. tomasellii* are suitable for cultivation in subtropical and warm temperate regions.
**Propagation:** From fresh seed; occasionally plants can be propagated by basal suckers.

## *Dioon tomasellii* var. *sonorense*

De Luca, Sabato & Vazquez-Torres
(from Sonoro, Mexico)

**Description:** Differs from *D. tomasellii* var. *tomasellii* by its narrower leaflets (10–18 cm x 0.45–0.65 cm), which are often glaucous. Seedlings are smaller, with strongly glaucous, very narrow leaflets.
**Distribution and Habitat:** Endemic to Sonora and northern Sinaloa in western Mexico, growing in mountainous areas between 800 m and 1100 m altitude.
**Notes:** Plants of this variety which originate from the Sierra Mazatan range in the northernmost parts of its distribution differ by having straight leaflets which are inserted on the rhachis at an acute angle.
**Cultivation and Propagation:** As for *D. tomasellii* var. *tomasellii*.

*Dioon tomasellii* in habitat.

A. MOF

GENUS

# *Encephalartos*

The genus *Encephalartos* is possibly the largest in the Cycadales consisting of about fifty-two described species and a number of others yet to be named. It is restricted to the African continent with more than half the species occurring in South Africa. A significant number of species occur in the tropical regions of Central Africa and East Africa and these have been the subject of much recent study, with new species still being discovered in remote areas. While some species of *Encephalartos* have a fairly wide distribution, others are remarkably localised and a few are known to be restricted to a single river system or even a solitary hill.

**Derivation:** The genus *Encephalartos* was described by the German botanist Johann Georg Christian Lehmann in 1834. The generic name, composed from the Greek words *en*, 'in', *cephale*, 'a head', and *artos*, 'bread', refers to the farinaceous, starchy material in the trunks of some species which is used for food by local native tribes.

**Generic Description:** Terrestrial cycads with stout, cylindrical trunks, which may be emergent or wholly subterranean, with contractile roots. Basal suckers are produced by the majority of species and some also develop offsets on the trunk. Leaf bases are usually retained at senescence. New leaves emerging in flushes, glabrous, hairy or with a powdery bloom. Mature leaves pinnate, oblong, elliptical or lanceolate in outline, flat or vee-ed in cross-section (inverted in some species), the older leaves often spreading or deflexing after a flush of growth. Petioles lacking spines or prickles, glabrous or hairy, swollen at the base and with a prominent, often colourful collar. Rhachis lacking prickles, straight, incurved, recurved or twisted in profile. Leaflets non-articulate at the base, opposite to nearly opposite, evenly spaced or becoming crowded towards the apex, straight or falcate, margins entire or beset with spines or with spine-tipped lobes, the margins themselves flat, revolute or rarely involute, veins inconspicuous, rarely raised on the upper surface, apex pungent, the lower leaflets sometimes reduced in size to spine-like processes which may be entire, bifurcate, trifurcate or multi-lobed. Cones arising from lateral buds, mostly markedly dissimilar in shape, size and colouration, rarely similar, solitary or clustered, sessile or pedunculate, the sporophylls lacking any horns or projections but often hairy. Male cones often narrow, erect, hairy. Female cones usually broad. Seeds radiospermic, ovoid, obovoid or oblong, often angular, the sarcotesta red, orange or yellow.

**Notable Generic Features:** Species of *Encephalartos* can be recognised by their non-articulate leaflets which are often spiny. Other useful features include:

- the suckering and clumping habit;
- unarmed petioles and rhachises;
- cataphylls with decurrent bases present on the cone peduncles;
- seeds attached directly to the sporophylls; and
- the sporophylls of both sexes lack any horns or projections.

**Recent Studies:** Whereas there have been a number of new species published over the last twenty-five years, these have been largely fragmentary, resulting from studies into species complexes or from collections in isolated areas. An overall revision of the genus is badly needed to clarify many problems and relationships. The pioneering studies of Robert Allen

Dyer and Inez Clare Verdoon should be singled out for special mention. See for example: 'New Species and Notes on Type Specimens of South African *Encephalartos*', *Journal of South African Botany* 31 (1965): 111–21; '*Encephalartos manikensis* and its Near Allies', *Kirkia* 7 (1969): 147–58; 'The Cycads of Southern Africa', *Bothalia* 8 (1965): 405–515. Two comprehensive books also provide up-to-date information and excellent illustrations. See Cynthia Giddy, *Cycads of South Africa*, C. Struik Publishers, Cape Town, 1986 and Douglas Goode, *Cycads of Africa*, Struik Winchester, Cape Town, 1989.

**Habitat:** Species of *Encephalartos* are frequently found growing among rocks in mountainous regions but are also known from the lowlands, including coastal and near-coastal scrubs. Many species grow in open situations among grass but some favour the shelter of forests which may be sparse to dense. Heavy frosts and snow, common occurrences in the mountainous districts where the southern species grow, are unknown in the tropical and subtropical regions further north where the summers are hot and humid and the winters mild. Thus, species of *Encephalartos* occupy a wide range of climatic regimes and habitats. Bushfires and grassfires are a very common phenomenon in many of the habitats and these may be of annual occurrence. The fires are often extremely hot and all above-ground parts of the cycads are usually destroyed. All species show growth adaptations to survive these fires with some species developing to such an extent that the fires have become a prerequisite for the induction of cones. A few species of *Encephalartos* from arid areas may become completely deciduous each year for a short period prior to the production of new leaves.

**Cultivation:** As a general rule most species of *Encephalartos* adapt very well to cultivation. The majority of species prefer to grow in conditions of bright light or full sun but there are a few which require shade. Air movement should be unimpeded as leaf fungal attack can result from stagnant conditions. Root-rotting fungi can cause plant deaths unless soil drainage is excellent. Humus-rich, acid soils are satisfactory for most species, with relatively few showing preference for neutral to slightly alkaline soils. Mulches, watering during dry periods and regular light fertiliser applications are beneficial for all species. Those species originating from high altitudes in the south of Africa are very tolerant of heavy frosts and even snow but species from tropical and subtropical regions are very sensitive to damage by cold.

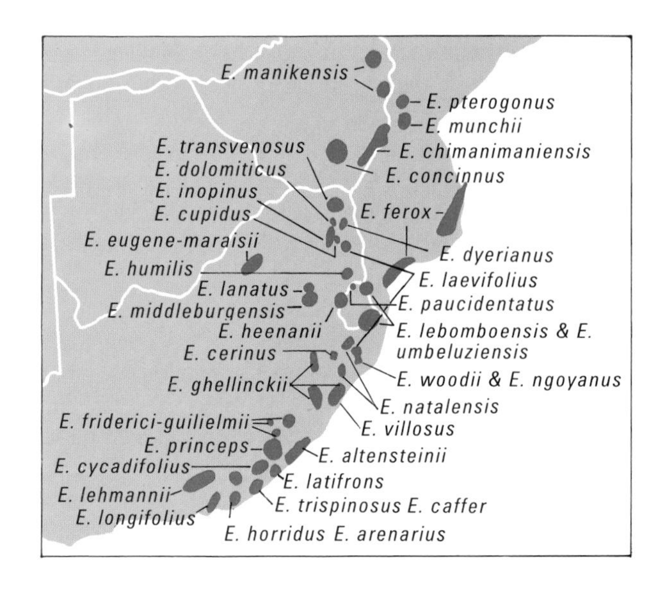

**Propagation:** The majority of species of *Encephalartos* are primarily propagated from seed which should be sown soon after maturity. These take about twelve to eighteen months to germinate. Most species produce basal suckers and these can be transplanted readily if due care is given to careful removal from the parent plant. A few species produce offsets from the trunk and these can also be used successfully for propagation. Although they lack roots these portions can be induced to grow if handled carefully.

## KEY TO SPECIES

Because of the size and complexity of this genus no key has been attempted.

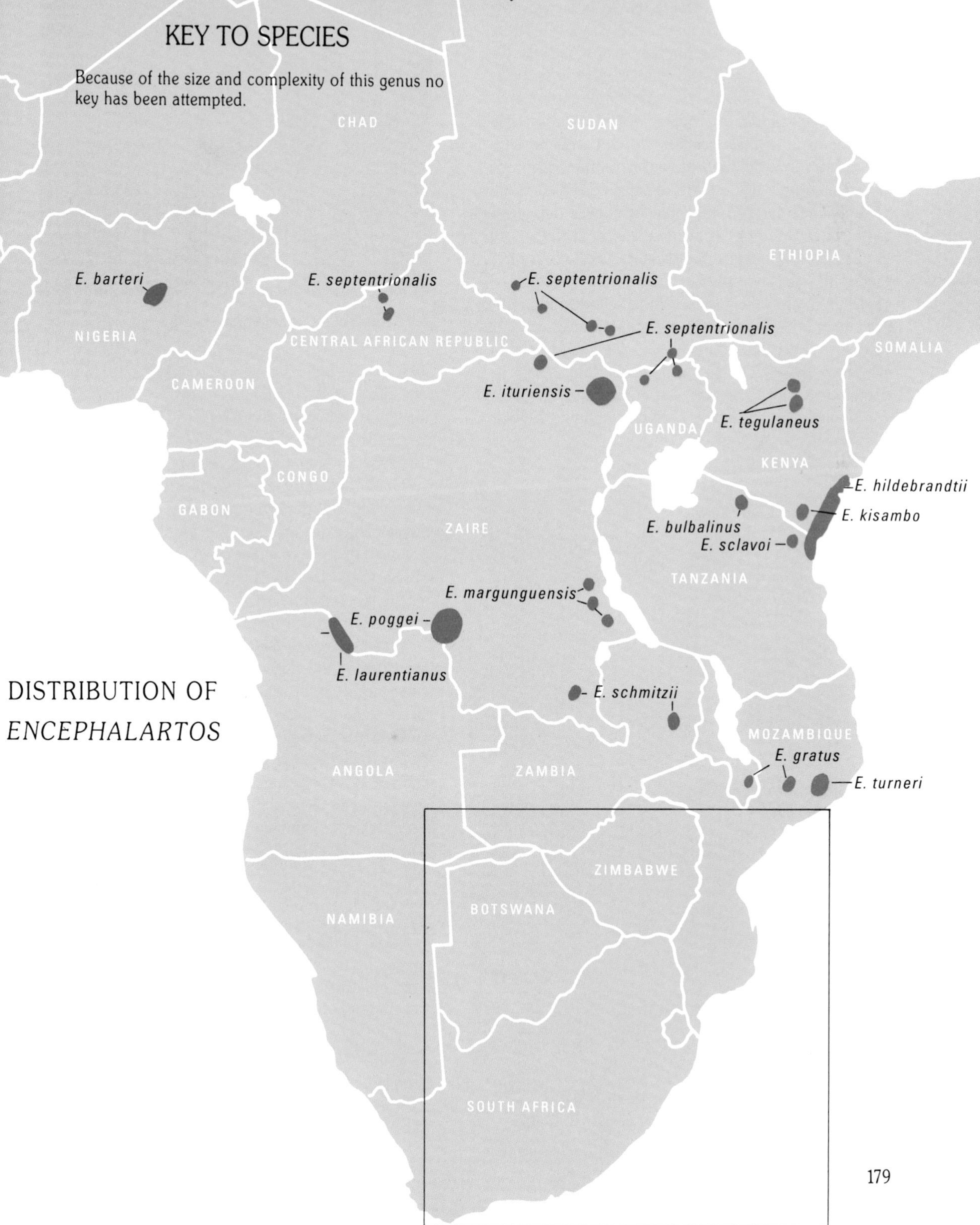


DISTRIBUTION OF *ENCEPHALARTOS*

# *Encephalartos aemulans*

Vorster

(rivalling, more or less equalling)

**Description:** A small to medium-sized cycad with an erect trunk to 3 m tall and 35 cm across, with new suckers produced from the base. **Mature leaves** numerous in an obliquely erect to rounded crown, 1.2–2 m x 20–30 cm, dark green and glossy, flat in cross-section; **petiole** 10–12 cm long; **leaflets** narrowly elliptical, widely spaced towards the base, those in the middle and upper parts of the leaf crowded and overlapping, inserted on the rhachis at about 35°, the **lower leaflets** reduced progressively into a series of small bifurcate or entire, spine-like processes; **median leaflets** 12.5–15 cm x 1.6–1.8 cm, slightly falcate, pungent-tipped, the upper and lower margins with one to three teeth. Cones nearly identical in shape, ellipsoid, covered with dense brown hairs, two to four on each stem, appearing sessile. **Male cones** 29–38 cm x 14–18 cm, narrowly ellipsoid, lemon yellow; **peduncle** 5–6 cm long, buried in the crown. **Female cones** 35–40 cm x 20–23 cm, green; **peduncle** 2–2.5 cm long, buried in the crown. **Seeds** 2.5–3 cm x 1.5–2 cm, bright red.
**Distribution and Habitat:** Apparently mostly restricted to a single hill in Natal where it grows at about 1000 m altitude. Two old male plants occur some 10 km away but there are no plants in the intervening localities. The geology is sandstone and the plants grow on south-facing cliffs and associated scree slopes beneath. The climate is hot in the summer and cool to cold in the winter with light frosts occurring. The rainfall is 600–800 mm per annum with a summer maximum.
**Notes:** The exact location of the locality of *E. aemulans*, described in 1990, was not given in the original description to protect the plants from poachers. Several hundred healthy plants are reported to occur and active seedling regeneration is occurring. The species is related to both *E. natalensis* and *E. lebomboensis* but can be distinguished by the close general similarity of the male and female cones, both of which are densely covered by hairs.
**Cultivation:** Suited to subtropical and warm temperate regions. Experience in cultivation is limited, although plants of this species are apparently grown, being collected prior to the description of the species. Requires sun and unimpeded drainage. Tolerates light frosts.
**Propagation:** From seed and by removal of suckers.

P. VORSTER

Male cones of *Encephalartos aemulans*.

P. VORSTER

*Encephalartos aemulans* in habitat.

# *Encephalartos altensteinii*

Lehm.

(after Altenstein, a nineteenth century German chancellor and statesman)

**Description:** A medium-sized to large cycad with an erect or reclining trunk to 5 m tall and 35 cm across, growing singly or in clumps of two or three stems, with new suckers produced from the base. **Young leaves**

*Encephalartos altensteinii.*

bright green, hairy. **Mature leaves** numerous in an obliquely erect to widely spreading crown, 1–3 m x 20–30 cm, bright yellowish green and glossy, more or less flat in cross-section, straight in profile (recurved in shaded plants); **petiole** 10–30 cm long; **leaflets** lanceolate, 200 or more on each leaf, moderately spaced and not crowded, evenly distributed throughout, inserted on the rhachis at 30–40°, the lower leaflets reduced but not spine-like; **median leaflets** 10–15 cm x 2–2.5 cm, rigid, pungent-tipped, entire or with one to five marginal spines. Cones markedly dissimilar, two to five on each stem. **Male cones** 40–50 cm x 12–15 cm, cylindrical but tapered from near the middle, yellow to yellowish green; **peduncle** 5–10 cm long, sturdy. **Female cones** 40–55 cm x 25–30 cm, oblong-ovoid, yellow with brown hairs, more or less sessile. **Seeds** 3.5–4 cm x 2–2.5 cm, oblong, bright red to scarlet, shiny.

**Distribution and Habitat:** Endemic to South Africa where widely distributed in coastal and near-coastal districts of the Eastern Cape Province nearly to the southern border of Natal. It grows in a range of habitats including rocky hillsides and exposed escarpments and shaded forests, sometimes being found in dense shade under tall trees. The climate is warm to hot in summer and cool in winter with frosts being a rarity. Rainfall ranges from 875 mm to 1000 mm per annum with a summer maximum.

**Notes:** *E. altensteinii*, described in 1834, is well-known because it is so common in cultivation. It is somewhat variable with specimens in the eastern parts of its range having less spiny leaflets than from elsewhere. This species is often erroneously recorded from Natal but these records apply to the closely related *E. natalensis*, which some authorities consider to be best treated as a subspecies of *E. altensteinii*. Cultivated plants of *E. altensteinii* may be difficult to identify due to subtle changes in characters. *E. altensteinii* can be distinguished from *E natalensis* and *E. lebomboensis* by the absence of reduced, spine-like leaflets on the leaf stalk and from *E. longifolius* by the bright yellowish green straight leaves and multiple cones. Natural hybrids with *E. trispinosus* are not unknown where the two grow in close proximity and a bewildering range of hybrids occurs. Natural hybrids, generally having intermediate characters between the parents, have also been recorded with *E. villosus*, *E. latifrons* and *E. arenarius*.

**Cultivation:** Suited to subtropical and temperate regions. Although frosts are rare in its natural habitat, cultivated plants will tolerate light to moderate frosts without damage. Highly ornamental and often planted in lawns as specimens. Adapts well to full sun but grows equally in shade. Large specimens transplant readily and it has been recorded in South Africa that trunks kept in a shed for four years grew vigorously after planting.

**Propagation:** From seed and by removal of suckers which transplant readily.

Female cones of *Encephalartos altensteinii*.

Mature female cones of *Encephalartos altensteinii*.

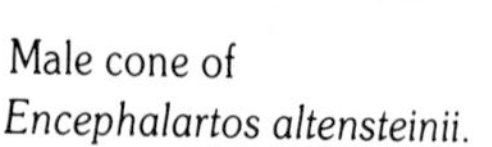
Male cone of
*Encephalartos altensteinii.*

*Encephalartos altensteinii* in habitat.

*Encephalartos arenarius.*

# *Encephalartos arenarius*

R. A. Dyer
(sandy, growing on sand)

**Description:** A small cycad with a subterranean or erect to reclining trunk to 1 m long and 30 cm across, growing singly or in clumps of up to three stems, with new suckers produced from the base. **Young leaves** light bluish green with a powdery bloom. **Mature leaves** numerous in a spreading crown, 1–1.5 m x 20–32 cm, dark dull green to blue green, flat or shallowly vee-ed in cross-section, straight in profile with a strongly recurved apex; **petiole** 10–20 cm long, with a swollen pale collar at the base; **leaflets** oblong-lanceolate, moderately crowded and evenly spaced throughout, obliquely erect to form a shallow vee, inserted on the rhachis at about 30°, the lower leaflets reduced, the lowest pair spine-like; **median leaflets** 12–16 cm x 2.5–4 cm, rigid, pungent-tipped, the lower margins with three to four large lobes each ending in a pungent spine, upper margins entire or with a solitary spine. Cones markedly dissimilar, solitary. **Male cones** 30–50 cm x 8–15 cm, cylindrical-ovoid, light green, when young with a powdery bloom; **peduncle** 4–8 cm long. **Female cones** 35–60 cm x 20–30 cm, broadly barrel-shaped, pale green; **peduncle** 3–6 cm long. **Seeds** 4–5 cm x 2–2.5 cm, narrowly oblong, angular, red, shiny.
**Distribution and Habitat:** Endemic to South Africa where it is found in a relatively limited area of the Cape Province in the vicinity of Alexandria. It grows in coastal scrub on stabilised sand dunes. The climate is hot in summer and mild and frost-free in winter. Rainfall ranges from 725 mm to 875 mm per annum with a summer

maximum although it is well distributed over most of the year.
**Notes:** *E. arenarius*, which was described in 1956, has suffered badly from clearing and illegal collection and it is now much less common in the wild than previously. It is closely related to *E. latifrons* but can be distinguished by its non-twisted leaflets which are not excessively crowded towards the tips of the leaves. Also *E. latifrons* has bright green leaves and the female cone scales are warty instead of smooth.
**Cultivation:** Suited to cool tropical and subtropical regions. This species is very frost-sensitive. For best appearance the plants should be grown in the shade or at least protected from excessive hot sun. Many growers advocate sandy soil for this species. Drainage must be excellent and plants respond to surface mulches and watering during dry periods.
**Propagation:** From seed and by removal of suckers which transplant readily.

Female cone of *Encephalartos arenarius*.

## *Encephalartos barteri* ssp. *barteri*

Carruth. ex Miq.
(after C. Barter, original collector of the species)

**Description:** A small cycad with a subterranean trunk to 30 cm long and 25 cm across, often forming clumps by multiple branching from below ground. **Young leaves** bright green, lightly hairy, arising in a flush. **Mature leaves** ten to twenty in an erect to arching crown, 1–1.8 m x 15–25 cm, dark green, dying in the dry season, flat in cross-section, straight in profile; **petiole** 5–10 cm long, the base swollen and covered with shaggy grey hairs; rhachis deeply grooved; **leaflets** about eighty pairs on each leaf, linear-lanceolate, moderately spaced and evenly distributed throughout, inserted on the rhachis at about 90°, the lower leaflets reduced to a series of weak spine-like processes; **median leaflets** 9–15 cm x 1–1.7 cm, rigid, pungent-tipped, the margins with up to six pungent spines. Cones markedly dissimilar, one to three on each stem. **Male cones** 8–23 cm x 3–5 cm, narrow cylindrical, green with brown tips on the **sporophylls**; **peduncle** 8–20 cm long, pale green. **Female cones** 12–35 cm x 8–15 cm, oblong-ovoid, dark olive green; **sporophylls** 4.5–6.5 cm, prominently ridged; **peduncle** 5–12 cm long. **Seeds** 2–3 cm x 1.8–2.3 cm, oblong-obovoid, crimson to scarlet.
**Distribution and Habitat:** Widely distributed in tropical West Africa where it occurs in Nigeria, Benin, Ghana and Togo (also erroneously reported from Uganda in Central Africa). This species grows in sparse to dense forests on sandstone and granite outcrops at about 400 m altitude. The climate is tropical with warm to hot, wet summers and mild dry winters.

Male cone of *Encephalartos arenarius*.

P. VORSTER

**Notes:** Plants of *E. barteri*, which was described in 1868, are completely deciduous in the wild with the leaves dying off annually in the dry season and new flushes appearing in the early wet season. The trunks have thick contractile roots which maintain the trunks in their underground position. Measurements of leaf remains on a

trunk about 30 cm x 25 cm indicate an estimated age of this trunk of about 100 years. Annual fires are common in the areas where this species grows. The typical subspecies (*E. barteri* ssp. *barteri*) occurs in western Nigeria, Benin, Tago and Ghana whereas those from an area in central Nigeria have been described as a distinct subspecies (see following entry).
**Cultivation:** Suited to tropical and warm subtropical regions. Mature plants transplant readily and seedlings establish quickly. Plants will grow in full sun, filtered sun and partial shade. Plants should be kept on the dry side when deciduous. Very senstive to cold weather, in particular frosts.
**Propagation:** From seed.

## *Encephalartos barteri* ssp. *allochrous*

L. E. Newton
(changing colour, in reference to the difference in colour between the new leaves and mature leaves)

**Description:** This subspecies is more robust with emergent trunks to 2.6 m long and 30 cm across, reclining with age; young leaves brown, maturing dark green; **median leaflets** longer (13–26 cm x 1.3–2.2 cm) and **male and female cones** slightly larger.
**Distribution and Habitat:** Endemic to the Jos Plateau in Central Nigeria where it is known from three localities close to the village of Tokkos. This subspecies grows on rocky slopes (granite) in grassland and sparsely forested areas at 1200–1400 m altitude. The climate is tropical but with cooler winters than experienced by plants of *E. barteri* ssp. *barteri*.
**Notes:** This subspecies was described in 1978 and the populations are reported to be regenerating strongly.
**Cultivation and Propagation:** As for *E. barteri* ssp. *barteri* although plants of *E. barteri* ssp. *allochrous* are probably more tolerant of cold periods.

## *Encephalartos bubalinus*

Melville
(buff-coloured, in reference to the close hairs of the scale leaves and leaf bases which impart a resemblance to an antelope's skin)

**Description:** A small cycad with an erect or reclining trunk to 2 m tall and 45 cm across, growing in clumps of up to six stems, with suckers produced from the base. **Young leaves** light green, with silky whitish hairs. **Mature leaves** numerous in an erect to obliquely erect crown, 0.6–1.6 m x 20–30 cm, dark green and glossy, widest above the middle, flat or shallowly vee-ed in cross-section, straight or slightly recurved in profile; **petiole** extremely short, 1–2 cm long, the swollen base covered with buff-coloured, felted hairs; **leaflets** narrow-linear, moderately spaced and evenly distributed throughout, obliquely erect, those towards the apex forming a deep vee, inserted on the rhachis at about 40°, the lower leaflets reduced to a series of bifurcate or trifurcate, spine-like processes; **median leaflets** 10–20 cm x 1–2 cm, rigid, pungent-tipped, the upper margin with one to four pungent spines, one on the lower margin. Cones more or less similar in shape but different in size, one to three on each stem, dark green. **Male cones** 11–22 cm x 5–6 cm, ovoid. **Female cones** 25–30 cm x 15–18 cm, barrel-shaped, nearly sessile. **Seeds** 3.5–4 cm x 2–2.5 cm, oblong, golden yellow.
**Distribution and Habitat:** Endemic to northern Tanzania where it grows on quartzite hills at about 1500 m altitude. Plants grow in open sunny positions among rocks and grass on slopes. The climate is tropical with hot, humid summers and mild winters. The rainfall is about 1000 mm per annum with a summer maximum.
**Notes:** *E. bubalinus* was first located in 1944 and described in 1957 but with details of the female cones lacking. The species can be recognised by the dense mat of buff-brown, felted hairs on the scale leaves and leaf bases.
**Cultivation:** Suited to tropical and subtropical regions. Uncommon in cultivation but apparently plants grow readily. Best in full sun and requires excellent drainage. Sensitive to damage by frosts.
**Propagation:** From seed and by removal of suckers.

## *Encephalartos caffer*

(Thunb.) Lehm.
(of Caffraria, a region in eastern Cape Province, South Africa)

**Description:** A small cycad with an entirely subterranean, rarely branched trunk to 40 cm long and 25 cm across, subtended by large, swollen fleshy roots. **Young leaves** light green and covered with brown wool. **Mature leaves** about ten in an obliquely erect, sparse crown, 0.4–1m x 15–20 cm, dark green, shallowly vee-ed in cross-section, straight in profile or slightly twisted; **petiole** 6–12 cm long, the swollen base covered with brown hairs; **leaflets** lanceolate, irregularly arranged in different planes to impart a plumose appearance, moderately crowded, sometimes twisted, evenly spaced throughout, inserted on the rhachis at 80–90°, the lower leaflets reduced and the lowest one or two pairs spine-like; **median leaflets** 8–10 cm x 0.8–1 cm, pungent-tipped, the margins lacking spines. Cones markedly dissimilar, greenish yellow, solitary. **Male cones** 20–30 cm x 6–12 cm, cylindrical; **peduncle** 12–15 cm long. **Female cones** 20–30 cm x 12–

*Encephalartos bubalinus.*

Male cone of *Encephalartos barteri.*

15 cm, cylindrical-ovoid; **peduncle** 4–7 cm long. **Seeds** 3.5–3.8 x 2–2.3 cm, bright red to scarlet, glossy.
**Distribution and Habitat:** Endemic to South Africa where it is found in the coastal and near-coastal districts of the Transkei and Eastern Cape Province. This species grows in open grassland often among rocks. The climate is hot in summer and mild to cool in winter with frosts being unknown. Rainfall ranges from 750 mm to 1000 mm per annum over its distribution with mainly a summer maximum.
**Notes:** *E. caffer*, possibly the southernmost occurring cycad in South Africa, was first described in the genus *Zamia* in 1775 and later transferred to *Encephalartos* in 1834. It shares many similarities with *E. ngoyanus* but can be distinguished by the densely crowded leaflets which lack marginal spines.
**Cultivation:** Suited to temperate regions. This species is well tried and known to be hardy and adaptable. Looks attractive when planted among rocks. Requires a sunny position and excellent drainage. Mature specimens transplant readily. Tolerates light frosts only.
**Propagation:** Solely from seed.

## *Encephalartos cerinus*

Lavranos & Goode
(waxy, in reference to the covering on the leaves)

**Description:** A small cycad with a subterranean or shortly emergent trunk to 30 cm long and 25 cm across, growing singly or in clumps of up to three stems, with very few basal suckers produced. **Young leaves** light bluish green. **Mature leaves** about fifteen in a stiffly erect, dense crown, 1–1.2 m x 35–40 cm, dark bluish green, flat or very shallowly V-shaped in cross-section, straight in profile; **petiole** 12–18 cm long; **leaflets** linear-lanceolate, moderately spaced and evenly distributed throughout, inserted on the rhachis at about 80°, the lower leaflets reduced but hardly spine-like; **median leaflets** 15–18 cm x 1–1.2 cm, rigid, pungent-tipped, the margins entire or the lower margin with one or two spines in the distal half. Cones markedly dissimilar, dark bluish green with a prominent waxy covering, becoming yellow at maturity, one on each stem. **Male cones** 55–60 cm x 9–10 cm, narrowly cylindrical; **peduncle** 6–8 cm long. **Female cones** 30–35 cm x 12–18 cm, ovoid, the **sporophylls**

with irregularly toothed or lobed margins. **Seeds** 2.5–3 cm x 1.6–1.8 cm, oblong, strongly angular, yellow to pale red.
**Distribution and Habitat:** Endemic to South Africa where it is restricted to two localities in northern Natal Province. It grows on cliffs and escarpments of river gorges with the plants often wedged in crevices, the roots growing in skeletal soil. The climate is very hot in summer and mild to cool in winter.
**Notes:** Although described as recently as 1989, *E. cerinus* has been reduced to great rarity in the wild as a result of illegal poaching. The species was known only from two colonies and it is estimated that less than fifty mature plants are left in the wild at one site, none at the other. *E. cerinus* is a very distinctive dwarf species which can be readily recognised by the prominent waxy bloom on the leaves and cones. Plants in nature usually shed their leaves completely and are deciduous prior to coning or the production of new leaves.
**Cultivation:** Suited to subtropical and warm temperate regions. Best grown in full sun but can be successful in partial shade. Requires excellent drainage and plants should be kept on the dry side when deciduous, although watering will delay leaf shed. Tolerates light frosts.
**Propagation:** From seed and less commonly by removal of suckers.

## *Encephalartos chimanimaniensis*

R. A. Dyer & I. Verd.
(from the type locality, the Chimanimani Mountains)

**Description:** A small to medium-sized cycad with an erect trunk to 1.8 m tall and 45 cm across, growing in clumps of up to three stems, with suckers produced from the base. **Young leaves** covered with white woolly hairs. **Mature leaves** about twenty in an obliquely erect crown, 1–1.5 m x 25–36 cm, bright green and glossy, flat in cross-section, straight in profile (sometimes twisted near the apex); **petiole** 2–6 cm long; rhachis relatively thin, ageing yellow; **leaflets** linear-lanceolate to lanceolate, irregularly twisted at the base and facing forwards, dark green above, yellowish beneath, moderately spaced and evenly distributed throughout, inserted on the rhachis at about 80°, the lower leaflets reduced to a series of spine-like processes; **median leaflets** 12–18 cm x 2–2.5 cm, pungent-tipped, the margins each with two to four spines near the base. **Cones** markedly dissimilar, bluish green, one to three on each stem. **Male cones** 50–70 cm x 8–10 cm, narrowly cylindrical, erect or lax; **peduncle** 10–15 cm long. **Female cones** 40–45 cm x 18–20 cm, barrel-shaped; **peduncle** 7–8 cm long. **Seeds** 2–3 cm x 1.5–2 cm, oblong, angular, red.
**Distribution and Habitat:** Restricted to the Chimanimani Mountain Range which forms the border between Zimbabwe and Mozambique. This species grows among rocks in high-rainfall grassland at about 1000 m altitude. The climate is subtropical, being hot and humid in summer and mild and wet in winter. The rainfall, which can be in excess of 1800 mm per annum, falls mainly in summer.
**Notes:** *E. chimanimaniensis* is a segregate species of the *E. manikensis* complex and was described in 1969. It is now extinct in the wild in Zimbabwe (from poaching), and has been reduced to great rarity in Mozambique. A concerted program is needed for the survival of natural populations of this species.
**Cultivation:** Suited to subtropical regions. Best grown in full sun and must be given excellent drainage. Mulching and regular watering, especially over summer, are of benefit. Not tolerant of frosts.
**Propagation:** From seed and by removal of suckers.

## *Encephalartos concinnus*

R. A. Dyer & I. Verd.
(neat, trim, elegant)

**Description:** A small to medium-sized cycad with erect trunks to 3 m long and 45 cm across, growing in clumps of up to eight stems, with suckers produced from the base and offsets along the trunk. **Young leaves** light green, covered with grey, woolly hairs. **Mature leaves** numerous in a rather untidy, obliquely erect crown, 1.5–2 m x 20–30 cm, dark green, flat in cross-section, straight in profile; **petiole** 6–10 cm long, woolly at the base; **leaflets** lanceolate, moderately spaced, those towards the base wider apart, inserted on the rhachis at about 40°, the lower leaflets reduced to a few multi-lobed, spine-like processes; **median leaflets** 10–15 cm x 2–2.5 cm, rigid, pungent-tipped, the margins each with one to three spines. **Cones** markedly dissimilar, green. **Male cones** 30–50 cm x 7–10 cm, narrowly fusiform, one to four on each stem; **peduncle** 5–12 cm long. **Female cones** 35–45 cm x 15–20 cm, barrel-shaped, one or two on each stem; **peduncle** 6–8 cm long. **Seeds** 3–3.5 cm x 1.8–2.3 cm, narrowly obovoid, red.
**Distribution and Habitat:** Endemic to south-central Zimbabwe where it grows among boulders in open deciduous forest and grassland and also on the steep sides of protected ravines and gorges. The climate is hot and humid in summer and mild in winter.
**Notes:** *E. concinnus*, part of the *E. manikensis* complex, was described in 1969. It is distributed over a relatively large area but is usually found growing in small, disjunct colonies.
**Cultivation:** Suited to subtropical and warm temperate

P. VORSTER

Female cone of
*Encephalartos cupidus*.

regions. Can be grown in full sun or filtered sun. Sensitive to damage by frosts.
**Propagation:** From seed and by removal of suckers.

## *Encephalartos cupidus*

R. A. Dyer
(desirable, attractive, in reference to its being sought after by poachers)

**Description:** A dwarf cycad with mostly subterranean trunks (rarely emergent to 15 cm tall) to 15 cm across, growing in much-branched clumps of up to twelve stems, with new suckers produced from the base. **Young leaves** erect, bluish green. **Mature leaves** about fourteen in a widely spreading to almost horizontal or arching stiff crown, 0.5–1 m x 20–30 cm, grey-green to yellowish grey, deeply vee-ed in cross-section, straight in profile, slightly twisted or recurved near the apex; **petiole** 8–12 cm long, the swollen base with a few brown hairs; **leaflets** linear-lanceolate, moderately spaced and evenly distributed throughout, steeply erect in a deep vee, inserted on the rhachis at about 40°, the lower leaflets reduced to a series of simple or bifurcate prickles; **median leaflets** 12–15 cm x 1–1.6 cm, rigid, pungent-tipped, with three or four spines on each margin. **Cones** dissimilar, one or two on each stem (females solitary), green. **Male cones** 18–20 cm x 5–8 cm, cylindrical, bluish green; **peduncle** 6–8 cm long. **Female cones** 18–20 cm x 12–14 cm, ovoid, light green to yellow green, sessile. **Seeds** 2–2.5 cm x 1.5–2 cm, ellipsoid, yellow to apricot coloured, ridged.
**Distribution and Habitat:** Endemic to South Africa where restricted to a limited region of the Eastern Transvaal Province. It grows on slopes, rocky ridges and gorges in full sun among grass. The climate is subtropical with hot, humid summers and mild to cool winters. The rainfall is about 650 mm per annum with a summer maximum.
**Notes:** *E. cupidus* is a distinctive species which was described in 1971 and since then its natural populations have suffered drastically from the activities of poachers. It is a small-growing species which is easily recognised by its widely spreading, blue-green leaves with stiffly erect leaflets which have a few marginal spines.
**Cultivation:** Suited to subtropical and warm temperate regions. Grows vigorously and suckers profusely. Best grown in full sun. Tolerates moderate to heavy frosts.
**Propagation:** From seed and by removal of suckers which transplant readily.

P. VORSTER

Male cone of
*Encephalartos cupidus*.

## *Encephalartos cycadifolius*

(Jacq.) Lehm.
(with leaves like the Sago Palm, *Cycas revoluta*)

**Description:** A small cycad with subterranean or emergent trunks to 1.5 m tall and 25 cm across, growing in clumps of up to twelve stems, with new suckers produced freely from the base. **Young leaves** light green, covered with fine woolly hairs. **Mature leaves** numerous in an erect, somewhat untidy crown, 0.5–1 m x 20–26 cm, dark green, shallowly vee-ed in cross-section, incurved or slightly twisted in profile; rhachis light green to yellow; **petiole** 10–14 cm long, the swollen base sparsely woolly; leaflets linear, moderately spaced and evenly distributed throughout, flat or obliquely erect to form a shallow vee, inserted on the rhachis at about 60°, the lower leaflets reduced but not spine-like; **median leaflets** 9–15 cm x 0.4–0.6 cm, pungent-tipped, the margins slightly recurved and lacking spines. Cones markedly dissimilar, one or two from a stem, yellow, densely covered with white to tawny wool. **Male cones** 13–22 cm x 4–7 cm, cylindrical, the apex broadly rounded, sessile. **Female cones** 20–30 cm x 16–18 cm, barrel-shaped, sessile. **Seeds** 2.5–3 cm x 1.5–2 cm, narrowly obovoid, orange to yellow or honey brown.

**Distribution and Habitat:** Endemic to South Africa where it is found in mountainous regions of the Eastern Cape Province. It grows on exposed shale slopes up to 1800 m altitude. The summers are hot and the winters cold with heavy frosts and snowfalls being common. The rainfall ranges from 500 mm to 800 mm per annum with a summer maximum.

**Notes:** *E. cycadifolius* was described in the genus *Zamia* in 1801 from a plant cultivated in the Imperial Botanic Garden at Vienna. It was later transferred to *Encephalartos* in 1834 and has been the subject of some confusion since it was also named as *E. eximius* by Inez Verdoon as recently as 1954 when it was mistakenly believed to be new. Superficially this species resembles *E. ghellinckii* which can be distinguished by the revolute margins of its leaflets.

**Cultivation:** Suited to temperate and cool subtropical regions. Seedlings of this species can be established read ily but large plants are very sensitive to disturbance and take many years to settle after transplanting. Best results are achieved in a sunny position in alkaline soils. Tolerates moderate to heavy frosts.

**Propagation:** From seed and by removal of small suckers. This is a very slow growing species — plants known to be thirty years old from seed have been reported as having trunks only 10 cm in diameter.

## *Encephalartos dolomiticus*

Lavranos & Goode
(growing on dolomite)

**Description:** A small to medium-sized cycad with an erect trunk to 2 m tall and 40 cm across, growing in

Female cones of *Encephalartos dolomiticus*.

P. VORSTER

clumps of two or three stems, with new suckers produced from the base. **Young leaves** silvery from a powdery bloom, also covered with long, silvery hairs. **Mature leaves** numerous in an obliquely erect crown, 0.6–0.8 m x 15–20 cm, glaucous, shallowly vee-ed in cross-section, straight or mostly twisted or flexuose in profile; **petiole** 5–12 cm long, with a red-brown collar; **leaflets** narrow-elliptical, slightly falcate, crowded in the distal two-thirds, the lower leaflets widely spaced, obliquely erect to form a shallow vee, inserted on the rhachis at about 45°, the lower leaflets reduced and spine-like; **median leaflets** 12–17 cm x 1–1.4 cm, pungent-tipped, margins entire or with one or two spines on the lower side. **Cones** markedly dissimilar, blue-green, one to four on each stem. **Male cones** 35–50 cm x 10 cm, narrowly ovoid to narrowly cylindrical; **peduncle** 8–10 cm long. **Female cones** 30–45 cm x 18–25 cm, broadly ovoid, the **sporophylls** strongly wrinkled and warty; **peduncle** 4–5 cm long, hidden by the cataphylls. **Seeds** 3–3.5 cm x 1.8–2 cm, ellipsoid, yellow.

**Distribution and Habitat:** Endemic to South Africa where found in the Drakensburg Mountains of the Transvaal Province. It grows on ridges and slopes of dolomite formation, among grass, at about 1200 m altitude. The summers are hot and the winters are cool to cold with occasional frosts. Rainfall is 600–800 mm per annum distributed mainly over summer.

**Notes:** *E. dolomiticus* is a segregate from *E. eugene-maraisii* and was described in 1988. Its natural populations have been drastically reduced and the species is now regarded as being seriously endangered. The total natural population may consist of only about fifteen plants, with many hundreds being removed for cultivation over the last two decades. Coning occurs rarely in nature. Hopefully protection and a replanting program will ensure its survival in the wild. The species is closely related to *E. eugene-maraisii* but can be distinguished by its twisted leaves and the strongly wrinkled sporophylls on the female cones. *E. verrucosus* Robbertse, Vorster & van der Westhuizen (see page 221) is synonymous.

**Cultivation:** Suited to temperate and cool subtropical regions. Requires sun and excellent drainage. Tolerates light to moderate frosts.

**Propagation:** From seed and by removal of suckers. Plants no longer set seed in nature and cones are seldom produced on cultivated plants.

# *Encephalartos dyerianus*

Lavranos & Goode

(after Dr R. Allen Dyer, twentieth century South African botanist who monographed the southern African species of *Encephalartos*)

**Description:** A medium-sized to large cycad with an erect or reclining trunk to 4 m tall and 60 cm across, growing in clumps of up to four stems, with suckers produced freely from the base. **Young leaves** silvery. **Mature leaves** numerous in an erect to spreading, rounded crown, 1.4–1.7 m x 30–40 cm, glaucous becoming yellowish green with age, shallowly vee-ed in cross-section, straight or slightly twisted towards the apex; **petiole** 4–6 cm long, glaucous, with a prominent brown collar; **leaflets** narrowly elliptical, slightly falcate, moderately crowded and evenly spaced throughout, obliquely erect to form a shallow vee, inserted on the rhachis at about 45°, the lower leaflets reduced and spine-like; **median leaflets** 17–24 cm x 1.3–1.8 cm, rigid, pungent-tipped, with one or two spines on both margins. **Male cones** 30–50 cm x 9–12 cm, narrowly ovoid, bluish green, ageing to yellow, five to eight on each stem; **peduncle** 10–17 cm long. **Female cones** 30–60 cm x 10–20 cm, cylindrical, bluish green ageing to yellow, up to five on each stem. **Seeds** 4–4.5 cm x 2.5–3 cm, narrowly oblong, yellow maturing to orange-brown.

**Distribution and Habitat:** Endemic to South Africa where it is known from a single granite hill in the northern part of the Transvaal Province. Plants grow in

P. VORSTER

*Encephalartos dyerianus* in habitat.

P. VORSTER

Male cones of *Encephalartos dyerianus*.

exposed situations among grass and large granite boulders at about 700 m altitude. The climate is subtropical with hot summers and mild winters with no frosts occurring. Rainfall is about 450 mm per annum all falling over summer.

**Notes:** *E. dyerianus*, a segregate from *E. eugene-maraisii*, was described in 1988. It can be distinguished by its extremely short petioles, the lower leaflets being reduced to short, spine-like processes and the smooth glabrous sporophylls of the cones. Although restricted to a single hill, the population of this species consists of about 600 plants which are reproducing freely and the site is a proclaimed nature reserve. *E. graniticolus* Robbertse, Vorster & van der Westhuizen (see page 195) is synonymous.

**Cultivation:** Suited to subtropical and warm temperate regions. Grows best in full sun and requires excellent drainage. Frost-tolerance is not high.

**Propagation:** From seed and by removal of suckers which transplant readily.

Female cones of *Encephalartos eugene-maraisii*.

P. VORSTER

# *Encephalartos eugene-maraisii*

I. Verd.

(after Eugene Marais, South African writer and naturalist)

**Description:** A medium-sized to large cycad with erect to sprawling trunks to 3 m tall and 45 cm across, growing in clumps of up to ten stems, with new suckers produced from the base. **Young leaves** silvery blue. **Mature leaves** numerous in an obliquely erect to spreading crown, 0.7–1.5 m x 30–40 cm, bluish green, deeply vee-ed in cross-section, straight in profile with an upcurved tip; **petiole** 5–16 cm long, circular in cross-section, with a relatively small, swollen base covered with grey wool; **leaflets** linear-lanceolate, crowded and evenly spaced throughout, obliquely erect to form a deep vee, inserted on the rhachis at about 50°, the lower leaflets reduced, the lowest one or two pairs spine-like; **median leaflets** 15–

20 cm x 1.3–1.8 cm, rigid, pungent-tipped, lacking marginal spines or occasionally with a single basal spine. **Cones** dissimilar, greenish grey with a dense covering of maroon hairs. **Male cones** 20–40 cm x 6–8 cm, cylindrical, tapered to the apex, giving off a strong odour, sessile, one to three on each stem. **Female cones** 20–50 cm x 16–20 cm, ovoid, usually borne in pairs; **peduncle** about 5 cm long. **Seeds** 3.5–4 cm x 2.3–3 cm, obovoid, light brown.

**Distribution and Habitat:** Endemic to South Africa where it occurs in scattered, disjunct mountainous districts of the Transvaal Province. Plants grow in sunny locations and among low shrubs at an altitude of about 1450 m. The climate is hot in summer and very cold in winter with severe frosts being common. Rainfall is 600–750 mm per annum with a summer maximum.

**Notes:** Populations of *E. eugene-maraisii*, which was described in 1945, are very disjunct and separated by large distances. These groups have been isolated for long periods and recent studies have shown that they have adapted genetically to differences in their habitat (see also *E. dolomiticus* and *E. dyerianus*).

Specimens from the Waterberg Region are representative of typical *E. eugene-maraisii*, whereas those from the Middelburg Region are treated by some authorities as a distinct species (see *E. middelburgensis*), and by others as a subspecies (*E. eugene-maraisii* ssp. *middelburgensis* Lavranos & Goode).

**Cultivation:** Suited to cool subtropical and temperate regions. Plants should be grown in full sun to encourage the attractive leaf colouration. Tolerates very heavy frosts.

**Propagation:** From seed and by removal of suckers which transplant readily.

*Encephalartos ferox.*

# *Encephalartos ferox*

G. Bertol.

(ferocious, fierce, in reference to the rigid, spiny leaves)

**Description:** A small cycad with a subterranean trunk (very rarely an emergent trunk to 2 m tall) to 35 cm across, usually solitary or rarely branched, with new suckers produced from the base. **Young leaves** dark green or coppery brown, hairy. **Mature leaves** numerous in an obliquely erect to arching, stiff to rigid crown, 1–2 m x 20–30 cm, dark green, shallowly vee-ed in cross-section, straight or slightly recurved; **petiole** 10–15 cm long; **leaflets** moderately spaced at the base, the rest crowded and often overlapping, obliquely erect to form a shallow vee, inserted on the rhachis at about 70°, the lower leaflets reduced to spine-like processes; **median leaflets** 12–15 cm x 4–8 cm, oblong-obovate, rigid, apex obtuse (often trifurcate), both margins with three to five, broadly triangular lobes which end in pungent spines, these lobes twisted out of the plane of the leaflet. **Cones** markedly dissimilar, pink to red or scarlet (occasionally yellow). **Male cones** 40–50 cm x 7–10 cm, cylindrical, one to ten on each stem; **peduncle** 2–3 cm long. **Female cones** 25–50 cm x 20–25 cm, one to five on each stem, sessile. **Seeds** 4.5–5 cm x 1.5–2 cm, narrowly oblong, angular, red, glossy.

**Distribution and Habitat:** Widely distributed in coastal districts of southern Mozambique and northern Natal. It occurs in evergreen forests and sparse scrub clothing sand dunes (to within 50 m of the beach, sometimes nearly to the high water mark). The climate is hot and humid in summer and mild in winter. The rainfall ranges from 1000 mm to 1250 mm per annum with a summer maximum. Frosts are unknown.

**Notes:** *E. ferox* was described in 1851 from material collected in Mozambique. It was redescribed in 1932 as *E. kosiensis* Hutchinson, based on material collected in Natal. After careful comparison of the type collection, the latter name has been reduced to synonymy. The starch from the stem was used by local people as a source of food. Its cones are among the most colourful in the genus. In some areas of Mozambique, plants with yellow

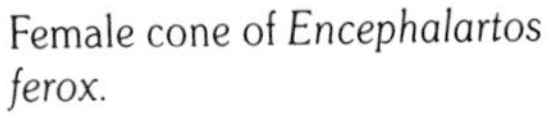

Female cone of *Encephalartos ferox.*

Male cone of *Encephalartos ferox.*

cones may predominate. The leaflets of some plants may have strongly revolute margins which impart an almost tubular appearance.
**Cultivation:** Suited to tropical and subtropical regions. Plants will grow in sun but have a better appearance if given partial protection. Tolerance of frosts is very low. This handsome species is found in the collection of many botanical gardens, especially those in the warmer regions. Plants should be sited where the highly colourful cones can be appreciated.
**Propagation:** From seed and by removal of suckers which transplant readily.

## *Encephalartos friderici-guilielmii*

Lehm.
(after Friedrich William, nineteenth century King of Prussia and patron of botany)
**Description:** A medium-sized to large cycad with sturdy, erect or sprawling trunks to 4 m tall and 60 cm across, growing in clumps of up to six stems, with new suckers produced freely from the base. **Crown** open, covered by brown wool. **Young leaves** light green or bluish green. **Mature leaves** numerous in an obliquely erect to widely spreading crown, 1–1.5 m x 20–35 cm, dull green to somewhat glaucous, deeply vee-ed in cross-section, straight in profile; **petiole** 15–25 cm long, slender, the swollen base covered with brown wool; **leaflets** linear, lower ones moderately spaced, distal ones crowded, obliquely erect and held at a steep angle to form a vee, inserted on the rhachis at about 45°, the lower leaflets reduced but not spine-like; **median leaflets** 10–17 cm x 0.7–0.8 mm, pungent-tipped, the margins lacking spines. **Cones** markedly dissimilar, yellow to brown, densely woolly. **Male cones** 20–40 cm x 6–10 cm, cylindrical, becoming curved, three to twelve per plant; **sporophylls** about 2.8 cm x 2 cm, with a beak about 0.5 cm long. **Female cones** 25–30 cm x 15–20 cm, barrel-shaped, three to six per plant. **Seeds** 2.5–3 cm x 1.5–2 cm, obovoid to oblong, yellow to orange.
**Distribution and Habitat:** Endemic to South Africa where it is common in the mountains of the Eastern Cape Province in the districts of Queenstown and Cathcart. It grows on rocky slopes and ridges in full sun among grass and low shrubs. The summers are hot and in the winter snow and frosts are common. Rainfall ranges from 375 mm to 500 mm per annum with a summer maximum.
**Notes:** *E. friderici-guilielmi*, which was described in 1834, has been confused with *E. cycadifolius* from which it can be distinguished by its taller, stouter trunk, its open, woolly crown with the leaves curving outwards and its longer, broader leaflets. Prior to the production of cones or a flush of new fronds, the old fronds become prostrate or may even droop. This species can produce more cones at any one time than any other in the genus.
**Cultivation:** Suited to temperate and cool subtropical regions. A very hardy species that requires an open sunny position, excellent drainage and is tolerant of heavy frosts. Very slow growing.
**Propagation:** From seed and by removal of suckers which transplant readily.

*Encephalartos friderici-guilielmi.*

P. VORSTER

Female cones of *Encephalartos friderici-guilielmi.*

# *Encephalartos ghellinckii*

Lem.
(after M. Ed de Ghellinck de Walle, Belgian amateur botanist and horticulturist)

**Description:** A medium-sized cycad with an erect or reclining trunk to 3 m tall and 40 cm across, growing in clumps of up to three stems, with new suckers produced from the base. **Young leaves** covered with dense grey hairs. **Mature leaves** erect and incurved in a relatively open crown, 0.8–1 m x 16–28 cm, yellowish green to dark green, flat, shallowly vee-ed or inversely vee-ed in cross-section, spirally twisted in profile; **petiole** 10–25 cm long, the swollen base covered with brown wool; **rhachis** yellow and twisted; **leaflets** linear with strongly revolute margins, crowded and evenly spaced throughout, obliquely erect to form a shallow vee, the arrangement being altered by the twisted rhachis, inserted on the rhachis at 45–50°, the lower leaflets reduced but not spine-like; **median leaflets** 8–14 cm x 0.2–0.4 cm, leathery, pungent-tipped, the margins lacking spines and strongly revolute. **Cones** markedly dissimilar, nearly sessile, one to five on each stem. **Male cones** 20–25 cm x 6–8 cm, cylindrical, pale yellow with a dense covering of cream to brown hairs. **Female cones** 20–25 cm x 12–15 cm, ovoid to ellipsoid, yellowish with a dense covering of cream wool. **Seeds** 2.5–3 cm x 1.5–2 cm, oblong, angular, yellow to golden yellow.

*Encephalartos friderici-guilielmii* From *L'Illustration Horticole* 29:123-4, (1882)

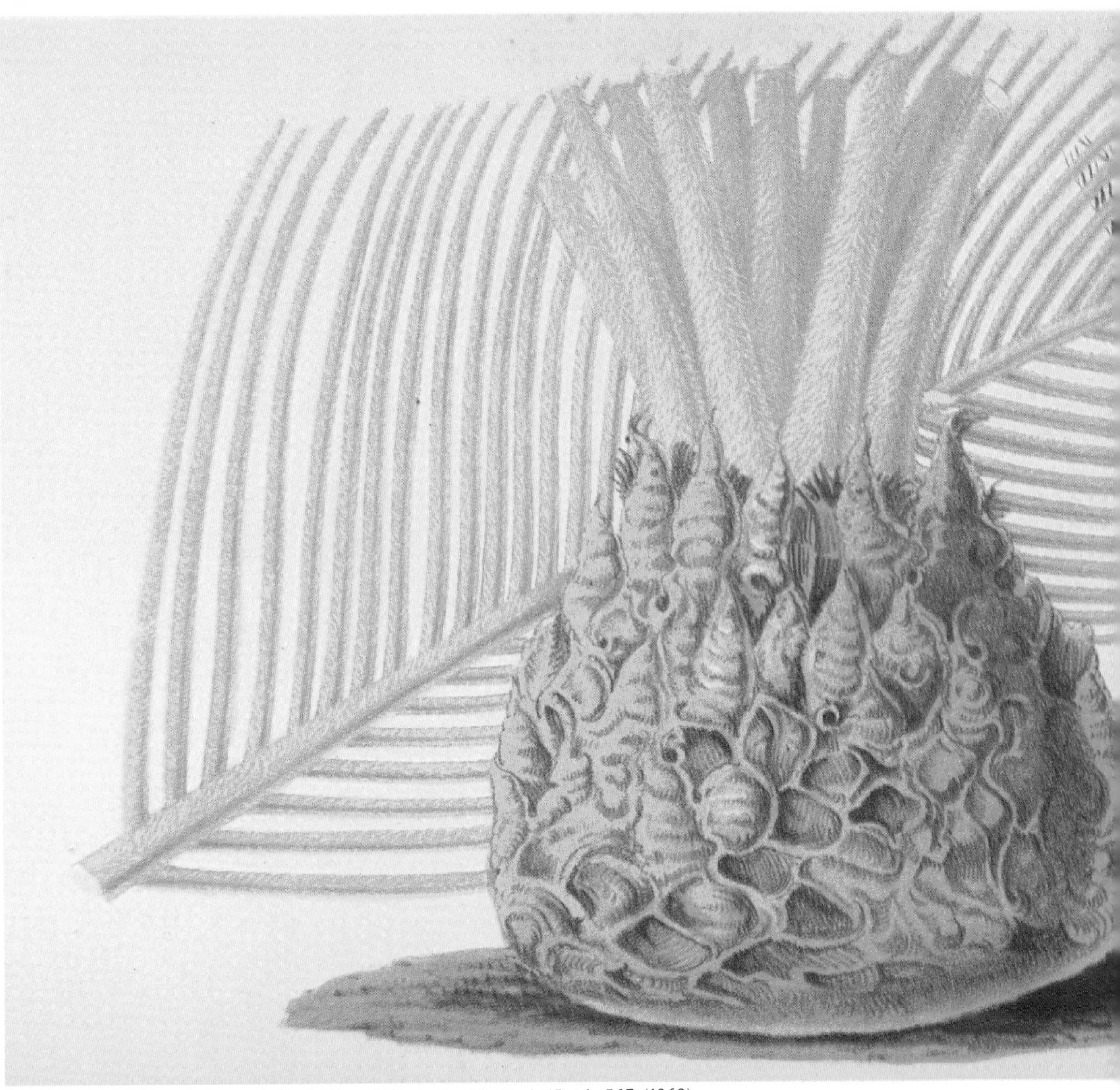

*Encephalartos ghellinckii* From *L'Illustration Horticole* 15, tab. 567, (1868)

**Distribution and Habitat:** Endemic to South Africa where widely distributed in the provinces of Transkei and Natal. This is a species of diverse habitats but is prominent in mountainous regions where it grows on sheltered rocky slopes and in grasslands. Its altitudinal range is amazing, extending as it does from about 700 m in Natal to about 2400 m in the Drakensburg Mountains. The climate is mild to hot in summer and cold to freezing in winter with heavy frosts and snowfalls being extremely common at the higher elevations.

**Notes:** *E. ghellinckii*, which was described in 1867, is easily recognised by its strongly revolute leaf margins which are unique in the genus. Superficially, plants may be confused with *Cycas revoluta* but that species can be readily distinguished by the prominent midribs on its leaflets and the outcurving or straight orientation of the fronds. Grassfires are common in the habitat of this species and may influence its cone production, although this phenomenon is reportedly infrequent in the wild and occurs every four to six years. Being widespread and with such disjunct altitudinal ranges it is not surprising that this species exhibits variation. Three main ecotypes can be recognised:

1. The high altitude variant from the Drakensburg Mountains is the most robust having the longest and stoutest trunks, the largest cones, an extremely woolly brown crown and the old leaves are retained as a skirt on the trunk.
2. The Transkei variant has slender, sinuous trunks which are often procumbent.

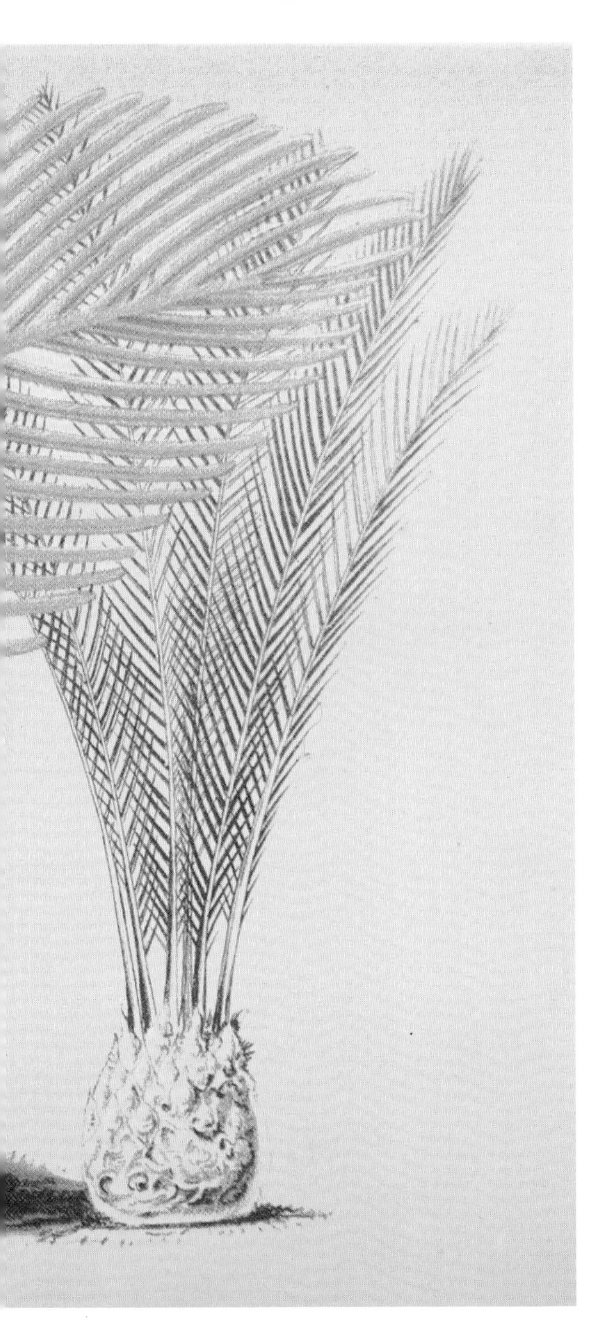

3. The dwarf variant from low altitudes has short stems and a grey, woolly crown.

**Cultivation:** Suited to temperate and cool subtropical regions. This species is generally regarded as being difficult to grow, because it dislikes high temperatures. Large specimens do not transplant readily and are commonly slow to re-establish. Seedlings adapt better to cultivation but are generally slow and appear to dislike humid conditions. Heavy frosts and snow present no problems to the cultivation of this species. It seems that cultivated plants rarely produced cones, this probably being a reflection of insufficient stimulus following lack of winter chilling.

**Propagation:** From seed and by removal of suckers.

# *Encephalartos graniticolus*

Robbertse, Vorster & van der Westhuizen
(growing on granite)

This species, described in 1988, is a synonym of *E. dolomiticus*. Although the latter was also described in 1988, its description was published prior to that of *E. graniticolus* and hence by botanical rules has priority.

# *Encephalartos gratus*

Prain
(pleasing, agreeable)

**Description:** A medium-sized cycad with an erect trunk to 2.5 m tall and 60 cm across although plants from higher altitudes are about 30 cm across, usually solitary but in clumps of up to eight stems, with suckers produced from the base. **Young leaves** light green, with loose fine white hairs. **Mature leaves** numerous in a spreading to rounded crown, 1.2–2 m x 34–44 cm, dark green, flat in cross-section, straight in profile or recurved; **petiole** 6–18 cm long, with some brownish hairs; **leaflets** lanceolate, rather soft in texture, moderately spaced and evenly distributed throughout, inserted on the rhachis at 70–90°,

*Encephalartos gratus.*

Young plant of *Encephalartos horridus*.

the lower leaflets reduced to simple, bifurcate or trifurcate, spine-like processes; **median leaflets** 18–26 cm x 2–3.5 cm, rigid, pungent-tipped, margins wth two to seven pungent spines towards the base. **Cones** dissimilar, pinkish brown, covered with brown felted hairs. **Male cones** 30–40 cm x 4.5–10 cm, cylindrical, fifteen to twenty on each stem, clustered or arising in groups; **peduncle** 15–20 cm long. **Female cones** 55–68 cm x 15–20 cm, broadly cylindrical, two to five on each stem, nearly sessile. **Seeds** 3–4 cm x 1.4–2 cm, narrowly obovoid, pinkish brown to reddish.

**Distribution and Habitat:** Occurring in south-eastern Malawi and north-western Mozambique. This species grows in rocky gorges and on slopes near streams at about 600 m altitude, often in deciduous, riverine forest. The climate is hot in summer and warm in winter. Rainfall is about 1500 mm per annum with a summer maximum.

**Notes:** Living plants of *E. gratus*, collected from Mulanje Mountain in Malawi, were growing at Kew Gardens in about 1900. The species was described in 1916 after male and female cones had been collected. This species is well established in nature and regenerating freely.

**Cultivation:** Suited to tropical, subtropical and perhaps warm temperate regions. Can be grown in full sun or partial shade. Tolerates light frosts only.

**Propagation:** From seed and by removal of suckers.

Female cone of *Encephalartos horridus*.

P. VORSTER

## Encephalartos heenanii

R. A. Dyer
(after Denis Heenan who discovered the species)

**Description:** A small to medium-sized cycad with trunks to 2.5 m tall and 35 cm across, growing in clumps of up to five stems, with numerous suckers produced from the base. Apex of trunk densely woolly. **Young leaves** densely covered with brown hairs. **Mature leaves** numerous in an erect, incurved crown, 1–1.25 m x 24–30 cm, dark green, hairy, flat or inversely vee-ed in cross-section, incurved in profile and often twisted near the apex; **petiole** 20–30 cm long, hairy; leaflets lanceolate to oblanceolate, moderately crowded, evenly spaced throughout, the leaflets in the upper part of the leaf reflexed, the veins on the lower side prominently raised, inserted on the rhachis at about 60°, the lower leaflets 12–15 cm x 1.2–1.5 cm, the margins usually lacking spines or with a single spine near the base. **Cones** alike, pale green, densely covered with brown wool, pimpled, one to three on each stem. **Male cones** 25–30 cm x 16–18 cm, ovoid; **peduncle** 8–9 cm long. **Female cones** 25–30 cm x 16–18 cm, ovoid; **peduncle** 5–6 cm long. **Seeds** 2.3–2.5 cm x 1.4–1.8 cm, narrowly obovoid, angular, orange.
**Distribution and Habitat:** Occurring in the south-east of the Transvaal Province of South Africa and across the border in adjacent areas of Swaziland. This species grows in open areas among grass on steep slopes at about 1500 m altitude. The climate is hot in summer and cold in winter. Rainfall is about 1200 mm per annum, falling mainly in summer.
**Notes:** *E. heenanii*, which was described in 1972, has been reduced to great rarity in the wild as the result of illegal poaching. A distinctive species, it apparently has as its closest relative *E. paucidentatus*. It can be distinguished from that species by its smaller stature, smaller leaflets and the presence of hairs, even on mature leaves. It shares two unique feature with *E. aemulans* — remarkably similar male and female cones and raised veins on the underside of the leaflets.
**Cultivation:** Suited to temperate and cool temperate regions. Best grown in full sun. Mature plants and seedlings establish readily. Tolerates light to moderate frosts.
**Propagation:** From seed and by removal of suckers.

## Encephalartos hildebrandtii

A. Br. & Bouche
(after J. M. Hildebrandt, German botanist, explorer and early collector of this species)

**Description:** A medium-sized to large cycad with an erect trunk to 6 m tall and 30 cm across, usually solitary with new suckers produced from the base. **Young leaves** coppery, brown or reddish, covered with fine white hairs. **Mature leaves** numerous in a rounded crown, 2–3 m x 30–60 cm, bright green and glossy, shallowly vee-ed in cross-section, straight in profile; **petiole** 2–7 cm long, the base swollen and woolly; **leaflets** linear-lanceolate to oblong, moderately crowded and evenly spaced throughout, obliquely erect to form a shallow vee, inserted on the rhachis at about 60°, the lower leaflets reduced progressively into a series of small, spine-like processes; **median leaflets** 20–26 cm x 2.8–3.6 cm, up to nine teeth on each margins, one or two subtending the apex to give a bifurcate or trifurcate appearance (not present in plants from Tanzania). **Cones** markedly dissimilar. **Male cones** 20–50 cm x 5–9 cm, narrowly cylindrical, greenish yellow, three to eight on each stem; **peduncle** 5–25 cm long. **Female cones** 28–60 cm x 15–25 cm, broadly oblong to barrel-shaped, yellow, two to four on each stem; **peduncle** 4–6 cm long. **Seeds** 3–3.8 cm x 1.5–2 cm, obovoid, angular, bright red or yellow to orange.
**Distribution and Habitat:** Occurs in Tanzania and Kenya, where it grows in near-coastal and coastal districts including the islands of Zanzibar and possibly Pemba. This species grows in sparse deciduous forests and savannah grassland from sea level to about 600 m altitude. The climate is hot and humid in the summer and mild and dry in winter. The rainfall is 1000–1400 mm per annum, falling mostly in summer.
**Notes:** *E. hildebrandtii* was described in 1874 from material collected by Hildebrandt near Mombasa in Kenya. It had been collected a few years earlier from the island of Zanzibar by the British Consul-General Sir John Kirk and was grown successfully in Kew Gardens but this material was not used for the description. Local natives prepare a flour from the seeds, after boiling and drying, and in times of scarcity the starchy part of the trunk may be prepared as a gruel or porridge. One variety has been named but its status has been questioned. The typical variety (*E. hildebrandtii* var. *hildebrandtii*) lacks teeth on the sporophylls of the male and female cones whereas *E. hildebrandtii* var. *dentatus* Melville has prominent teeth on these organs.
**Cultivation:** Suited to tropical and warm subtropical regions. This species is found in the collections of many botanic gardens thanks to the wide dispersal of plants in the late 1800s. Plants require a sunny location, free and unimpeded drainage and protection from frost.
**Propagation:** From seed or by removal of suckers which transplant readily.

## Encephalartos horridus

(Jacq.) Lehm.
(dreadful, horrible, in reference to its heavily armed leaves)

**Description:** A small cycad with a trunk to 0.8 m long and 30 cm across, the trunks branching freely and form-

ing dense clusters of overlapping or entangled foliage. **Young leaves** silvery blue, contrasting markedly with the old leaves. **Mature leaves** numerous in a rigid, entangled, heavily glaucous to blue-green crown, 0.6–1 m x 10–20 cm, flat or vee-ed in cross-section, straight in the proximal half, strongly recurved to almost coiled near the apex; **petiole** 12–22 cm, with a red-brown collar; **leaflets** ovate-lanceolate, moderately spaced but appearing crowded because of long lobes, inserted on the rhachis at about 80°, the lower leaflets reduced and lacking lobes but not spine-like; **median leaflets** 8–12 cm x 2.5–3.5 cm, blue-green, rigid, pungent-tipped, the lower margins with one to three prominent pungent-tipped lobes to 4 cm long which are twisted out of the plane of the leaflets and form an impenetrable barrier. **Cones** markedly dissimilar, bluish green, covered with black hairs when young, solitary on each stem. **Male cones** 30–40 cm x 10–12 cm, narrowly cylindrical; **peduncle** 6–8 cm long, thick. **Female cones** 30–40 cm x 18–20 cm, ovoid to barrel-shaped, smooth; **peduncle** 4–8 cm long, thick. **Seeds** 3–3.5 cm x 2–2.5 cm, oblong, angular, pale red.
**Distribution and Habitat:** Endemic to South Africa where restricted to the Eastern Cape Province. This species grows in a range of habitats including rocky ridges in shallow, infertile soil and on gentle slopes in deeper, fertile soils. The climate is hot in summer and cool to mild in winter with frosts a rarity. Rainfall ranges from 250 mm to 650 mm per annum, distributed more or less evenly throughout the year.
**Notes:** *E. horridus* was described in the genus *Zamia* in 1801 from a plant cultivated in Vienna and the species was then transferred to the genus *Encephalartos* in 1834. A very distinctive species which can form impenetrable thickets due to its suckering habit and rigid, extremely spiny leaves. Two variants of this species can be distinguished in nature: the robust variant as described above and a dwarf variant, which has shorter trunks (to 20 cm x 15 cm) which sucker more freely, shorter leaves and smaller male and female cones and seeds. Some of these characters appear to be retained in cultivation.
**Cultivation:** Suited to temperate and subtropical regions. Very hardy and adaptable although best not planted too close to paths. Requires full sun and excellent drainage. Tolerates light to moderate frosts.
**Propagation:** From seed and by removal of suckers which transplant readily.

# *Encephalartos humilis*

I. Verd.
(lowly, low growing, dwarf)

**Description:** A small cycad with a subterranean trunk (rarely emergent), to 50 cm long and 30 cm across (usually smaller), growing in clumps of up to six stems, with offsets produced along the trunk and suckers arising from the base. **Young leaves** bluish, covered with fine silky hairs. **Mature leaves** about fifteen in an erect to spreading, sparse, untidy crown, 0.3–0.5 m x 16–25 cm, bluish green, deeply vee-ed in cross-section, recurved and twisted in profile, arising at about 70° from the trunk; rhachis yellow; **petiole** 10–16 cm long, the swollen base sparsely woolly; **leaflets** linear, moderately crowded, the lower leaflets well spaced, deeply vee-ed, inserted on the rhachis at about 30°, the lower leaflets reduced but not spine-like; **median leaflets** 10–14 cm x 0.4–0.6 cm, pungent-tipped, the margins lacking spines, juvenile leaves with a single basal spine. **Cones** dissimilar, solitary on a stem, yellow, covered densely with grey wool. **Male cones** 15–20 cm x 4–5 cm, cylindrical; **peduncle** 10–12 cm long. **Female cones** 25–30 cm x 8–10 cm, barrel-shaped, appearing sessile; **peduncle** 3–5 cm long. **Seeds** 2.5–3 cm x 2–2.5 cm, oblong, bright yellow, shiny.
**Distribution and Habitat:** Endemic to South Africa where it grows in the catchments of the Crocodile River in the Eastern Transvaal Province. It grows on grassy mountain slopes and is often associated with outcrops of sandstone. The climate is cool and temperate in summer and cold in winter. Rainfall is about 800 mm per annum, with a summer maximum.
**Notes:** *E. humilis*, the smallest species in the genus, was separated from *E. lanatus* and described as new in 1951. It can be immediately recognised by its dwarf habit and subterranean stems. Plants usually become completely deciduous before the onset of a flush of new leaves or cones. The numbers of this species in the wild have been greatly reduced by habitat clearing and poaching. It has also been observed that cone production is very limited and may only occur in years of heavy rain and dense mists.
**Cultivation:** Suited to subtropical and temperate regions. Best grown in a sunny position. Cultivated plants may become deciduous each year and should be then kept on the dry side until new growth appears. If watered and mulched, plants may not shed their leaves between flushes. It is recorded that the main stems of cultivated plants may die when about 10 cm tall.
**Propagation:** From seed and by removal of suckers.

# *Encephalartos inopinus*

R. A. Dyer
(unexpected)

**Description:** A medium-sized cycad with erect or more usually pendulous trunks (which are upcurved near the apex) to 4 m long and 25 cm across, growing in clumps of up to eight stems, with new suckers produced freely from the base. **Young leaves** covered with white silky hairs. **Mature leaves** numerous in an obliquely erect to

spreading crown, 0.8–1.5 m x 28–40 cm, soapy green, flat in cross-section, straight in profile with a characteristic twist near the apex; rhachis yellow; **petiole** 15–25 cm long, with two distinctive ridges near the base; **leaflets** linear-lanceolate, moderately spaced and evenly distributed throughout, the lower leaflets deflexed the upper ones obliquely erect, inserted on the rhachis at about 90°, the lower leaflets sometimes reduced and spine-like. **Cones** dissimilar, one to three on each stem. **Male cones** 18–25 cm x 6–8 cm, narrowly ovoid, green with a white powdery bloom; **peduncle** 6–8 cm long. **Female cones** 30–40 cm x 15–20 cm, ovoid, soapy green; **peduncle** 4–5 cm long. **Seeds** 2–2.5 cm x 1.5–2 cm, narrowly oblong, orange to yellowish red.
**Distribution and Habitat:** Endemic to South Africa where restricted to a few small areas in the Lydenburg district of the Transvaal Province. It grows among sparse to dense deciduous vegetation on escarpments and on the margins of gorges of dolomite rock. The summers are very hot and the winters mild. The rainfall is low (about 375 mm per annum) mostly falling in summer.
**Notes:** *E. inopinus* was discovered as recently as 1955 and was not described until 1964 when the second collection was made. The plants commonly hang off the sides of steep gorges with the stems reclining or almost pendulous and the apex upturned. This species can be readily distinguished from all others in the genus by the silvery green leaves, the lower leaflets of which are deflexed and the upper ones obliquely erect. This species is uncommon to rare in the wild and cultivated plants will play an important role in its conservation.
**Cultivation:** Suited to temperate and subtropical regions. Requires filtered sun or full sun and excellent drainage.
**Propagation:** From seed or by removal of suckers which transplant readily.

Emerging leaves of *Encephalartos inopinus*.

# *Encephalartos ituriensis*

Bamps & Lisowski
(from the Ituri Forest in Zaire)

**Description:** A large cycad with an erect or reclining trunk to 6 m tall and 50 cm across, usually solitary, with suckers produced from the base. **Young leaves** light green. **Mature leaves** numerous in an erect crown, 2–3 m x 35–40 cm, dark green, flat in cross-section, straight in profile or recurved near the apex; **petiole** 6–12 cm long; **leaflets** lanceolate to oblong, leathery, moderately crowded and evenly spaced throughout, inserted on the rhachis at about 80°, the lower leaflets reduced, the lowest three to nine pairs spine-like; **median leaflets** 20–25 cm x 2.5–3 cm, pungent-tipped, the margins with three to nine spines on each side. **Cones** markedly dissimilar, green ageing to light brown or yellow. **Male cones** 20–26 cm x 5–7 cm, fusiform, one to four on each stem, semi-pendulous; **peduncle** 20–26 cm x 7 cm, sturdy. **Female cones** 18–20 cm x 10–12 cm, ovoid to cylindrical, one or two on each stem, erect; **peduncle** 3–4 cm long. **Seeds** 2.5–3.5 cm x 1.5–2.5 cm, narrowly oblong, angular, red.
**Distribution and Habitat:** Occurring in north-eastern Zaire and possibly also in Uganda. The species grows on granitic domes which are emergent from surrounding rainforest. It grows on steep slopes in grassland and among granite boulders and along the margins of rainforests. Altitude is about 1000 m. The climate is subtropical with hot humid summers and a cool, mild dry winter. Frosts are unknown.
**Notes:** *E. ituriensis*, described in 1990, was previously confused with *E. hildebrandtii*. It can be distinguished from that species by its smaller leaflets and smaller cones.
**Cultivation:** Suited to subtropical and warm temperate regions. Little known in cultivation. Will grow in full sun or semi-shade. Frost-tolerance is unlikely to be very high.
**Propagation:** From seed and by removal of suckers.

# Encephalartos kisambo

Faden & Beentje
(the local native name for the species)

**Description:** A medium-sized cycad with erect trunks to 2.5 m tall and 70 cm across, growing in clumps of up to twelve stems, with suckers produced from the base. **Young leaves** bluish green, covered with fine white hairs. **Mature leaves** numerous in an obliquely erect to spreading crown, 2–4 m x 50–90 cm, soapy green, flat in cross-section, straight in profile; **petiole** 10–20 cm, swollen and hairy at the base; **leaflets** linear to oblanceolate, crowded and evenly spaced throughout, leathery, inserted on the rhachis at about 80°, the lower leaflets reduced to trifurcate, bifurcate or simple spine-like processes; **median leaflets** 25–40 cm x 3–4 cm, rigid, pungent-tipped, with three to four long, pungent spines on the basal part of the upper margins, the lower margins lacking spines or with a single spine. **Cones** dissimilar, yellow to orange, two or three on each stem. **Male cones** 40–60 cm x 10–12 cm, narrowly cylindrical; **peduncle** 5–15 cm long. **Female cones** 40–60 cm x 15–16 cm, cylindrical; **peduncle** 11–15 cm long, hairy. **Seeds** 2–3 cm x 2–2.5 cm, narrowly oblong, yellow to orange.
**Distribution and Habitat:** Endemic to southern Kenya where restricted to isolated mountains in the Voi district. It grows among rocks on steep slopes at altitudes of between 800 m and 1200 m. Most plants grow in mist forest or cloud forest but some grow in open, exposed situations. The climate is hot and humid in summer and mild and dry in winter. Frosts are unknown.
**Notes:** *E. kisambo* was described in 1989 after being discovered in 1973. It has a very restricted distribution and has suffered badly at the hands of poachers. Its closest relative is *E. hildebrandtii* from which it can be distinguished by being shorter and with fewer marginal spines on the leaflets. *E. voiensis* Moretti, D. Stevenson & Sclavo is a synonym.
**Cultivation:** Suited to tropical and subtropical regions. Can be grown in full sun but best appearance is obtained with some protection. Sensitive to damage frcm frosts.
**Propagation:** From seed and by the removal of suckers.

P. VORSTER

*Encephalartos laevifolius* in habitat showing male cones.

P. VORSTER

Female cones of *Encephalartos laevifolius*.

# Encephalartos laevifolius

Stapf & Burtt-Davy
(with smooth leaves)

**Description:** A medium-sized cycad with a sturdy, erect or sprawling trunk to 4 m tall and 30 cm across, growing in clumps of up to six stems, with new suckers produced freely from the base. **Crown** open, lacking wool, the long cataphylls exposed. **Young leaves** bluish green and hairy. **Mature leaves** numerous in an attractive, spreading to rounded crown, 1–1.3 m x 24–30 cm, blue-green, deeply vee-ed in cross-section, straight or recurved in profile; **petiole** 6–25 cm long, yellow, the swollen base covered with short, dense wool; **leaflets** linear, moderately crowded and evenly spaced throughout, obliquely erect and held at a steep angle to form a vee, inserted on the rhachis at about 45°, the lower ones reduced but not spine-like; **median leaflets** 10–15 cm x 0.5–0.7 cm, pungent-tipped, the margins lacking spines. **Cones** dissimilar, densely hairy, one to five on each stem, sessile. **Male cones** 30–40 cm x 8–10 cm, cylindrical, curved when mature, at first covered with white hairs, becoming brown. **Female cones** 20–30 cm x 10–15 cm, barrel-shaped, with white to yellowish hairs, usually borne in threes. **Seeds**

*Encephalartos lanatus* in habitat.

P. VORSTER

2.5–2.7 cm x 2–2.3 cm, broadly oblong, orange-yellow.
**Distribution and Habitat:** Endemic to South Africa where occurring in the mountains of the Eastern Transvaal Province and Swaziland. It grows on rocky slopes and ridges in full sun among grass and low shrubs at 1300—1800 m altitude. The summers are mild and severe frosts are common in winter. Rainfall is more than 1000 mm per annum, with a summer maximum.
**Notes:** The stems of *E. laevifolius*, which was described in 1926, have a characteristic pattern formed by the densely packed, small scars left by the fallen leaves. New fronds are produced in a spectacular flush. Prior to this event the old leaves in the crown become horizontal or even droop. This species has similarities to *E. lanatus* but can be distinguished by its smooth stems and non-woolly crowns.
**Cultivation:** Suited to temperate and cool subtropical regions. Generally uncommon in cultivation although plants grow readily, are hardy and adaptable. Plants may become nearly deciduous prior to a flush of new leaves and at this stage should be kept on the dry side. Tolerant of heavy frosts. Requires full sunlight.
**Propagation:** From seed and by removal of suckers which transplant readily.

# *Encephalartos lanatus*

Stapf & Burtt-Davy
(woolly)

**Description:** A medium-sized cycad with an erect or reclining trunk to 2 m tall and 30 cm across, growing singly or in clumps of two or three, with suckers produced from the base. **Young leaves** densely hairy and covered with a silvery bloom. **Mature leaves** numerous in an obliquely erect crown, 0.5–1 m x 20–26 cm, blue-green, deeply vee-ed in cross-section, straight or slightly twisted in profile with a strongly recurved apex; **petiole** 6–10 cm long, yellowish, the swollen base covered with white or grey wool; **leaflets** linear, crowded and evenly spaced throughout, obliquely erect to form a deep vee, inserted on the rhachis at about 90°, the lower leaflets reduced but not spine-like; **median leaflets** 11–13 cm x 0.6–0.8 cm, rigid, pungent-tipped, the margins lacking spines. **Cones** markedly dissimilar, densely woolly, one to four on each stem. **Male cones** 25–30 cm x 5–6 cm, more or less cylindrical,

P. VORSTER

Female cones of *Encephalartos lanatus*.

*Encephalartos latifrons.* P. VORSTER

Leaflets of *Encephalartos latifrons.*

P. VORSTER

yellow to creamy grey; **peduncle** 3–5 cm long. **Female cones** 25–35 cm x 12–15 cm, barrel-shaped, yellow to creamy grey; **peduncle** 2–3 cm long. **Seeds** 2.5–3 cm x 2–2.5 cm, ellipsoid, yellow.

**Distribution and Habitat:** Endemic to South Africa where restricted to the catchment areas of the Wilge, Olifants and Little Olifants Rivers in the Transvaal Province. The cycads grows among grass in sheltered rocky valleys at about 1500 m altitude. The climate is warm to moderately hot in summer and extremely cold in winter with heavy frosts being common. Rainfall ranges from 660 mm to 770 mm per annum with a summer maximum falling mainly during thunderstorms.

**Notes:** *E. lanatus*, described in 1926, has similarities with *E. laevifolius* but can be distinguished immediately by the dense wool which covers the leaf bases and areas in between and the leaflets being held in a deep vee formation. Fires occur almost annually in its natural habitat and the plants are regularly stimulated into cone production as a result of the burning. The seeds of this species have a very thin, fleshy covering.

**Cultivation:** Suited to cool subtropical and temperate regions. This species is more difficult than most to grow successfully. Large plants are extremely sensitive to disturbance and in most cases die. Seedlings can be established readily but are generally very slow growing. New leaves are sensitive to damage by strong winds and excessively hot sun, although plants should be grown in full sun. Tolerates very heavy frosts.

**Propagation:** From seed.

## *Encephalartos latifrons*

Lehm.
(with broad leaves)

**Description:** A medium-sized cycad with erect trunk to 3 m tall and 35 cm across, growing in clumps of up to eight stems, with occasional offsets produced on the trunk and numerous suckers arising at the base. **Young leaves** covered with fine hairs. **Mature leaves** numerous in an obliquely erect to widely spreading crown, 1–1.5 m x 20–30 cm, usually with a skirt of dead fronds, bright green and glossy, stiff and rigid, deeply vee-ed in cross-section, recurved towards the apex; **petiole** 10–18 cm long; **leaflets** ovate-lanceolate, stiff, rigid, glossy above, dull beneath, surface roughened, crowded, obliquely erect, inserted on the rhachis at about 40°, the lower leaflets reduced but not spine-like; **median leaflets** 10–15 cm x 4–6 cm, rigid, pungent-tipped, the distal margin usually entire, the proximal margin with two to four, rigid, pungent-tipped lobes, these twisted out of the plane of the leaf. **Cones** dissimilar, one to three on each stem, dark green. **Male cones** 30–50 cm x 10–17 cm, obovoid, each scale with a well-developed drooping beak, appear-

ing sessile. **Female cones** 50–60 cm x 20–25 cm, barrel-shaped, the surfaces of the **sporophylls** warty or wrinkled. **Seeds** 2.5–3.3 cm x 1.6–1.8 cm, narrowly obovoid, angular, red.
**Distribution and Habitat:** Endemic to South Africa where restricted to a couple of localities in the Eastern Cape Province. This species grows among rocks and on slopes among dense scrub. The climate is hot in summer and cool to cold in winter. Rainfall is about 600 mm per annum, falling mainly over summer.
**Notes:** *E. latifrons*, which was described in 1837–38, has been reduced to a desperate state in the wild as the result of illegal poaching. Active regeneration is no longer occurring since the male and female plants are now too widely scattered for pollination to be effective. An active program by enthusiasts to ensure pollen collection and hand pollination has resulted in the production of seedlings for replanting in the wild. Plants of this species are extremely slow growing, often not producing new leaves for two or three years and coning even more sporadically. *E. latifrons* is a very distinctive species which can be recognised by its stiff, arching leaves recurved near the apex and its broad leaflets having pungent-tipped lobes on the margins. It is easily confused with *E. arenarius* but can be distinguished by its bright green instead of dull to somewhat glaucous fronds, male cone scales with well-developed drooping tips and female cone scales being warty instead of smooth.
**Cultivation:** Suited to temperate and cool subtropical regions. An easily grown species which will grow in full sun or partial shade. Tolerates light to moderate frosts.
**Propagation:** From seed and by removal of suckers which transplant readily.

# *Encephalartos laurentianus*

De Wild.
(after Emile Laurent, Belgian collector who introduced the first plants to cultivation)

**Description:** A very large cycad which forms clumps of erect or decumbent, whitish trunks to 15 m long and about 1 m across, with new suckers arising from the base. **Young leaves** bright green, covered with reddish brown wool. **Mature leaves** numerous in a large spreading crown, 4–6 m x 50–90 cm, dark green above, paler beneath, flat in cross-section, straight or gently recurved in profile; **petiole** 4–10 cm long, more or less trigonous in cross-section; **leaflets** linear-lanceolate, moderately spaced and evenly distributed throughout, inserted on the rhachis at about 90°, the lower leaflets reduced to a series of bifurcate or simple, spine-like processes; **median leaflets** 35–50 cm x 4–7 cm, rigid, pungent-tipped (sometimes with three tips), broad at the base, the margins each with ten to fifteen pungent spines. **Cones** markedly dissimilar, two to six on each stem, greenish yellow to yellow with reddish hairs. **Male cones** 17–35 cm x 6–10 cm, cylindrical; **peduncle** 25–30 cm long. **Female cones** 35–40 cm x 18–20 cm, barrel-shaped; **peduncle** 6–9 cm long, stout. **Seeds** 4–5 cm x 2.5–3 cm, more or less ovoid, angular, bright red to vermillion red.
**Distribution and Habitat:** Endemic to northern Angola and south-western Zaire where it is restricted to valleys of the Kwango River and its tributaries on both sides of the border between the two countries. It commonly grows in large colonies in the deep shade of dense riverine forests but plants are also found fully exposed on sandstone cliffs. The climate is tropical with hot, humid summers and mild, dry winters.
**Notes:** *E. laurentianus* is one of the most robust species in the genus with massive, long (although usually reclining) trunks and large fronds in an impressive crown. Apart from its extra vigour it can be distinguished from the related *E. hildebrandtii* by the wider leaves and more numerous marginal spines on the leaflets.
**Cultivation:** Suited to tropical and subtropical regions. Will grow in shade or sun and responds to mulches and watering during dry periods. Plants are not tolerant of frosts.
**Propagation:** From seed and by removal of suckers.

# *Encephalartos lebomboensis*

I. Verd.
(from Lebombo, in reference to the type locality in the Lebombo Mountains)

**Description:** A medium-sized to large cycad with an erect trunk to 4 m tall and 30 cm across, growing singly or in clumps of up to eight stems, with numerous suckers from the base and occasional offsets on the trunks. **Young leaves** light green, covered with light golden hairs. **Mature leaves** numerous in an obliquely erect crown, 1–2 m x 20–35 cm, bright green and glossy, flat or shallowly vee-ed in cross-section, straight or slightly recurved in profile; **petiole** 5–10 cm long, the base sometimes covered with golden wool; **leaflets** linear-lanceolate, crowded and overlapping on the lower side, inserted on the rhachis at about 30°, the lower leaflets reduced progressively to a series of small, bifurcate or trifurcate, spine-like processes; **median leaflets** 12–18 cm x 1.2–2.2 cm, rigid, pungent-tipped, usually with one to four spines on each margin. **Cones** markedly dissimilar, yellow, one to three on each stem. **Male cones** 30–45 cm x 10–12 cm, cylindrical, yellow to apricot; **peduncle** 1–3 cm long. **Female cones** 40–45 cm x 20–25 cm, broadly ovoid, sparsely hairy, sessile, the scales smooth. **Seeds** 3–4 cm x 1.8–2.2 cm, more or less oblong, angular, bright red to scarlet.
**Distribution and Habitat:** Occurring in southern Moz-

P. VORSTER

*Encephalartos lebomboensis.*

Male cones of *Encephalartos lebomboensis.*

Female cone of *Encephalartos lebomboensis.*

P. VORSTER

ambique, and Swaziland and the provinces of northern Natal and south-eastern Transvaal in South Africa. This species grows on exposed rocky cliffs and steep slopes among sparse, low shrubs and grass. The climate is hot in the summer and mild to cool in winter. Frosts are rare. Rainfall ranges from 600 mm to 750 mm per annum, with a summer maximum. Clouds and mists are frequent.
**Notes:** *E. lebomboensis*, which was described in 1949, is closely related to *E. natalensis*. *E. lebomboensis* has shorter, narrower leaflets which are crowded and overlap on the basal margins and the female cone scales are smooth instead of warty. Plants of this species and *E. aemulans* are vegetatively very similar. More than 5000 specimens of *E. lebomboensis* were removed from the wild during a rescue dig in 1963 prior to the construction of a large dam.
**Cultivation:** Suited to subtropical and warm temperate regions. Easily grown and hardy in full sun. Requires excellent drainage but may be damaged by heavy frosts.
**Propagation:** From seed or by removal of suckers which transplant readily.

Encephalartos lehmannii.

Female cone of *Encephalartos lehmannii.*

# *Encephalartos lehmannii*

Lehm.
(after Professor J. G. C. Lehmann, a nineteenth century German botanist who studied cycads and described the genus *Encephalartos*)

**Description:** A small to medium-sized cycad with an erect trunk to 2 m tall and 40 cm across, growing in clumps of up to ten stems, with suckers produced from the base. **Young leaves** an attractive blue-green, covered with powdery bloom. **Mature leaves** numerous in an erect to obliquely erect crown, 1–1.5 m x 24–36 cm, heavily bluish, deeply vee-ed in cross-section, straight in the proximal half, recurved in the distal half; **petiole** 10–15 cm long, the base surrounded by a brown or yellow collar; **leaflets** linear-lanceolate moderately crowded and more or less evenly spaced throughout (more widely spaced towards the base), obliquely erect to form a deep vee, inserted on the rhachis at about 60°, the lower leaflets widely spaced, the basal pair spine-like; **median leaflets** 12–18 cm x 1.6–2 cm, rigid, pungent-tipped, the margins entire, when juvenile with a single basal spine. **Cones** markedly dissimilar, green with fine black hairs, solitary on each stem. **Male cones** 25–35 cm x 8–10 cm, cylindrical-ovoid; **peduncle** 2–4 cm long. **Female cones** 45–50 cm x 20–25 cm, ovoid, sessile. **Seeds** 4.7–5 cm x 1.5–2 cm, more or less oblong, slightly angular, bright red.

**Distribution and Habitat:** Endemic to South Africa where it is found in the Eastern Cape Province, growing on cliffs and mountain slopes of sandstone formation. This species grows in semi-arid regions in very sparse vegetation. The climate is very hot in summer and cold in winter with frosts being common. Rainfall (250–350 mm per annum) falls mainly in summer but is very erratic and droughts are frequent.

**Notes:** *E. lehmannii* was described in the genus *Zamia* in 1833 and transferred to *Encephalartos* in 1834. It is a very drought-resistant species which grows naturally in semi-arid habitats. Although related to *E. trispinosus* and *E. horridus*, *E. lehmannii* can be readily distinguished by its entire leaflets. Natural populations of this highly attrac-

P. VORSTER

*Encephalartos longifolius.*

tive species have suffered drastically from a range of factors including illegal collection, droughts which have reduced cone production and animal predation.
**Cultivation:** Suited to temperate regions including those with a semi-arid climate. Requires excellent drainage, must not be overwatered and achieves its best appearance if planted in a sunny position. Tolerates alkaline soils and moderate to heavy frosts. A very attractive species with excellent ornamental qualities.
**Propagation:** From seed and by removal of suckers which transplant readily.

## *Encephalartos longifolius*

(Jacq.) Lehm.
(with long leaves)

**Description:** A medium-sized cycad with stout, erect trunks to 4.5 m tall and 45 cm across, growing in clumps of up to ten stems, with numerous suckers produced from the base. **Young leaves** light green to bluish green, covered with fine hairs. **Mature leaves** numerous in an obliquely erect to umbrella-shaped crown, 1–2 m x 30–40 cm, dark green or bluish green, shallowly to deeply vee-ed in cross-section, straight in the proximal half, recurved in the distal half; **petiole** 30–35 cm long, pale green to yellowish; rhachis yellowish; **leaflets** ovate-lanceolate, moderately spaced in the proximal half, crowded and overlapping towards the apex, inserted on the rhachis at about 50°, the lower leaflets reduced but not spine-like; **median leaflets** 15–20 cm x 2–3 cm, rigid, pungent-tipped (blunt and rounded in one area), the margins usually lacking spines but one to three spines sometimes present on the lower side near the base (possibly indicating hybridisation). **Cones** markedly dissimilar, olive green to brownish green, solitary on each stem. **Male cones** 40–60 cm x 15–20 cm, cylindrical; **peduncle** 3–5 cm long. **Female cones** 50–60 cm x 25–30 cm, broadly ovoid, the **sporophylls** warty or pimply, sessile. **Seeds** 4–5 cm x 2–3 cm, linear-oblong, angular, bright red.
**Distribution and Habitat:** Endemic to South Africa where it occurs in the Eastern Cape Province. This species grows in a range of habitats but principally on mountain slopes and ridges in sclerophyll vegetation (Fynbos), at altitudes of about 1500 m. It also occurs on flatter sites in broad, near-coastal valleys at altitudes as low as 200 m. The soils are invariably acid with a pH of about 5.5. The climate is hot in the summer with cool to cold winters. This species grows more or less in the zone between summer rainfall and winter rainfall regions with annual falls ranging from 300 mm (in inland districts) to

Male cone of *Encephalartos longifolius.*

Female cone of *Encephalartos longifolius.*

*Encephalartos longifolius* From the *Botanical Magazine* 82, tab. 4903, (1856)

1250 mm per annum near the coast.

**Notes:** *E. longifolius* was first described in the genus *Zamia* in 1775 and later transferred to *Encephalartos* in 1834. Coincidentally a specimen of this species, which is still living, arrived at the Kew Gardens in England in 1775, and is believed to be the oldest greenhouse plant in the garden. Mature female cones of this species are among the largest of any species of *Encephalartos* and can weigh up to 40 kg. The local natives used the pith of the stems of this species to prepare a form of bread; preparation included burying for up to two months prior to baking. The species is closely related to *E. altensteinii* but has much darker green leaves the upper leaflets of which are crowded and overlap. *E. longifolius* forms natural hybrids with *E. horridus* where the two grow in close proximity.

**Cultivation:** Suited to temperate and cool subtropical regions. Easily grown and adaptable. Best in full sun in well-drained, moderately acid sandy soils. Tolerates light frosts only. Mature specimens transplant readily.
**Propagation:** From seed and by removal of suckers which transplant readily.

Old male cone of *Encephalartos manikensis*.

## *Encephalartos manikensis*

(Gilliland) Gilliland
(from the Manica district in western Mozambique)

**Description:** A small to medium-sized cycad with an erect trunk to 1.5 m tall and 30 cm across, growing singly or in pairs, with new suckers produced from the base. **Young leaves** dark green, covered with fine white wool. **Mature leaves** numerous in an erect crown, older leaves drooping downwards, 1–2 m x 20–34 cm, dark green, glossy, flat in cross-section, straight in profile; **petiole** 4–8 cm long, the swollen base covered with fawn, cottony hairs; rhachis ageing to yellow; **leaflets** lanceolate, moderately spaced and evenly distributed throughout, bright yellow at the base where attached to the rhachis, inserted on the rhachis at about 90°, the lower leaflets reduced to a series of simple to trifurcate, spine-like processes; **median leaflets** 12–15 cm x 2–2.5 cm, rigid, pungent-tipped, apex curved forwards, the margins each with two or four pungent spines. **Cones** dissimilar, one to four on each stem. **Male cones** 22–60 cm x 7–15 cm, narrow conical to cylindrical, dark green to yellow green or blue green; **peduncle** 10–15 cm long. **Female cones** 32–50 cm x 14–18 cm, cylindrical, dark green to yellowish with a powdery bloom; **peduncle** 4–6 cm long. **Seeds** 3–3.6 cm x 1.8–2.5 cm, more or less oblong, dark red.
**Distribution and Habitat:** Occurring in north-eastern Zimbabwe and central Mozambique. It grows on the ridges and slopes of mountainous regions, often among granite boulders, at altitudes of about 1400 m. Plants can grow in open situations among grass or in shady deciduous forests. The climate is subtropical with hot, humid summers and cold winters in which frosts occur.
**Notes:** *E. manikensis* is part of a complex which includes *E. chimanimaniensis*, *E. pterogonus*, *E. munchii* and *E. concinnus*. It was first described in 1938 as a variety of *E. gratus* and was raised to specific rank in the following year. It can be recognised by the pale woolly hairs which clothe the scale leaves on the trunk and frond bases and the shape and colouration of the cones. Disjunct populations of this species exhibit a fair degree of variation and are in need of further study.
**Cultivation:** Suited to cool tropical and subtropical regions. Can be grown in sun or shade. Tolerates light frosts. Grown on a limited scale in Zimbabwe.
**Propagation:** From seed and by removal of suckers.

Female cones of *Encephalartos manikensis*.

# *Encephalartos marunguensis*

Devred
(from the Marungu Mountains in south-eastern Zaire)

**Description:** A small cycad with a subterranean trunk (rarely shortly emergent) to 40 cm long and 16 cm across, growing singly or in clumps of up to four stems, with suckers produced from the base. **Young leaves** bluish green, covered with white silky hairs. **Mature leaves** about fifteen in an erect to obliquely erect crown, 0.5–0.85 m x 20–25 cm, bluish green, deeply vee-ed in cross-section, mostly straight in profile but recurved near the apex; **petiole** about 2 cm long; **leaflets** linear-lanceolate, moderately spaced and evenly distributed throughout, erect to form a deep vee, inserted on the rhachis at about 50°, the lower leaflets reduced to a series of small spine-like processes; **median leaflets** 10–12 cm x 1.2–1.3 cm, the margins lacking spines (rarely with a single spine). **Cones** dissimilar, bluish green to green, one to three on each stem. **Male cones** 18–25 cm x 5–7.5 cm, narrowly ovoid; **peduncle** 2–4 cm long. **Female cones** 20–30 cm x 10–15 cm, ovoid, appearing sessile. **Seeds** 2–3 cm x 2–2.5 cm, oblong or narrowly ovoid, red or yellow.
**Distribution and Habitat:** Endemic to south-eastern Zaire where restricted to a few mountainous regions at about 1500 m altitude. The species grows among rocks in grassland and sparse forests. The climate is hot and humid in summer and mild and dry in winter.
**Notes:** *E. marunguensis*, described in 1959, can be difficult to locate in the wild because the plants are usually overgrown by grass for much of the year and then fires destroy all of the above-ground growth. In the absence of fire, plants are deciduous during the dry season. This species is very closely related to *E. schmitzii* but lacks any marginal spines on the leaflets. Further study is needed to evaluate the relationships between these two taxa.
**Cultivation:** Suited to tropical and subtropical regions. Will grow in full sun or filtered sun but is sensitive to damage from frosts. Plants should be kept on the dry side if they become deciduous.
**Propagation:** From seed and by removal of suckers.

# *Encephalartos middelburgensis*

Robbertse, Vorster & Van der Westhuizen
(from the Middleburg district)

**Description:** A large cycad with an erect trunk to 7.5 m tall and 40 cm across, patterned with large, rounded leaf scars, growing in clumps of up to twelve stems, with suckers produced freely from the base. **Young leaves** silvery due to a dense covering of powdery bloom. **Mature leaves** numerous in an obliquely erect to spreading

P. VORSTER

*Encephalartos middelburgensis* in habitat.

crown, 1–1.8 m x 30–38 cm, bluish green with persistent powdery bloom, deeply veed in cross-section, straight in profile, arising at about 60° from the trunk; **petiole** 20–40 cm long, triangular in cross-section; **leaflets** narrowly elliptical, slightly falcate, moderately crowded and slightly overlapping, obliquely erect to form a deep vee, inserted on the rhachis at about 50°, the lower leaflets widely spaced and reduced progressively but not spine-like; **median leaflets** 18–20 cm x 1.4–2 cm, pungent-tipped, the margins usually lacking spines. **Cones** markedly dissimilar, bright green with patches of brown hairs, four to eight on each stem. **Male cones** 30–35 cm x 8–12 cm, narrowly ovoid; **peduncle** 5–17 cm long. **Female cones** 35–45 cm x 17–20 cm, cylindrical; **peduncle** 10–15 cm long, buried in the cataphylls and appearing sessile. **Seeds** 3–3.5 cm x 2–2.5 cm, oblong, yellow to light brown.
**Distribution and Habitat:** Endemic to the Middelburg District of the Transvaal Province of South Africa. This species grows in open grassy habitats on sheltered slopes at between 1000 m and 1400 m altitude. The climate is

Leaf of *Encephalartos munchii*.

Male cones of *Encephalartos natalensis*.

*Encephalartos natalensis* in habitat.

P. VORSTER

hot in summer and cold in winter with severe frosts being common. Rainfall is about 600 mm per annum with a summer maximum.

**Notes:** *E. middelburgensis*, described in 1988, is a segregate of *E. eugene-maraisii*. It is more robust than that species with a petiole triangular in cross-section, the tips of the leaves not curving upwards and the lowest leaflets not spine-like. This species is regarded as being endangered in the wild as a direct result of overcollecting combined with poor natural regeneration. It is treated by some authorities as *E. eugene-maraisii* ssp. *middelburgensis* Lavranos & Goode.

**Cultivation:** Suited to temperate and cool subtropical regions. Best grown in full sun. Tolerates heavy frosts.

**Propagation:** From seed and by removal of suckers which transplant readily.

## *Encephalartos munchii*

R. A. Dyer & I. Verd.
(after R. C. Munch, twentieth century cycad enthusiast from Zimbabwe)

**Description:** A small cycad with an erect or leaning trunk to 1 m tall and 35 cm across, solitary or in pairs, with suckers produced from the base. **Young leaves** powdery blue, with a covering of white woolly hairs. **Mature leaves** numerous in an obliquely erect to spreading crown, 1–1.3 m x 20–30 cm, soapy green, shallowly vee-

ed in cross-section, straight in profile or recurved near the apex, arising at about 45° from the trunk; **petiole** 7–14 cm long, swollen and hairy at the base; **leaflets** lanceolate, moderately crowded and evenly spaced throughout, obliquely erect to form a shallow vee, inserted on the rhachis at about 70°, the lower leaflets reduced to a series of spine-like processes; **median leaflets** 12–15 cm x 1.8–2 cm, rigid, pungent-tipped, the margins each with two to six pungent spines. **Cones** markedly dissimilar, bluish green, one to six on each stem. **Male cones** 40–65 cm x 7–9 cm, narrowly cylindrical; **peduncle** 15–20 cm long. **Female cones** 40–50 cm x 15–20 cm, narrowly ovoid, appearing sessile; **peduncle** 8–10 cm long. **Seeds** 2.5–3 cm x 2–2.3 cm, oblong, red.
**Distribution and Habitat:** Endemic to central Mozambique where restricted to a single colony on Zembe Mountain. It grows among rocks in sparse scrub. The climate is hot in summer and cool in winter with a mainly summer rainfall.
**Notes:** *E. munchii*, originally included as part of the *E. manikensis* complex, was described in 1969. It can be distinguished by its blue-green leaves with the toothed leaflets being held obliquely erect in a shallow vee. The species has been reduced to great rarity by illegal poaching.
**Cultivation:** Suited to cool tropical and subtropical regions. This species is very difficult to transplant with most such plants declining or dying. Seedlings can be established readily. Will grow in filtered sun or full sun. Frost-tolerance is not high.
**Propagation:** From seed and by removal of suckers.

# *Encephalartos natalensis*

R. A. Dyer & I. Verd.
(from Natal)

**Description:** A large cycad with an erect or reclining trunk to 6 m tall and 40 cm across, usually solitary but sometimes growing in clumps of up to eleven stems, with occasional offsets produced along the trunks and numerous basal suckers. **Young leaves** light green, with silky white hairs. **Mature leaves** numerous in a rounded crown, 1–3 m x 30–40 cm, bright green and glossy, flat in cross-section or obliquely erect to form a shallow or steep vee, inserted on the rhachis at about 90°, straight in profile or gently recurved near the apex, arising at about 45° from the trunk; **petiole** 10–20 cm long, expanded at the base; **leaflets** broadly lanceolate, moderately crowded to crowded, flat to obliquely erect to form a vee, inserted on the rhachis at about 70°, the lower leaflets reduced into a series of small, spine-like processes; **median leaflets** 15–25 cm x 2.5–4 cm, rigid, pungent-tipped, the margins lacking spines or with up to five spines on each margin. **Cones** markedly dissimilar, yellow, three to five

P. VORSTER

Female cones of *Encephalartos natalensis*.

on each stem. **Male cones** 40–50 cm x 10–12 cm, narrowly fusiform; **peduncle** 2–4 cm long. **Female cones** 50–60 cm x 25–30 cm, ovoid, appearing sessile. **Seeds** 2.5–3 cm x 1.6–1.8 cm, narrowly obovoid, bright red.
**Distribution and Habitat:** Endemic to South Africa where widely distributed in Natal and Zululand. It grows in inland mountainous regions in open situations among rocks on cliffs and escarpments. The climate is hot in summer and cold in winter, with frosts a regular occurrence.
**Notes:** *E. natalensis* is a large species which was segregated from *E. altensteinii* in 1951, it being readily distinguished by the lower leaflets being reduced to a series of spine-like processes. *E. natalensis* is a variable species.
**Cultivation:** Suited to temperate and subtropical regions. A hardy and adaptable species which grows best in full sun. Tolerates moderate frosts.
**Propagation:** From seed or by removal of suckers which transplant readily.

# *Encephalartos ngoyanus*

I. Verd.
(from Ngoya Mountain, the type locality)

**Description:** A small cycad with a solitary, unbranched, subterranean trunk to 30 cm long and 20 cm across, subtended by swollen, fleshy roots. **Young leaves** light green

with silky or woolly hairs. **Mature leaves** five to ten in an obliquely erect, somewhat sparse crown, 0.5–1.25 m x 14–21 cm, mid green to dark green, flat in cross-section, straight or slightly twisted in profile; **petiole** 10–15 cm long, the swollen base covered with brown hairs; **leaflets** linear, moderately crowded but not overlapping, evenly spaced throughout, inserted on the rhachis at 60–90°, the lower ones reduced but not spine-like; **median leaflets** 7–12 cm x 0.9–1.3 cm, pungent-tipped, the lower margins usually with one to three spines. **Cones** markedly dissimilar, solitary on each plant. **Male cones** 20–25 cm x 5–7 cm, cylindrical to narrowly ovoid, pale olive green to yellow; **peduncle** 2–6 cm long. **Female cones** 20–25 cm x 10–12 cm, ovoid-cylindrical to pear-shaped, yellow, the **sporophylls** sometimes with a fringed lower margin; **peduncle** extremely short. **Seeds** 2.5–3 cm x 1.5–2 cm, oblong-narrowly ovoid, scarlet.
**Distribution and Habitat:** Endemic to southern Africa where found in Zululand, Swaziland and the Transvaal. It grows on rocky slopes, in open grassland and along forest margins. The summers are hot and the winters are mild. Rainfall is 750–1000 mm per annum with a summer maximum.
**Notes:** *E. ngoyanus*, described in 1949, has many similarities with *E. caffer* but can be distinguished by the less crowded **leaflets** which usually have a few spines on the lower margins. It is common for plants to shed their leaves and be deciduous for a period prior to the production of new leaves or cones. *E. ngoyanus*, which does not reproduce prolifically in nature, is endangered by overcollection of plants from the wild, overgrazing of its habitat and annual grass fires.
**Cultivation:** Suited to subtropical and warm to mild temperate regions. A slow growing but hardy species which requires a sunny position with excellent drainage. In cold areas plants often shed leaves in the autumn and should then be kept on the dry side. Tolerates light to moderate frosts.
**Propagation:** From seed.

## *Encephalartos paucidentatus*

Stapf & Burtt Davy
(with few teeth, in reference to the leaflets)

**Description:** A large cycad with a mostly erect trunk to 6 m tall and 70 cm across, thickened towards the base, growing in clumps of up to five stems, with numerous suckers produced from the base. **Young leaves** light green, covered with brown hairs. **Mature leaves** numerous in an erect, incurved, almost rounded crown, 1–2 m x 30–50 cm, bright green and glossy, flat in cross-section, incurved in profile or occasionally twisted, arising at about 80° from the trunk; **petiole** 35–45 cm long, swollen at the base; rhachis ageing to yellow; **leaflets** lanceolate, crowded, evenly spaced throughout, the leaflets from the middle of the leaf to the apex drooping prominently, the veins on the lower side prominently raised, inserted on the rhachis at about 90°, the lower leaflets reduced to a series of multilobed, spine-like processes; **median leaflets** 15–25 cm x 2–3 cm, in juveniles the margins each with up to three spines. **Cones** very similar, yellow but appearing brown from a dense covering of brown wool, one to five on each stem. **Male cones** 40–60 cm x 12–15 cm, ovoid-cylindrical; **peduncle** 4–6 cm long. **Female cones** 35–50 cm x 20–25 cm, ovoid-cylindrical, appearing sessile. **Seeds** 3–3.5 cm x 1.8–2 cm, oblong-narrowly ovoid, red.
**Distribution and Habitat:** Occurring in the south-east of the Transvaal Province in South Africa and across the border in adjacent areas of Swaziland. It grows in mountainous regions on heavily forested, steep slopes at about 1500 m altitude. The climate is hot in summer and cold in winter. Rainfall is about 1400 mm per annum falling mainly in summer.
**Notes:** *E. paucidentatus* is a very handsome species which was described in 1926 and has been now reduced to rarity in the wild as the result of illegal collecting. It is a distinctive species which can be recognised by its erect and incurved leaves, with the leaflets having few or no spines on each margin. It can be confused with *E. heenanii* which is much woollier in its crown and leaves. It is also easily confused with the Soutpansberg form of *E. transvenosus*, from which it can be distinguished by the raised veins on the underside of the leaflets.
**Cultivation:** Suited to temperate and cool temperate regions. Best grown in shade or filtered sun and should be mulched and watered during dry periods. Mature plants resent disturbance and are slow to re-establish often taking a number of years before producing new leaves.
**Propagation:** From seed and by removal of suckers.

## *Encephalartos poggei*

Ascherson
(after Paul Pogge, nineteenth century German collector in Central Africa)

**Description:** A small cycad with a swollen subterranean or emergent trunk to 2 m long and 30 cm across, growing in clumps of two or three stems, with a few suckers produced from the base. **Young leaves** covered with white hairs. **Mature leaves** about twelve in an erect to obliquely erect crown, 0.7–1.5 m x 15–27 cm, somewhat glaucous, flat in cross-section, straight or arched in profile, sometimes with an apical twist; **petiole** 7–20 cm long, the base swollen and with fine, white hairs; **leaflets** linear-lanceolate, well spaced and evenly distributed throughout,

*Encephalartos paucidentatus* in habitat.

*Encephalartos poggei*.

Male cones of *Encephalartos paucidentatus*.

inserted on the rhachis at about 30°, the lower leaflets abruptly reduced to a series of spine-like processes; **median leaflets** 8–15 cm x 0.7–1.3 cm, rigid, pungent-tipped, striate beneath, the margins entire or with one to four pungent spines near the base. **Cones** markedly dissimilar, two or three on each stem, green or pinkish maturing to yellow. **Male cones** 10–30 cm x 3–7.5 cm, cylindrical, erect at first, drooping with age; **peduncle** 8–10 cm long. **Female cones** 17–23 cm x 9–12 cm, oblong

Female cone of *Encephalartos poggei*.

to barrel-shaped, appearing sessile; **peduncle** 4–6 cm long. **Seeds** 2–3.3 cm x 1.7–2.3 cm, narrowly obovoid, pink to red.
**Distribution and Habitat:** Occurring in southern Zaire and north-eastern Angola. It grows in open grassland, often as solitary individuals but sometimes forming extensive colonies. The climate is hot and dry in summer and mild in winter. Fires occur annually in the grassland where this species grows.
**Notes:** *E. poggei*, although described in 1878, appears to have been poorly studied in the field. It is reported that elephants eat the fleshy sporophylls of the cones and seeds and are probably the main dispersal agent for the species. Plants are deciduous during the dry season.
**Cultivation:** Suited to tropical and subtropical regions. Best grown in sun in sandy soil. Sensitive to frost damage.
**Propagation:** From seed and by removal of suckers.

# *Encephalartos princeps*

R. A. Dyer
(the first, alluding to its evolutionary appearance being a forerunner of related species)

**Description:** A medium-sized to large cycad with sturdy, erect or sprawling trunk to 4 m tall and 40 cm across, growing in clumps of up to fifteen stems, with new suckers produced freely from the base. **Young leaves** silvery blue, hairy. **Mature leaves** numerous in a spreading to rounded crown, 1–1.3 m x 26–30 cm, strongly blue-green, shallowly to deeply vee-ed in cross-section, straight in profile or recurved near the apex; **petiole** 12–20 cm long, swollen and pale brown at the base; **leaflets** linear, glaucous, moderately spaced in the proximal half, becoming crowded and overlapping towards the apex, obliquely erect and held at a steep angle to form a vee, inserted on the rhachis at about 45°, the lower ones reduced and one or two becoming spine-like; **median leaflets** 12–15 cm x 1–1.3 cm, rigid, pungent-tipped, the margins lacking spines or with a single one on the lower margin. **Cones** markedly dissimilar, dull green to olive green, one to three on each stem. **Male cones** 25–30 cm x 8–10 cm, cylindrical to ovoid-cylindrical, sessile; **sporophylls** with a prominent, beak-like projection to 1.5 cm long. **Female cones** 30–40 cm x 20–25 cm, ovoid to barrel-shaped, sparsely hairy, sessile; **sporophylls** warty on the outer facet. **Seeds** 3.5–4 cm x 1.5–2 cm, oblong, red, slightly warty.
**Distribution and Habitat:** Endemic to South Africa where found in the catchment area of the Great Kei River and its tributaries in the Eastern Cape Province. It grows on dolerite escarpments and rocky outcrops usually in exposed situations, rarely in the shade of trees. The summers are mild to hot and the winters are cool to cold with heavy frosts being frequent. Rainfall is 400–500 mm per annum with a summer maximum.
**Notes:** *E. princeps* was included with *E. lehmannii* until separated and described in 1965. Plants of *E. princeps* are more robust, grow in a different habitat and the leaflets in the distal third of the leaf are crowded and overlap. The main differences lie in the cones, the males of *E. princeps* having a prominent, beak-like projection on the scales and those of the female are warty.
**Cultivation:** Suited to subtropical and temperate regions. Plants require full sun and excellent drainage. Tolerates heavy frosts.
**Propagation:** From seed and by removal of suckers which transplant readily.

# *Encephalartos pterogonus*

R. A. Dyer & I. Verd.
(with winged seeds, in reference to the unusual wing-like structures which protrude from the sporophylls of the male cone)

**Description:** A medium-sized cycad with an erect trunk to 1.5 m tall and 40 cm across, growing in clumps of three or four stems, with suckers produced from the base. **Young leaves** densely covered with grey woolly hairs. **Mature leaves** numerous in an obliquely erect crown, 1–1.5 m x 30–36 cm, light green, flat in cross-section, straight in profile; **petiole** 4–8 cm long, hairy at the base; **leaflets** lanceolate, moderately crowded and evenly spaced throughout, inserted on the rhachis at about 75°, the lower leaflets reduced to a series of trifurcate, bifurcate or simple spine-like processes; **median leaflets** 15–18 cm x 2–2.5 cm, rigid, leathery, pungent-tipped, the margins each with one or two spines towards the base. **Cones** markedly dissimilar, bright green and shiny, two or three on each stem. **Male cones** 30–38 cm x 9–11 cm, narrowly ovoid, the **sporophylls** with prominent yellow or orange toothed lobes protruding from the outer margins, appearing sessile. **Female cones** 30–40 cm x 16–18 cm, ovoid to barrel-shaped; **peduncle** 2–4 cm long. **Seeds** 3–3.5 cm x 2–2.3 cm, oblong, scarlet or orange.
**Distribution and Habitat:** Endemic to central Mozambique where restricted to Mt Mruwere. The species was restricted to a single colony growing among boulders in or near forested areas. The climate is hot and humid in summer and cool and dry in winter.
**Notes:** The sad history of *E. pterogonous* can be summarised in a few words: discovered in 1949, described in 1969 and apparently extinct in the wild by about 1975. The single known colony consisted of only about 200 individuals and these were removed to cultivation by the depredations of poachers. Cultivated plants apparently cone very infrequently and a concerted program is needed for the survival of this species. Although related

to *E. manikensis*, the species is readily recognised by the unusual lobes which protrude from among the sporophylls of the male cone.
**Cultivation:** Suited to cool tropical and subtropical regions. Will grow in sun or partial shade. Frost-tolerance is not high.
**Propagation:** From seed and by removal of suckers.

## *Encephalartos schmitzii*

Malaisse
(after Andre Schmitz, the original collector of this species)

**Description:** A small cycad with a subterranean trunk to 30 cm long and 15 cm across, growing in dense clumps of up to fifty stems, with suckers produced freely from the base. **Young leaves** bluish green, covered with white hairs. **Mature leaves** about nine leaves in an obliquely erect crown, 0.4–0.6 m x 20–30 cm, bluish green, deeply vee-ed in cross-section, straight in profile but recurved near the apex; **petiole** about 2 cm long; **leaflets** linear to linear-oblong, moderately spaced and evenly distributed throughout, erect to form a deep vee, inserted on the rhachis at about 50°, the lower leaflets reduced to a series of bifurcate or simple, spine-like processes; **median leaflets** 10–14 cm x 0.8–1 cm, margins each with one or two (rarely absent) basal spines. **Cones** markedly dissimilar, green or blue-green. **Male cones** 8–10 cm x 3–4 cm (when dry), obovoid to cylindrical, one to three on each stem; **peduncle** 3–5 cm long. **Female cones** 20–25 cm x 10–12 cm, ovoid, solitary on each stem, appearing sessile. **Seeds** 2–2.5 cm x 2–2.5 cm, oblong, angular, orange.
**Distribution and Habitat:** Occurring in south-eastern Zaire and also north-eastern Zambia where known by a single female plant. It grows among rocks in sparse woodland and grassland in mountainous regions at altitudes of between 1500 m and 2000 m. The climate is hot and humid in summer and mild to cool and dry in winter.
**Notes:** A male plant of *E. schmitzii* was collected in 1955 by Andre Schmitz but the species was not described until 1969 following the collection of a female plant during an expedition to the same plateau. Then in 1987 a solitary clump of the species was located in Zambia. This clump was a single female plant which consisted of more than forty stems. Much information is lacking about this species. Plants apparently become deciduous during the dry season.
**Cultivation:** Suited to tropical and subtropical regions. Will grow in full sun or filtered sun but is sensitive to damage from frosts. Requires excellent drainage and should be kept bone dry in winter. This species is rarely cultivated and apparently produces cones infrequently.
**Propagation:** From seed and by the removal of suckers.

## *Encephalartos sclavoi*

Moretti, D. Stevenson & De Luca
(after Jean Pierre Sclavo who discovered this species)

**Description:** A small cycad with a subterranean or emergent trunk to 1 m long and 35 cm across, growing singly or in clumps of up to three stems, with new suckers produced from the base. **Young leaves** covered with white silky hairs. **Mature leaves** numerous in an erect to obliquely erect crown, 1.75–2 m x 36–56 cm, dark green, flat in cross-section, straight in profile or with the tip slightly twisted; **petiole** 10–15 cm long, swollen and hairy at the base; **leaflets** narrow-oblong to elliptical, blue green to dark green, yellow at the point of attachment to the rhachis, leathery, apex recurved, crowded and overlapping, evenly spaced throughout, the lower leaflets reduced progressively to a series of spine-like processes; **median leaflets** 18–28 cm x 3–4 cm, pungent-tipped, the apex recurved and hook-like, the margins each with up to three spines near the base. **Cones** markedly dissimilar, greenish yellow to yellow, one or two on each stem. **Male cones** 20–25 cm x about 10 cm, more or less conical; **peduncle** 2–4 cm long. **Female cones** 30–40 cm x 15–20 cm, ovoid-cylindrical, sometimes light brown when mature; **peduncle** 3–4 cm long. **Seeds** 2.5–3 cm x 2 cm, oblong, golden yellow.
**Distribution and Habitat:** Endemic to the Usambara Mountains in north-eastern Tanzania. This species grows on steep rocky hillsides and ridges among grass and low shrubs. The climate is tropical with hot humid summers and mild winters.
**Notes:** *E. sclavoi*, which may have been cultivated by enthusiasts as 'Species A', was described in 1990 having previously been confused with *E. hildebrandtii*. It can be distinguished from that species by its male and female cones being about half the size.
**Cultivation:** Suited to tropical and subtropical regions. Will grow in full sun or semi-shade. Needs protection from frosts.
**Propagation:** From seed and by removal of suckers.

## *Encephalartos septentrionalis*

Schweinf.
(northern)

**Description:** A small to medium-sized cycad with an erect or reclining trunk to 2 m tall and 30 cm across, growing in clumps of two or three stems, with suckers produced sparsely from the base. **Young leaves** yellowish green, covered with fine white hairs. **Mature leaves** numerous in an obliquely erect although straggly crown, 1–1.5 m x 30–45 cm, light green, flat in cross-section,

straight in profile, arising at about 70° from the trunk; **petiole** very short or absent; **leaflets** oblong to lanceolate, moderately spaced and evenly distributed throughout, inserted on the rhachis at about 80°, the lower leaflets reduced to a series of spine-like processes; **median leaflets** 7.5–18 cm x 1.5–3.5 cm, rigid, pungent-tipped, broad at the base, the margins each with one to seven pungent spines. **Cones** markedly dissimilar green becoming yellowish brown. **Male cones** 20–22 cm x 6–8 cm, cylindrical to ellipsoid, drooping, eight to ten on each stem; **peduncle** 20–30 cm x 2–3 cm. **Female cones** 30–35 cm x 18–20 cm, ovoid, drooping or pendulous, the inner surface of the **sporophylls** pink; **peduncle** 25–30 cm x 4–5 cm. **Seeds** 2.3–3.5 cm x 1.6–2.5 cm, oblong, angular, dark red.
**Distribution and Habitat:** Occurs in northern Uganda and southern Sudan where it grows among rocks on slopes and plateaux of mountain ranges at about 1200 m altitude. The climate is tropical with hot, humid summers and mild winters.
**Notes:** Although described as early as 1871, *E. septentrionalis* still remains poorly known and its biology, variation and relationships with other species are in need of further study. Annual fires are common in the grassland habitat where this species grows.
**Cultivation:** Suited to tropical and subtropical regions. Plants will grow in full sun or filtered sun. Requires excellent drainage. Frost-tolerance is unlikely to be very high.
**Propagation:** From seed and by removal of basal suckers.

## *Encephalartos tegulaneus*

Melville
(overlapping like roof tiles, in reference to the sporophylls of the male cone)

**Description:** A large cycad with an erect trunk to 10 m tall and 60 cm across, usually solitary, with suckers produced freely from the base. **Young leaves** light green, covered with brown hairs. **Mature leaves** numerous in a stiff, obliquely erect crown, 1.2–1.8 m x 30–40 cm, soapy green, flat in cross-section, straight in profile, arising at about 60° from the trunk; **petiole** about 2 cm long or absent; **leaflets** oblong-lanceolate, crowded and overlapping, evenly spaced throughout, inserted on the rhachis at about 70°, the lower leaflets reduced to a series of trifurcate, bifurcate or simple spine-like processes; **median leaflets** 16–22 cm x 1.6–2.8 cm, leathery, rigid, pungent-tipped, the margins revolute and each with one to three spines towards the base. **Cones** markedly dissimilar, green maturing to yellow, three to six on each stem. **Male cones** 35–50 cm x 10–14 cm, cylindrical; **peduncle** 25–30 cm long. **Female cones** 40–70 cm x 25–30 cm, oblong to barrel-shaped, appearing sessile. **Seeds** 3–3.7 cm x 2–2.3 cm, narrowly oblong, scarlet.
**Distribution and Habitat:** Endemic to Central Kenya where it grows on the ranges around Mt Lolokwe at about 2000 m altitude. The species grows on slopes among rocks and also in shady evergreen, montane forests. The climate is tropical with hot humid summers and mild, dry winters.
**Notes:** *E. tegulaneus*, described in 1957, is a distinctive, robust species which can be recognised by its large size and the leaflets having revolute margins and one to three spines on each margin near the base.
**Cultivation:** Suited to tropical and warm subtropical regions. Can be grown in full sun, semi-shade or filtered sun. Must be protected from frosts.
**Propagation:** From seed and by removal of suckers.

## *Encephalartos transvenosus*

Stapf & Burtt Davy
(with a network of cross-veins — an apparent misnomer)

**Description:** A large cycad with erect trunks to 13 m tall and 65 cm across, growing singly or in clumps of up to three stems, with new suckers produced freely from the base and offsets from buds on the trunk. **Young leaves** light green, densely covered with brown hairs. **Mature leaves** numerous in an erect to spreading crown, 1–2.5 m long, bright green and glossy, flat in cross-section, straight in profile; **petiole** 20–28 cm long, the base covered with brown woolly hairs; **leaflets** asymmetrically lanceolate, reflexed from the rhachis, bright green above, yellow-green beneath, crowded and overlapping on the upper side, inserted on the rhachis at about 90°, the lower leaflets moderately spaced and reduced to multilobed, spine-like processes; **median leaflets** 16–25 cm x 2.5–4.5 cm, rigid, pungent-tipped, one to three teeth on each margin. **Cones** dissimilar, yellow to orange, two to five on each stem. **Male cones** 40–60 cm x 13–15 cm, cylindrical, woolly at first but the hairs rapidly shed; **peduncle** 3–6 cm long. **Female cones** 50–80 cm x 20–30 cm, broadly ovoid, hairy, sessile. **Seeds** 4–5 cm x 2–2.7 cm, oblong, red, glossy, occasionally yellow.
**Distribution and Habitat:** Endemic to South Africa where distributed in parts of the northern and north-eastern Transvaal Province. The species grows in large colonies on mountain slopes in short grassland at 600–1000 m altitude. The climate is warm to hot and humid in the summer and mild and frost-free in winter. Rainfall ranges from 500 mm to 1000 mm per annum with a summer maximum.
**Notes:** *E. transvenosus*, the tallest and possibly most majestic species of *Encephalartos*, was described in 1926. These cycads were of major interest to the local tribes and they have been protected over the centuries by hereditary rulers known as the 'rain queens'. They are lo-

cally common and grow in large colonies which dominate the vegetation and are reproducing freely. Many of these areas are now statutory nature reserves and the dominant vegetation is literally a cycad forest. The best known of these, the Modjadji Nature Reserve, about 30 km north-east of Tzaneen in the self-governing State of Lebowa, contains an estimated 15000 mature plants in about 308 ha. Mature female cones of this species are among the largest of any species of *Encephalartos* and can weigh in excess of 40 kg. The female cones mature on the plant for a longer period than other species (about eighteen months) and it is not uncommon for cones from two successive seasons to occur on a trunk simultaneously.
**Cultivation:** Suited to subtropical and warm temperate regions. This species must be given ample room to develop to its potential. In hot climates protection from excessive sun is advisable since leaf scorching can result. Young leaves can also be damaged by strong winds. Needs excellent drainage, full sunlight and responds to regular watering during dry periods. Protection from frost is essential. One of the fastest growing of all species of *Encephalartos*, plants have been known to produce cones after just eleven years from seed.
**Propagation:** From seed, which germinates quickly because of the highly advanced state of embryo development, and by the removal of suckers which transplant readily.

Male cone of *Encephalartos trispinosus*.

# *Encephalartos trispinosus*

(Hook.) R. A. Dyer
(with three spines, in reference to the leaflet tips)

**Description:** A small cycad with erect or reclining trunks to 1 m long and 30 cm across, growing in clumps of up to six stems, with numerous suckers arising from the base. **Young leaves** bluish green. **Mature leaves** numerous in a stiff, obliquely erect crown, 0.7–1.3 m x 20–36 cm, strongly blue-green, deeply vee-ed in cross-section, straight or shallowly recurved in profile; **petiole** 10–25 cm long; **leaflets** lanceolate, moderately spaced, crowded towards the apex of the leaf, more widely spaced towards the base, obliquely erect to form a deep vee, inserted on the rhachis at about 80°, the lower leaflets reduced but not spine-like; **median leaflets** 10–18 cm x 1.5–2.5 cm, rigid, pungent-tipped, the margins each with one or two large, broad, pungent-tipped lobes which are twisted out of the plane of the leaf. **Cones** markedly dissimilar, covered with brown hairs, one on each stem. **Male cones** 25–40 cm x 6–10 cm, narrowly cylindrical, yellow; **peduncle** 2–3 cm long. **Female cones** 25–45 cm x 15–20 cm, broadly ovoid, bluish green maturing to yellow, appearing sessile. **Seeds** 3–3.5 cm x 1.8–2 cm, narrowly obovoid, angular, pale red to orange.

Young plant of *Encephalartos trispinosus*.

*Encephalartos trispinosus* From the *Botanical Magazine* 89, tab. 5371, (1863)

**Distribution and Habitat:** Endemic to South Africa where it occurs in the catchments of the Bushmans River and the Fish River in the Eastern Cape Province. It grows among rocks in open situations and in dry forests and scrub. The climate is hot in summer and cold in winter with frosts a frequent occurrence. The rainfall is about 600 mm per annum falling mainly in summer.
**Notes:** First described as a variety of *E. horridus* by J. D. Hooker in 1863, *E. trispinosus* was raised to specific rank in 1965. *E. horridus* has intensely blue leaves and the basal leaflets are broad and lobed. By contrast the lower leaflets of *E. trispinosus* are relatively narrow and lack the pungent lobes so prominent on the median leaflets. The cones also differ in colour between the two species. *E. trispinosus* is extremely variable and ecological or geographical variants are known.
**Cultivation:** Suited to temperate and cool subtropical regions. A hardy species which takes on its best appearance when grown in full sun. Tolerates moderate frosts.
**Propagation:** From seed and by the removal of suckers which transplant readily.

P. VORSTER

Female cone of *Encephalartos trispinosus*.

# *Encephalartos turneri*

Lavranos & Goode
(after Ian S. Turner, original collector of the species and noted cycad enthusiast)

**Description:** A medium-sized cycad with an erect or reclining trunk to 3 m tall and 80 cm across, growing in clumps of up to four stems, with numerous suckers produced from the base. **Young leaves** light green, covered with fine white wool. **Mature leaves** numerous in an obliquely erect to spreading crown, 1–1.5 m x 30–40 cm, dark green and glossy, shallowly vee-ed in cross-section, straight in profile, arising at 50–60° from the trunk; **rhachis** yellowish; **petiole** 1–2 cm long, swollen at the base and covered with soft grey wool; **leaflets** lanceolate, thick and leathery, crowded and overlapping in the distal

Large clump of *Encephalartos trispinosus*.

half, moderately spaced in the proximal half, folded upwards and with recurved to revolute margins, the apex hooked, inserted on the rhachis at about 40°, the lower leaflets reduced to a series of bifurcate or trifurcate, spine-like processes; **median leaflets** 15–20 cm x 2–3 cm, pungent-tipped and strongly hooked, the margins of leaflets in the proximal half of the leaf each with up to three spines, those in the distal half entire. **Cones** dissimilar, one to three on each stem, greenish yellow to yellow with a pink bloom. **Male cones** 25–30 cm x 6–8 cm, cylindrical; **peduncle** to 12 cm x 2 cm. **Female cones** 26–30 cm x 14–16 cm, ovoid; **peduncle** 4–5 cm long. **Seeds** 4–6 cm x 2–2.5 cm, oblong, angular, yellow ripening to scarlet.
**Distribution and Habitat:** Endemic to north-eastern Mozambique where it grows on the rocky slopes of low hills, usually in full sun among boulders or in crevices but sometimes in light shade. The climate is hot and humid in summer and mild in winter. Frosts are unknown. Rainfall is between 800 mm and 1000 mm per annum with the majority falling in summer.
**Notes:** *E. turneri*, described in 1985, can be immediately recognised by its leaflets which have its margins turned up like the sides of a boat and a strongly hooked apex. The pink bloom on the cones is also distinctive.
**Cultivation:** Suited to tropical and subtropical regions. Will grow in full sun or filtered sun and requires excellent drainage. Sensitive to frost damage.
**Propagation:** From seed or by removal of suckers.

*Encephalartos umbeluziensis.*

Male cone of *Encephalartos umbeluziensis.*

## *Encephalartos umbeluziensis*

R. A. Dyer
(from the type locality of the Umbeluzi River)

**Description:** A small cycad with an unbranched, spherical, subterranean trunk to 30 cm long and 25 cm across. **Young leaves** light green and silky hairy. **Mature leaves** five to eight in a stiffly erect crown, 1–2 m x 20–40 cm, dark green, flat in cross-section, straight or recurved in profile; **petiole** 15–25 cm long, circular in cross-section, the swollen base slightly woolly; **leaflets** linear-lanceolate, moderately crowded and evenly spaced throughout, twisted near the base so they are not held in a flat plane, inserted on the rhachis at about 90°, the lower ones reduced but not spine-like; **median leaflets** 10–25 cm x 0.8–1.5 cm, pungent-tipped, one or two spines on each margin. **Cones** markedly dissimilar, green, up to four on each plant. **Male cones** 20–30 cm x 6–10 cm, narrowly cylindrical, maturing pink-orange; **peduncle** 10–12 cm x 2.5–3 cm. **Female cones** 25–30 cm x 12–15 cm, broadly cylindrical, maturing yellowish or apricot; **peduncle** 10–15 cm x 3–3.5 cm. **Seeds** 3–3.2 cm x 1.8–2.3 cm, narrowly obovoid, deep red, glossy.
**Distribution and Habitat:** Occurring in the Kingdom of Swaziland — in the valley of The Mbeluzi (Umbeluzi)

River — and Mozambique. It grows in sparse to dense, deciduous low forests (thornveld), along flood plains at low altitudes (no higher than 120 m). The summers are hot and often humid and the winters are mild and frost-free. Rainfall is about 500 mm per annum with a summer maximum.
**Notes:** *E. umbeluziensis*, which was described in 1951, was originally included with *E. villosus* by some authors. It can be immediately distinguished from that species by the long bare petiole lacking any spinose, reduced lower leaflets. *E. umbeluziensis* is reported to be locally common but in recent years the numbers have been reduced by illegal collecting and there is a need to ensure cultivated plants are pollinated for seed collection.
**Cultivation:** Suited to tropical, subtropical and warm temperate regions. Rarely cultivated and best grown in a shady or filtered sun situation. Frost-tolerance is very low.
**Propagation:** Solely from seed.

# *Encephalartos verrucosus*

Robbertse, Vorster & Van der Westhuizen
(warty)

This species described in 1988, is a synonym of *E. dolomiticus*. Although the latter was also described in 1988 its description was published prior to that of *E. verrucosus* and hence by botanical rules has priority.

# *Encephalartos villosus*

Lem.
(covered with soft hairs)

**Description:** A small cycad with a subterranean trunk to 30 cm tall and 20 cm across, with new suckers produced from the base. **Young leaves** densely covered with white hairs. **Mature leaves** erect or arching in an attractive crown, 1–3 m x 30–50 cm, dark green and glossy, flat in cross-section, recurved in profile; **petiole** 1–3 cm long, the swollen base covered with white wool; **leaflets** lanceolate, spreading or recurved, moderately crowded but not overlapping, evenly spaced throughout, inserted on the rhachis at about 70°, the lower leaflets reduced to a series of sharp, often trifurcate, spine-like processes; **median leaflets** 15–30 cm x 1.5–2 cm, pungent-tipped (often with three spines), the margins with a few small spines towards the apex. **Cones** markedly dissimilar, one to four on each stem. **Male cones** 50–70 cm x 6.5–12 cm, narrowly cylindrical, pale yellow to yellow green, emitting an unpleasant odour; **peduncle** 6–10 cm long. **Female cones** 30–50 cm x 12–15 cm, barrel-shaped, bright yellow to apricot-coloured, shiny; **sporophylls** usually with a well-developed basal fringe (absent in some provences); **peduncle** 6–10 cm long. **Seeds** 2.5–3 cm x 1.8–2 cm, narrowly obovoid, dark red to scarlet.
**Distribution and Habitat:** Endemic to southern Africa where widespread, occurring in Swaziland and the South African provinces of the Eastern Cape, Natal and Trans-

Female cones of *Encephalartos villosus*.

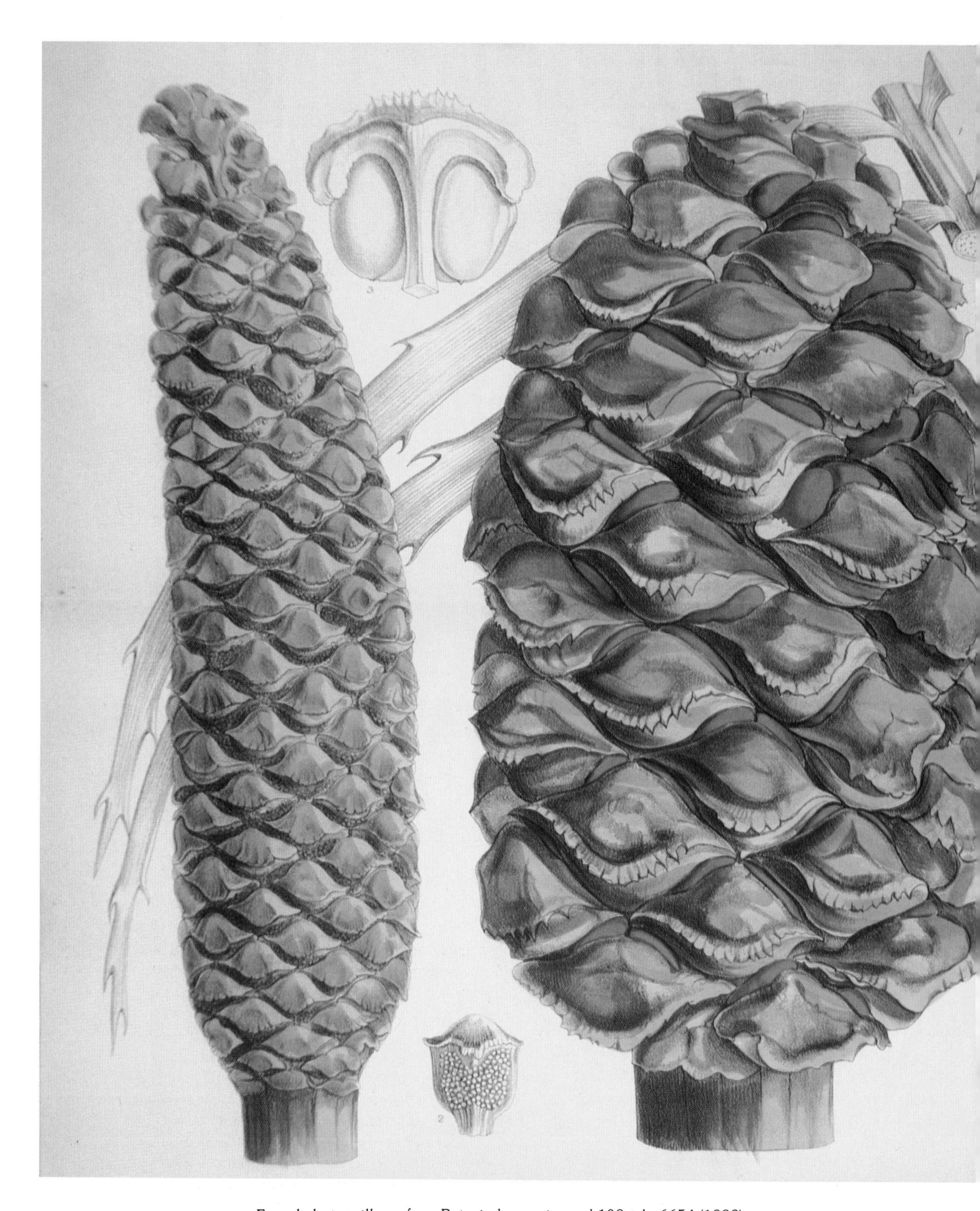

*Encephalartos villosus* from Botanical magazine, vol 108 tab. 6654 (1882)

vaal. This locally common species grows in moist, shaded, evergreen or deciduous forests. The climate is hot and wet in summer, mild and dry in winter. Rainfall ranges from 1000 mm to 1250 mm per annum with a summer maximum.
**Notes:** *E. villosus*, an attractive species described in 1867, was collected and exported in significant numbers in the late 1800s, but it is still common in its native habitat. It is closely related to *E. umbeluziensis* but can be immediately recognised by its arching leaves with the basal leaflets reduced to sharp spine-like processes. Natural hybrids occur with *E. lebomboensis* and *E. altensteinii* where the species grow in close proximity. Plants of *E. villosus* exhibit some variation in the leaflets. Those originating in the Eastern Cape have shorter, heavily spinose leaflets whereas those from northern areas have longer, nearly entire leaflets.
**Cultivation:** Suited to subtropical and warm temperate regions. An excellent species for landscaping. Best grown in shade or filtered sun. Seedlings grow fast and establish readily. Plants respond well to surface mulches and watering during dry periods. Intolerant of cold weather and is sensitive to frosts.
**Propagation:** From seed and by removal of suckers which transplant readily.

# *Encephalartos voiensis*

Moretti, D. Stevenson & Sclavo
(from the vicinity of Voi in Kenya)

This species, described in 1989, is a synonym of *E. kisambo*. Although the latter was also described in 1989, its description was published prior to that of *E. voiensis* and, hence, by botanical rules, has priority.

# *Encephalartos woodii*

Sander
(after Medley Wood who discovered the species)

**Description:** A large cycad with a massive, erect trunk to 6 m tall and 60 cm across, the base swollen and buttress-like, growing in clumps of up to four stems, numerous suckers produced from the base, offsets developing on the trunk. **Young leaves** bright green and densely hairy. **Mature leaves** numerous in a dense, gracefully arching crown which is umbrella-shaped, 2–2.5 m x 40–50 cm, bright glossy green, flat in cross-section, shallowly curved in profile; **petiole** 13–20 cm long; **leaflets** narrowly ovate, moderately crowded towards the base, middle and upper leaflets crowded and overlapping, attached at an oblique angle to the rhachis and arranged similar to the slats of a venetian blind, inserted on the

*Encephalartos voiensis.*

rhachis at about 40°, the lower leaflets reduced to a series of spine-like processes; **median leaflets** 20–25 cm x 4–5 cm, bright glossy green above, paler beneath, rigid, pungent-tipped, the margins usually lacking spines. **Male cones** 40–90 cm x 15–20 cm, oblong-obovoid, yellow to orange, covered with minute reddish brown hairs, usually four to six on each stem but up to twenty-one being recorded. Female plants are unknown.
**Distribution and Habitat:** Endemic to South Africa where known only from a single collection in Natal. This was on a steep, south-facing slope. The summers are hot and the winters are mild. Rainfall is 750–1000 mm per annum with a summer maximum.
**Notes:** *E. woodii* must be considered enigmatic since only a single clump has ever been found in the wild and it is a male plant. No female plants are known to exist and thus sexual reproduction is not possible. The solitary clump of the species was discovered in 1895 and subsequently described in 1908 from a plant in the collection of the English nurserymen Sander & Sons. This, apparently, was one of the basal offsets taken from the original clump. Interestingly, and despite its rarity, *E. woodii* has unique characters which distinguish it from closely related species such as *E. natalensis*. These include the stout trunk prominently swollen at the base, its arching fronds, ovate to narrowly ovate leaflets tapered to an acute tip and the blunt lobes on the leaflets of suckers.

One of the original trunks of *Encephalartos woodii* thriving in the Durban Botanic Gardens.

R. OSBORNE

The original clump of *E. woodii* consisted of four stems with a few offsets at the base. An offset was received at Kew Gardens in 1899 and further basal offsets were removed in 1903 and planted in the Durban Botanic Gardens. Subsequently two of the larger trunks were transplanted to the same location in 1907 and basal offsets had also by then been distributed to various botanic gardens throughout the world. No other specimens of *E. woodii* have ever been discovered in the wild, despite sporadic rumours to the contrary. To this day no female plants are known to exist.

Despite its sexual limitations, *E. woodii* reproduces very freely by basal suckers and to a lesser extent offsets on the trunk. Because of this asexual proclivity the species is now well represented in the collections of botanical institutions and private individuals in various countries around the world. While this male clone is hardly endangered, long-term research is underway on two fronts in the hope of developing a female plant. The first involves repeated back-crossing of *E. woodii* pollen on to female cones of *E. natalensis*. The theory is that after five successive generations of such hybridisation the resultant progeny will be genetically almost pure *E. woodii*. The second process involves the chemical manipulation of tissue-cultured plantlets grown under sterile conditions.

**Cultivation:** Suited to temperate and subtropical region. This species has proved to be extremely vigorous and adaptable to a range of soils and conditions. Tolerates light to moderate frosts and will grow in dappled shade to full sun.

**Propagation:** By the removal of suckers which transplant readily.

Young plant of *Encephalartos woodii* showing characteristic leaves.

GENUS

# *Lepidozamia*

The genus *Lepidozamia* consists of four species (two are fossils), with both of the extant species being endemic to eastern Australia (for fossil species see page 29).

**Derivation:** The genus *Lepidozamia* was described by the Russian botanist Eduard August von Regel in 1857 and was based on a living plant of *L. peroffskyana* cultivated in the St Petersburg Botanic Garden. The generic name is composed from the Greek *lepis*, *lepidis*, 'a scale', and *Zamia*, another genus of cycads, and refers to the frond-like scale bases which clothe the stems of these plants.

**Generic Description:** Terrestrial cycads with a stout, cylindrical, erect trunk. Leaf bases retained at senescence. New leaves emerging in flushes, covered with short hairs which are shed with age. Mature leaves pinnate, oblong or lanceolate in outline, flat in cross-section, the older leaves spreading or deflexing after a flush of growth. Cataphylls fleshy, shortly hairy, produced in alternating flushes with the leaves. Petioles lacking spines or prickles, swollen at the shortly hairy base. Rhachis lacking prickles, straight in profile. Leaflets not articulate at the base, alternate to nearly opposite, evenly spaced except for the lower leaflets which are more widely spaced, inserted on the upper midline of the rhachis, straight or falcate, margins entire, lacking a callous base, on emerging leaves there is a yellowish glandular area on the upper margins of the leaflet base, the lower leaflets never spine-like. Cones markedly dissimilar in shape and size, both sexes sessile or nearly so, appearing as if terminal but actually being pseudo-axillary among the cataphylls. Sporophylls shortly hairy on the outer face, lacking any spine-like appendages or ornamentation, the outer face of the female sporophylls prominently deflexed. Male cones narrow, opening in a prominent spiral at maturity. Female cones broad, massive. Seeds radiospermic, oblong, somewhat angular, the sarcotesta red, shiny.

**Notable Generic Features:** Species of *Lepidozamia* can be recognised by the leaflet bases being inserted on the upper midline of the rhachis. Other useful features include:

- an emergent, cylindrical trunk clothed with persistent leaf bases;
- an absence of suckers and offsets;
- young parts covered with short hairs;
- alternate flushes of leaves and cataphylls;
- petioles and rhachises lacking spines and prickles;
- leaflets without a midrib and decurrent at the base;
- lower leaflets not spine-like;
- cones sessile or nearly so;
- male cones opening in a prominent spiral at maturity;
- sporophylls non-peltate, shortly hairy on the outer surface, lacking spines, appendages or ornamentation; and
- outer face of female sporophylls prominently deflexed.

**Recent Studies:** The most recent study published on the genus *Lepidozamia* was by L. A. S. Johnson in 'The Families of Cycads and the Zamiaceae of Australia', *Proceedings of the Linnaean Society of New South Wales 84* (1959): 83–7.

**Habitat:** Species of *Lepidozamia* grow in sheltered, forested situations such as in rainforests, along rainforest margins or on protected shady slopes and in the vegetation along stream margins. The soils where they occur are generally well-structured loams. Fires occur in some localities but are virtually unknown in most areas where these cycads grow.

**Miscellaneous:** The leaves of both extant species of *Lepidozamia* are highly ornamental and are excellent for indoor decoration after cutting.

No cases of stock poisoning are recorded for either species of *Lepidozamia*, but this would appear to be due to lack of access by stock because the lethal compound macrozamin is found to be experimentally present at potentially toxic levels.

**Cultivation:** Both species of *Lepidozamia* have a highly decorative appearance and are popular subjects for cultivation. When young they are excellent in containers and can be used for indoor decoration, glasshouses and conservatories. They have proved to be adaptable to a range of soils and climates and have the following basic requirements: warmth, humidity, air movement and moisture. Humus-rich acid loams are ideal and bleaching due to nutrient deficiency may be a problem on alkaline soils. Mulching, watering during dry periods and regular, light fertiliser applications produce strong growth. Best appearance is attained when the plants are given some protection from excessive hot sun and in fact both species will grow well in quite shady situations.

**Propagation:** Species of *Lepidozamia* are propagated solely from seed. This has a limited period of viability and for best results should be sown fresh. The seeds have an after-ripening period of six to twelve months and hence the period of germination is usually twelve to twenty-four months (see also page 84).

## KEY TO SPECIES

Leaflets 0.7–1.3 cm wide ............................ ***L. peroffskyana***
Leaflets 1.5–3 cm wide ............................................ ***L. hopei***

## DISTRIBUTION OF *LEPIDOZAMIA*

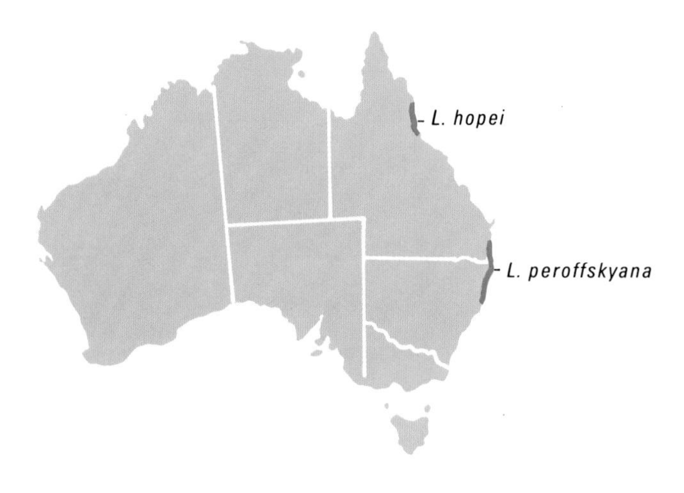


# *Lepidozamia hopei*

Regel
(after Captain L. Hope)

**Description:** A large cycad with an erect trunk to 20 m tall and 50 cm across. **Young leaves** brownish, bearing short, soft hairs. **Mature leaves** numerous in an obliquely erect to widely spreading crown, 2–3 m x 30–50 cm, dark green and glossy, flat in cross-section, straight to slightly curved in profile; **petiole** 30–60 cm long, grey-green, sparsely hairy, swollen and shortly hairy at the base; rhachis straight, not twisted, round in cross-section, sparsely hairy; **leaflets** 150–250 on each leaf, moderately crowded, evenly distributed throughout, widely spreading, arching or drooping towards the apex, recurved and falcate, leathery, dark green and glossy above, the lower leaflets reduced in size but not spine-like; **median leaflets** 30–40 cm x 1.5–3 cm, broadly ligulate, falcate, apex not pungent. **Cones** markedly dissimilar, shortly stalked but appearing sessile, appearing as if terminal on the trunk, usually solitary. **Male cones** 40–70 cm x 10–15 cm, more or less cylindrical, strongly curved or somewhat contorted; **sporophylls** 6–8 cm x 4–6 cm, broadly wedge-shaped, usually reflexed with upcurved, acute tips. **Female cones** 40–80 cm x 20–30 cm, broadly ovoid,



A Female plant of *Lepidozamia peroffskyana*.

Old male cone of *Lepidozamia peroffskyana*.

grey-green; **sporophylls** 5–8 cm x 6.5 cm, broadly wedge-shaped, the outer face with short hairs, deflexed and upcurved near the tip. **Seeds** 4–6 cm x 2.5–4 cm, oblong, bright red.

**Distribution and Habitat:** Endemic to Australia where it occurs in north-eastern Queensland from near Cooktown to near Ingham. This species commonly grows in rainforests, along the fringes of rainforests, in sheltered gullies close to streams and on shady, protected slopes in wet sclerophyll forest. Altitudinal range is from near sea level to about 1000 m altitude on the tablelands. The climate is tropical with summer temperatures to 33°C on the coast and to 29°C on the tablelands; winter day temperatures to 26° on the coast and to 19°C on the tablelands with the night temperatures to 14°C and 5°C respectively. Frosts are very rare. The rainfall ranges from 1700 mm to 2500 mm per annum falling mainly in summer.

**Notes:** *L. hopei* was described in 1876 from a living plant cultivated in Europe. It is the tallest of the Australian cycads and one of the tallest (if not the tallest) cycad species in the world. The seeds were collected and eaten by the Aborigines and in some areas (e.g. near Ingham) old plants are to be found with notches cut in their trunks, no doubt to aid climbing for seed harvesting. In

Cataphylls of *Lepidozamia peroffskyana*.

some localities branched plants are common, these possibly resulting from the destructive effects of cyclones. *L. hopei* is similar in most respects to *L. peroffskyana* but is taller growing and with wider leaflets.
**Cultivation:** Suited to tropical, subtropical and warm temperate regions (those provenances originating from highland regions are more cold-tolerant). A very attractive cycad which in humid areas can be grown in full sun but in areas with a drier climate needs some protection or else the fronds bleach severely. Likes warm, humid conditions and responds to mulching, watering during dry periods and regular light applications of fertiliser. Also an excellent container subject which can be used for decorative purposes in conservatories and well-lit positions indoors. Tolerates light frosts only.
**Propagation:** From seed.

## *Lepidozomia peroffskyana*

Regel
(after Count Peroffsky, an Imperial Russian Minister and benefactor of the St Petersburg Botanic Garden where the type plant was cultivated)

**Description:** A medium-sized to large cycad with an erect trunk to 7 m tall and 40 cm across. **Young leaves** brownish, sparsely hairy. **Mature leaves** numerous in an obliquely erect to widely spreading crown, 2–3 m x 20–40 cm, dark green and glossy, flat in cross-section, straight to slightly curved in profile; **petiole** 30–60 cm long, grey-green, sparsely hairy, swollen and shortly hairy at the base; rhachis straight, not twisted, round in cross-section, sparsely hairy; **leaflets** 200–250 on each leaf, moderately crowded, evenly distributed throughout, widely spreading, arching or drooping towards the apex, recurved and falcate, leathery, dark green and glossy above, the lower leaflets reduced in size but not spine-like; **median leaflets** 20–30 cm x 0.7–1.3 cm, broadly linear, falcate, tapered to an acute apex. **Cones** markedly dissimilar, shortly stalked but appearing sessile, appearing as if terminal on the trunk, usually solitary. **Male cones** 40–60 cm x 10–12 cm, more or less cylindrical, strongly curved or somewhat contorted; **sporophylls** 6–8 cm x 4–6 cm, broadly wedge-shaped, usually reflexed with up-curved tips. **Female cones** 40–80 cm x 25–30 cm, broadly ovoid, grey-green; **sporophylls** 6–8 cm 4–6 cm, broadly wedge-shaped, the outer face with short hairs, deflexed and upcurved near the tip. **Seeds** 5–6 cm x 3–3.5 cm, oblong, bright red.
**Distribution and Habitat:** Endemic.to Australia where it occurs in south-eastern Queensland and north-eastern New South Wales, roughly between Gympie and Taree. This species grows on protected slopes and gullies in wet sclerophyll forest, along rainforest margins and in littoral rainforests. Altitudinal range is from near sea level to

Multiple female cone development of *Lepidozamia peroffskyana*.

about 1000 m altitude. Soils are usually well-structured, humus-rich loams. The climatic range is from subtropical to warm temperate with summer temperatures to 32°C by day and 19°C and winter temperatures to 22°C by day and 6°C. Rainfall averages about 1500 mm per annum with heavy falls in summer and lighter winter rains.
**Notes:** *L. peroffskyana* was described in 1857 from a living plant cultivated in the St Petersburg Botanic Garden in Russia. A very distinctive cycad which often grows in loose colonies. The species is very similar to *L. hopei* but is shorter growing and with narrower leaflets.
**Cultivation:** Suited to subtropical and temperate regions. A very attractive species which is frequently planted in the public gardens of temperate Australia. Also makes an excellent tub plant for conservatories and well-lit situations indoors. Can be grown in full sun and filtered sun and also performs well in complete shade. Plants respond to fertiliser applications, mulches and watering during dry periods. Tolerates moderate to heavy frosts.
**Propagation:** From seed.

GENUS

# *Macrozamia*

The genus *Macrozamia* consists of about twenty-five species, all of which are restricted to Australia where they occur in subtropical and temperate regions, ranging from close to the coast to inland districts. One fossil species has also been described (see page 30).

**Derivation:** The genus *Macrozamia* was described by Friedrich Anton Wilhelm Miquel in 1842. The generic name is composed of the Greek words *macros*, 'large' and *Zamia*, another genus of cycads.

**Generic Description:** Terrestrial cycads with a trunk which is either ovoid and subterranean or cylindrical and subterranean or emergent. Leaf bases retained at senescence. New leaves emerge singly or in flushes, sparsely hairy. Mature leaves pinnate, oblong or lanceolate in outline, flat or vee-ed in cross-section, the older leaves spreading or deflexing after a flush of growth. Cataphylls leathery, hairy when young. Petioles lacking spines or prickles, expanded at the woolly base. Rhachis lacking prickles, straight or spirally twisted up to eight times. Leaflets not articulate at the base, alternate to nearly opposite, more or less evenly spaced except the lower leaflets which are more widely spaced, inserted on the margin of the rhachis, straight or falcate, simple or dichotomously forked one to many times, margins entire, apex usually pungent, base with a coloured area of callous tissue, in one group the lower leaflets reduced in size and spine-like. Cones markedly dissimilar in shape and size, prominently stalked, usually green, arising among the leaf bases. Peduncle with prominent short cataphylls attached by decurrent bases. Sporophylls peltately attached, with a prominent apical, spine-like appendage which is much longer on the distal sporophylls than the proximal ones. Male cones narrow, usually curved at maturity. Female cones broad, erect or spreading. Seeds radiospermic, oblong or obovoid, often angular, the sarcotesta red, orange or yellow.

**Notable Generic Features:** Species of *Macrozamia* can be recognised by their non-articulate leaflets, which are attached to the lateral margins of the rhachis, having a small, often colourful, callous area at the base. Other features include:

- a cylindrical or ovoid, subterranean or emergent trunk clothed with persistent leaf bases;
- an absence of suckers and offsets;
- young parts sparsely hairy;
- new leaves produced singly or in flushes;
- leaflets without a midrib and decurrent at the base;
- cones basically green or glaucous-green at maturity;
- peduncle with short cataphylls attached by decurrent bases; and
- sporophylls with an apical, spine-like appendage, the longest on the distal sporophylls of the cones.

**Infrageneric Features:** Two natural groups can be recognised in the genus *Macrozamia* and these have been treated as sections.

**Section Macrozamia:** This includes *M. communis, M. diplomera, M. dyeri, M. douglasii, M. johnsonii, M. macdonnellii, M. miquelii, M. moorei, M. mountperriensis, M. riedlei* and

*M.* species 'Eneabba'. Members of this section are large plants, often with cylindrical, emergent trunks and with numerous spreading leaves in the crown. The leaflets are undivided (except in *M. diplomera*), spread widely from the non-twisted rhachis in a flat plane, contain mucilage canals and the lower leaflets are reduced in size and spine-like. Cones are produced on a regular basis and often many mature plants in a colony will cone at the same time. These large-growing species frequently grow in extensive communities which may dominate the understorey vegetation.

**Section Parazamia:** This includes *M. fawcettii*, *M. fearnsidei*, *M. flexuosa*, *M. heteromera*, *M. lomandroides*, *M. lucida*, *M. pauli-guilielmi*, *M. platyrachis*, *M. plurinervia*, *M. secunda*, *M. spiralis*, *M. stenomera*, *M.* species 'Southern Pilliga' and *M.* species 'Northern Pilliga'. These are generally small plants, usually with an ovoid, subterranean trunk and with very few leaves in the crown. The leaflets, which may be entire or forked one to three times, are often erect and curved, lack mucilage canals and the lower leaflets are never reduced and are not spine-like. Cones are produced sporadically often with many years between coning events and usually very few plants in a colony will cone at the same time (often those growing close together). These small-growing species either occur in small, disjunct groups or in large but scattered communities.

## DISTRIBUTION OF *MACROZAMIA*

Macrozamia communis in habitat.

**Recent Studies:** The most recent studies published on the genus *Macrozamia* were by L. A. S. Johnson, 'The Families of Cycads and the Zamiaceae of Australia', *Proceedings of the Linnaean Society of New South Wales* 84 (1959): 87–116.

A number of species complexes are currently being investigated by David Jones, Australian National Botanic Gardens, Canberra.

**Habitat:** Species of *Macrozamia* are distributed from the coast to inland gorges and from the lowlands to moderate altitudes in the ranges. The majority of species grow in open forest and woodland with those from inland sites gaining protection by growing in gullies, on sheltered slopes and in deep valleys and gorges. Some species are frequently found growing among rocks. Bushfires and grassfires are common in some of the habitats (and may influence cone production), but in other habitats fires are rare and have limited effects on the biology of the cycads.

**Cultivation:** As a general rule, species of *Macrozamia* are easy to grow and adapt to a variety of soils and conditions. Most species are adaptable to situations of full sun but for best appearance may do better in partial shade or filtered sun. Surprisingly, many species will grow well in shady conditions but may suffer damage from attacks by leaf fungi unless the air movement is adequate. Soil drainage needs to be free and unimpeded or else damage from root-rotting fungi can be a significant problem. Humus-rich, acid soils are satisfactory for most species of *Macrozamia*. Watering during dry periods, mulches and regular, light fertiliser applications produce strong growth. Most species of *Macrozamia* are very tolerant of cold and are able to withstand exposure to moderate or heavy frosts.

**Propagation:** All species of *Macrozamia* are propagated mainly from seed. Seeds have a limited period of viability and for best results should be sown fresh. The seeds have an after-ripening period of six to twelve months and the period of germination is usually twelve to eighteen months (see also page 84).

Some species, such as *M. lomandroides* and *M. pauli-guilielmi*, have been successfully propagated by slicing off pieces of stem containing a leaf base. This section, if treated as a cutting, can develop into a separate plant. Sections of stem which lack a leaf can also sometimes be induced to grow.

## KEY TO SPECIES

**1** Rhachis with a gentle spiral twist less than 180° ......... **2**
Rhachis with at least one spiral twist of 360°, usually several such twists per frond ......... **23**

**2** Leaves held more or less horizontal, rhachis suddenly curved near the base, leaflets more or less erect ......... **3**
Leaves erect to obliquely erect, rhachis straight or gently curved ......... **4**

**3** Leaflets to 40 cm x 2 cm, concave, not crowded ......... ***M. platyrachis***
Leaflets to 20 cm x 0.8 cm, flat, crowded ......... ***M. secunda***

**4** Proximal pinnae reduced and spine-like ......... **5**
Proximal pinnae slightly shorter than median pinnae but not reduced and spine-like ......... **11**

**5** Reduced spine-like pinnae extending nearly to the base of the leaf ......... **6**
Reduced spine-like pinnae stopping well short of the leaf base ......... **7**

**6** Leaflets green ......... ***M. johnsonii***
Leaflets blue-green ......... ***M. moorei***

**7** Leaflets markedly glaucous ......... **8**
Leaflets green ......... **9**

**8** Spine-like appendages on sporophylls to 6 cm long, seeds to 5 cm x 3.5 cm ......... ***M. riedlei***
Spine-like appendages on sporophylls to 2 cm long, seeds to 8 cm x 5.5 cm ......... ***M. macdonnellii***

**9** Stomata on both surfaces of leaflets ......... ***M. riedlei***
Stomata on lower leaflet surface only ......... **10**

**10** Leaflets shiny, thin-textured ......... ***M. miquelii***
Leaflets dull, thick-textured ......... ***M. communis***

**11** Stomata on lower leaflet surface only ......... **12**
Stomata on both surfaces of leaflets ......... **14**

**12** Median leaflets attached to the rhachis at about 90° ......... ***M. lucida***
Median leaflets attached to the rhachis at about 40° ......... **13**

**13** Leaflets about 1 cm wide, dark green ......***M. douglasii***
Leaflets about 0.5 cm wide, light green ........................................................***M. mountperriensis***

**14** At least some leaflets forked ........................................ **15**
All leaflets entire ........................................................ **16**

**15** Leaflets steel blue, 2-4 mm across ........ ***M. stenomera***
Leaflets green to yellow-green, 5-10 mm across ..............................................................***M. diplomera***

**16** Trunk less than 0.5 m tall ............................ ***M. riedlei***
Trunk to more than 1 m tall ........................................ **17**

**17** Leaves erect to obliquely erect, forming a crown shape like a shuttlecock ............................................... ***M. dyeri***
Leaves spreading in a rounded crown ........................................................***M. species* 'Eneabba'**

**18** Leaflets all entire ........................................................ **19**
At least some leaflets forked ........................................ **21**

**19** Leaflets with a single point ...................***M. platyrachis***
Leaflets with 3 or more spine-like teeth .....................**20**

**20** Leaflets held stiffly erect ................... ***M. lomandroides***
Leaflets recurved or lax, widely spreading ..............................................................***M. fawcettii***

**21** Leaflets strongly glaucous at least when young ........................................***M. species* 'Northern Pilliga'**
Leaflets dull green ........................................................**22**

**22** Most leaflets on a leaf forked ..............***M. heteromera***
Few leaflets forked, sometimes all entire ........................................***M. species* 'Southern Pilliga'**

**23** Leaflets forked, steel blue ...................... ***M. stenomera***
All leaflets entire ........................................................**24**

**24** Leaflets 2-4 mm across .................. ***M. pauli-guilielmi***
Leaflets 6 mm across or more ........................................**25**

**25** Leaves held more or less horizontally, leaflets erect ..............................................................***M. platyrachis***
Leaves erect to obliquely erect, leaflets spreading .....**26**

**26** Leaflets flat in cross-section .........................***M. spiralis***
Leaflets concave in cross-section ...............................**27**

**27** Leaflets dull green to glaucous, cones glaucous ..............................................................***M. plurinervia***
Leaflets dark green and shiny, cones green ................**28**

**28** Leaflets to 28 cm x 0.7 cm ........................ ***M. flexuosa***
Leaflets to 60 cm x 1.1 cm ..................... ***M. fearnsidei***

Disintegrating cone and seeds of *Macrozamia communis*.

# *Macrozamia communis*

L. Johnson
(common in communities)

**Description:** A medium-sized cycad with a subterranean or emergent trunk to 2 m tall and 60 cm across. **Young leaves** light green. **Mature leaves** up to 100 in a graceful rounded crown, 0.7–2 m x 30–50 cm, bright green to yellowish, flat in cross-section, arching in profile; **petiole** 10–40 cm long, swollen and woolly at the base; rhachis not or only slightly twisted, flat above, convex and angular beneath; **leaflets** 70–130 on each leaf, moderately crowded in the middle of the leaf, closely packed towards the apex, becoming widely spaced towards the base, inserted on the rhachis at about 45°, pungent-tipped, the lower ones reduced and spine-like; **median leaflets** 15–35 cm x 0.5–1.2 cm, linear, bright green to yellowish, contracted to a whitish callous base. **Cones** markedly dissimilar. **Male cones** 20–45 cm x 8–14 cm, cylindrical, usually curved, green, one to five on each plant; **sporophylls** 2–4 cm x 1.5–2.5 cm, obovate, with an erect, apical, spine-like appendage 0.2–5 cm long; **peduncle** 10–15 cm x 3 cm. **Female cones** 20–45 cm x 10–20 cm, ovoid to barrel-shaped, glaucous green, one to three on each plant; **sporophylls** 4–7 cm x 4–8 cm, obovate, with an erect, apical, spine-like appendage 1–8 cm long. **Seeds** 2.5–3 cm x 1.5–2 cm, oblong to ovoid, yellow, orange or bright red.

**Distribution and Habitat:** Endemic to Australia where widely distributed in eastern New South Wales between the Macleay River on the north coast and Bega on the south coast. *M. communis* forms colonies (often dense and extensive) on ridges and slopes of inland and near-coastal ranges in gravelly loams under tall, sclerophyll forests and on tertiary sands and stabilised dunes in coastal districts. The climate where this species grows is temperate, being warm to hot in summer with a day/night temperature regime to 28°C and 12°C respectively. Winters are cool to cold with a day/night temperature regime to 12°C and 4°C respectively. Frosts occur in some areas. Rainfall ranges from 1000 mm to 1500 mm per annum spread fairly evenly throughout the year.

Mature female cone of *Macrozamia communis*.

Old male cone of *Macrozamia communis*.

Seeds of *Macrozamia communis* showing sarcotesta gnawed by rats.

**Notes:** This species was described in 1959 from material collected near Mossy Point in southern New South Wales. Previously it had been confused with *M. spiralis* by numerous authors. A distinctive species which probably has *M. miquelii* as its closest relative and from which it can be distinguished by its thicker textured, dull leaflets and broader cones. *M. communis* ranges over about 900 km in a north-south distribution and studies show a strong north-south clinal variation. Northern populations have fewer leaves in the crown, smaller cones and smaller seeds than plants from southern areas. Plants in some areas have yellow seeds. *M. communis* hybridises sporadically with *M. flexuosa* and *M. secunda* where the two species grow in close proximity.
**Cultivation:** Suited to subtropical and temperate regions. One of the most popular species of *Macrozamia* for cultivation both in Australia and overseas. Prized for its graceful crown of pleasant green fronds. Makes an excellent specimen for large containers. Will grow in full sun or filtered sun to shade and tolerates heavy frosts. Mature specimens transplant readily. Plants are sometimes retained for their ornamental qualities in urban areas where it occurs naturally.
**Propagation:** From seed.

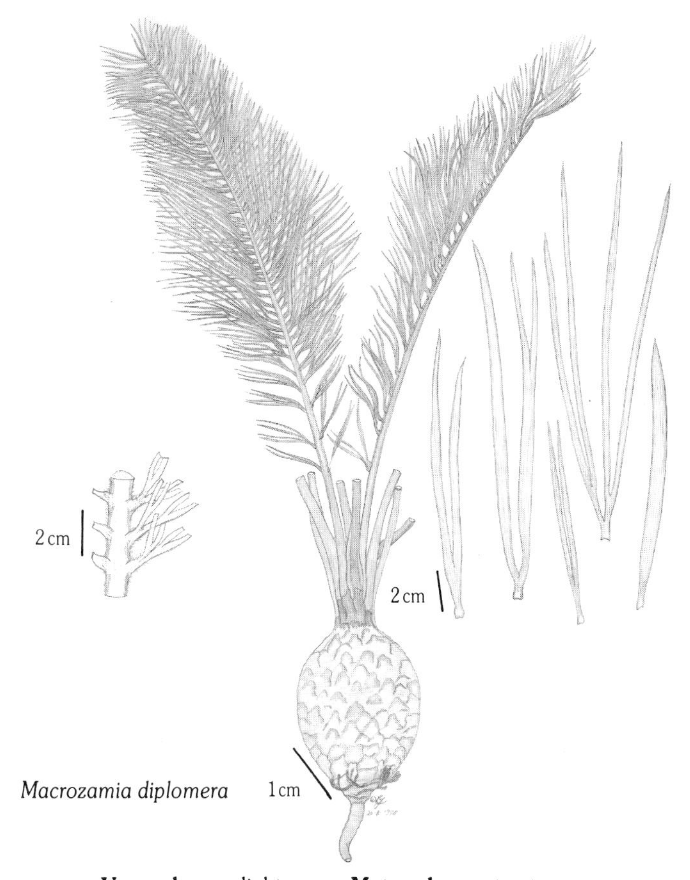

*Macrozamia diplomera*

# *Macrozamia corallipes*

J. D. Hook.
(coral-like, in reference to the coralloid roots)

This species, described in 1872, is a synonym of *M. spiralis*, the latter being first described in the genus *Zamia* in 1796. The original illustration of the species can be seen on page 258.

# *Macrozamia denisonii*

C. Moore & F. Muell.
(after Sir William Denison)

Described in 1858 from a collection made on the Manning River in northern New South Wales, this species is a synonym of *Lepidozamia peroffskyana*. Plants of the latter species, however, may still be wrongly labelled as *M. denisonii* in some Australian botanic gardens.

# *Macrozamia diplomera*

(F. Muell.) L. Johnson
(twice divided)

**Description:** A small to medium-sized cycad with an ovoid, subterranean trunk to 60 cm long and 40 cm across. **Young leaves** light green. **Mature leaves** ten to about fifty in a stiff, obliquely erect crown, 0.6–1.2 m x 26–40 cm, dark green, dull, flat in cross-section, straight in profile; **petiole** 10–20 cm long, swollen and woolly at the base; rhachis not or slightly twisted, flat above, concave to keeled beneath; **leaflets** 50–120 on each leaf, moderately spaced in the middle of the leaf, crowded towards the apex, more widely spaced towards the base,

*Macrozamia diplomera* in habitat.

inserted on the rhachis at about 40°, most of the leaflets forked once into two segments, the angle between them acute, rarely all the leaflets entire, sometimes the upper segment forked again or with a tooth near the apex, the lower leaflets reduced and spine-like; **median leaflets** 15–25 cm x 0.5–1 cm, the lobes linear, rigid, tapered to the pungent tip, contracted to orange, red or pale yellowish callous bases. **Cones** markedly dissimilar. **Male cones** 15–20 cm x 4–7 cm, cylindrical, curved, green, one or two on each plant; **sporophylls** 2–3 cm x 1.2–1.8 cm, narrowly wedge-shaped, with an erect, apical, spine-like appendage 0.5–2 cm long; **peduncle** 8–12 cm x 2 cm. **Female cones** 15–30 cm x 6–12 cm, cylindrical, green, one to four on each plant; **sporophylls** 3–4 cm x 3–4 cm, obovate, with an erect apical spine-like appendage 0.5–3.5 cm long; **peduncle** 10–15 cm x 2 cm. **Seeds** 2–3 cm x 2–2.5 cm, ovoid to oblong, red to reddish brown or yellow.
**Distribution and Habitat:** Endemic to Australia where it is restricted to the southern parts of the north-western slopes of New South Wales. This species grows in sandy or gravelly soils under low forests which are usually dominated by bloodwoods, ironbarks or species of *Callitris*.

Leaf of *Macrozamia diplomera*.

The climate is temperate having hot, dry summers with a day/night temperature regime to 36°C and 17°C. Winters are cool to cold with a day/night temperature regime to 18°C and zero, with frosts being common. The rainfall is about 500 mm per annum falling mainly in winter.
**Notes:** This species was described in the genus *Encephalartos* in 1866 (as a variety of *E. spiralis*), and transferred to *Macrozamia* in 1959. It is based on material collected in the Warrumbungle Ranges near the Castlereagh River. It is readily distinguished by its numerous, obliquely erect leaves (the crown has the appearance of a shuttlecock), and the usually forked leaflets with the lobes held close together. Often the apical 15 cm or so of the frond is extremely slender and appears drawn out. This species grows in loose colonies (often extensive), and exhibits considerable variation. Specimens which are apparently mature but lack any forked leaflets are not uncommon in some colonies and may grow right alongside others in which nearly every leaflet is forked. Puzzling (but uncommon) specimens, again apparently mature because of the size of the underground stem, may have only four or five leaves. Occasional specimens have lax leaves and rarely the leaflets are widely spaced on the rhachis. These variations have been noted by the author in areas where *M. diplomera* is the only cycad present and hybridisation is unlikely to be the cause.
**Cultivation:** Suited to temperate and subtropical regions. Will grow in partial shade or full sun. Hardy to heavy frosts. Requires free and unimpeded drainage. Can make an attractive specimen for a large container.
**Propagation:** From seed.

## *Macrozamia douglasii*

W. Hill & Bailey
(after John Douglas)

**Description:** A medium-sized cycad with an erect, subterranean trunk to 1 m tall and 70 cm across. **Young leaves** light green. **Mature leaves** to ninety in an erect to obliquely erect or spreading crown, 2–2.5 m x 40–50 cm, bright green to dark green, flat in cross-section, arching in profile; **petiole** 40–60 cm long, swollen and densely woolly at the base; rhachis not or slightly twisted, pale green, flat or grooved above, convex and angular beneath; **leaflets** numerous, moderately crowded, those towards the apex closely packed, those towards the base becoming more widely spaced, inserted on the rhachis at about 40°, pungent-tipped, the lower leaflets gradually reduced to one or two pairs of rigid, fairly long, yellowish, spine-like processes; **median leaflets** 25–35 cm x 1–1.2 cm, linear, dark bright green and shiny above, paler beneath, tapered to a pungent apex, contracted to a prominent white callous base. **Cones** markedly dissimilar. **Male cones** 20–35 cm x 5–7 cm, cylindrical, straight or

Macrozamia douglasii in habitat.

Coralloid roots of *Macrozamia douglasii* exposed naturally on the soil surface.

Mature female cones of *Macrozamia douglasii*.

Leaf of *Macrozamia douglasii* showing prominent white callus bases.

curved, green, one to three on each plant; **sporophylls** 2–3.5 cm x 1.5–2.5 cm, wedge-shaped, with an erect, apical, spine-like appendage 0.5–4 cm long; **peduncle** 20–35 cm x 2–3 cm. **Female cones** 35–45 cm x 10–15 cm, cylindrical to barrel-shaped, green; **sporophylls** 3–4 cm x 4–6 cm, broadly wedge-shaped, with an erect, apical, spine-like appendage to 1.5 cm long; **peduncle** 30–45 cm x 2–3 cm. **Seeds** 2.5–3.5 cm x 2–2.5 cm, oblong, orange to red.
**Distribution and Habitat:** Endemic to Australia where it occurs on Fraser Island and adjacent coastal districts north of Noosa in eastern Queensland. This species grows in tall open forests, along streams and on the fringes of rainforest, with some specimens occurring within the rainforest canopy. The soils are deep grey to white sands with a permanent water table. The climate is subtropical and has hot, humid summers with a day/night temperature regime to 31°C and 20°C respectively. Winters are mild to cool with a day/night temperature regime to 22°C and 8°C respectively. The rainfall is about 2000 mm per annum with summer rains heaviest.
**Notes:** This species was described in 1883 from material collected on Fraser Island. It has been wrongly treated as a synonym of *M. miquelii* in most recent botanical texts but the distinctiveness of the two has been well known to regular users of the bush in Queensland and by cycad enthusiasts. The species is locally abundant and often grows in scattered colonies. The roots of these cycads can probably tap the ground water of the area where they grow. *M. douglasii* can be distinguished from *M. miquelii* by its larger cones and more numerous, longer, lusher looking leaves in which only the lowest one or two pairs of leaflets are reduced to spine-like processes. It also has the largest and most conspicuous callous bases of any *Macrozamia*. The male cones of this species decay extremely rapidly after shedding pollen, leaving only the peduncle present within about six months. Often this peduncle is still green even though most traces of the remainder of the cone have vanished.
**Cultivation:** Suited to subtropical and warm temperate regions. Best grown in filtered sun or shade. Appreciates mulches and regular watering during dry periods. Seedlings and small plants must be protected from frosts, as they are very sensitive to cold damage.
**Propagation:** From seed.

Disintegrating female cone of *Macrozamia douglasii*.

*Macrozamia dyeri* in habitat.

# *Macrozamia dyeri*

(F. Muell.) C. A. Gardner
(after Sir William Turner Thistelton Dyer)

**Description:** A large cycad with an emergent, barrel-shaped trunk to 6.5 m tall and 1 m across, usually black-

*Macrozamia fawcettii* in habitat.

ened by fire. **Young leaves** bright green to yellowish green. **Mature leaves** numerous in an obliquely erect crown shaped like a shuttlecock, 1–2.2 m x 30–50 cm, dark green, flat in cross-section, straight to arching in profile; **petiole** 20–35 cm long, domed above, convex and ridged beneath, yellowish green, swollen and woolly at the base; rhachis not twisted, domed above, ridged beneath; **leaflets** 60–90 per leaf, dull green above, paler and with prominent veins beneath, crowded, overlapping like the slats of a venetian blind, inserted on the rhachis at about 25°, lower leaflets slightly reduced but not spine-like; **median leaflets** 30–40 cm x 1.5–2 cm, narrowly linear-lanceolate, tapered to a stiff, yellow, pungent apex, contracted to a whitish, raised callous base. **Cones** dissimilar. **Male cones** 20–40 cm x 8–10 cm, broadly cylindrical, curved, one to five on each plant; **sporophylls** 4–6 cm x 2–2.5 cm, narrowly wedge-shaped, with an erect, straw-coloured, apical, spine-like appendage 0.5–6 cm long; **peduncle** 10–18 cm long. **Female cones** 20–40 cm x 15–20 cm, yellowish green, one to three on each plant; **sporophylls** 7–9 cm x 4–5 cm, broadly wedge-shaped, with an erect, apical, spine-like appendage 2–5 cm long; **peduncle** 10–15 cm long. **Seeds** 3–3.5 cm x 2–2.5 cm, obovoid, red.

**Distribution and Habitat:** Endemic to Australia where widely distributed between Stokes Inlet and Cape Arid to the east and west of Esperance. The species grows in sparse woodland, low scrub and in heathland. Soils range from sandy clays to sands and laterites. The climate is temperate having hot, dry summers with a day/night temperature regime to 25°C and 18°C respectively. Winters are cool to cold with a day/night temperature regime to 16°C and 7°C respectively. The rainfall ranges from 600 mm to 800 mm per annum, falling mainly in winter.

**Notes:** *M. dyeri* was described in the genus *Encephalartos* in 1885 and transferred to *Macrozamia* in 1930. The type material was collected from Esperance Bay. Despite its very distinctive features, *M. dyeri* has been sunk under *M. riedlei* almost since it was described. It can be distinguished readily by its massive trunk with the obliquely erect leaves imparting the impression of a shuttlecock. It also has the broadest leaflets of any of the western members of the genus. In some areas of its natural habitat, populations of this species are being decimated by root-rotting fungi of the genus *Phytophthora*.

**Cultivation:** Suited to temperate regions. Best in filtered sun or full sun. Requires excellent drainage and is very sensitive to root-rotting fungi. Tolerates light to moderate frosts.

**Propagation:** From seed.

# *Macrozamia fawcettii*

C. Moore

(after C. Fawcett, original collector of the species)

**Description:** A small cycad with an ovoid subterranean trunk to 30 cm long and 20 cm across. **Young leaves** light green to bronze and shiny. **Mature leaves** two to twelve in an obliquely erect to spreading crown, 0.6–1 m x 20–30 cm, dark green and shiny, the leaflets appearing whorled or plumose because of the strongly twisted rhachis, straight in profile; **petiole** 15–30 cm long, slender, brownish green, swollen and very woolly at the base; rhachis twisted in one or more strong spirals, rounded to angular beneath; **leaflets** 50–120 on each leaf, not appearing crowded but those towards the apex more closely placed, twisted at the base and facing forwards, straight, recurved or lax and drooping in the distal half, thin-textured, the lower leaflets not reduced and not spine-like; **median leaflets** 18–30 cm x 0.7–1.7 cm, broadly linear, with two to seven spine-like teeth near the apex, especially on the basiscopic margin, the basal callous red. **Cones** markedly dissimilar. **Male cones** 15–25 cm x 4–6 cm, cylindrical, erect, straight or curved, green, one to three on each plant; **sporophylls** 1.5–2.5 cm x 1–1.5 cm, wedge-shaped, with an erect, apical, spine-like appendage 0.2–2 cm long; **peduncle** 8–12 cm x 2 cm. **Female cones** 20–30 cm x 10–12 cm, cylindrical to barrel-shaped, green to glaucous, one or two on each plant; **sporophylls** 3–4 cm x 4–5 cm, obovoid, with an erect, apical, spine-like appendage 0.3–2.5 cm long;

Leaf of *Macrozamia fawcettii*.

Young female cone of *Macrozamia fawcettii*.

**peduncle** 10–15 cm x 2 cm. **Seeds** 2.5–3 cm x 2–2.5 cm, ovoid to oblong, bright red.
**Distribution and Habitat:** Endemic to Australia where it occurs on the north coast of New South Wales. This species grows on sheltered slopes and gullies in dry sclerophyll forest and in flat areas and close to streams in wet sclerophyll forest. The climate is temperate having hot summers with a day/night temperature regime to 28°C and 19°C respectively. Winters are cool to cold with a day/night temperature regime of 19°C and 4°C respectively, with frosts occurring in some localities. The rainfall ranges from 1000 mm to 1700 mm per annum with most falling in summer and autumn.
**Notes:** *M. fawcettii* was described in 1884 from material collected in the hills around the upper reaches of the Richmond River in north-eastern New South Wales. It is a very distinctive species readily recognised by its broad, thin-textured leaflets presented in a whorled or plumose appearance because of the strong, spiral twists in the rhachis. Often this species grows in loose, scattered groups but in some localities extensive patches occur. Coning is somewhat irregular and seems to take place every three to five years, although in one area cones were noticed in three successive years.
**Cultivation:** Suited to temperate and subtropical regions. This is one of the most attractive species in the genus. Grows best in a position sheltered from excessive hot sun, otherwise the leaves and leaflets become much reduced and have a stunted appearance. Responds to mulches, watering during dry periods and light fertiliser applications. Tolerates moderate frosts.
**Propagation:** From seed.

## *Macrozamia fearnsidei*

D. Jones
(after Geoff Fearnside, strongly conservation-minded cattle farmer on whose property the type collection was made)

**Description:** A small cycad with an ovoid, subterranean trunk to 30 cm long and 20 cm across. **Young leaves** light green. **Mature leaves** five to twenty in an erect to obliquely erect crown, 0.7–1.4 m x 30–60 cm, dark green, the leaflets appearing whorled or plumose because of the strongly twisted rhachis, straight or recurved in profile; **petiole** 15–40 cm long, flat or slightly concave above, convex beneath; rhachis twisted strongly in more than two spirals, rounded beneath; **leaflets** 55–120 on each leaf, more or less crowded, twisted at the base and usually with other twists along their length, inserted on the rhachis at about 50°, widely spreading and drooping in the distal third, linear, dark green, shiny, strongly concave in cross-section, the lower leaflets not reduced and not spine-like; **median leaflets** 40–60 cm x 0.6–1.1 cm, tapered to a more or less pungent apex, contracted to a cream to pale yellow callous base. **Cones** markedly dissimilar, green. **Male cones** 15–27 cm x 4.5–6.5 cm, cylindrical, usually curved, one to four on each plant; **sporophylls** 1.6–2.2 cm x 1.5–2 cm, wedge-shaped, with an apical spine-like appendage 0.1–1.4 cm long; **peduncle** 18–30 cm x 1.5 cm. **Female cones** 12–18 cm x 8–10 cm, ovoid, one or two on each plant; **sporophylls** 1.7–2.8 cm x 1.5 cm, broadly ovate, with an erect, apical, spine-like appendage 0.2–2 cm long; **peduncle** 18–27 cm x 2 cm. **Seeds** 2.3–2.7 cm x 1.9–2.3 cm, ovoid, orange to scarlet.

*Macrozamia fearnsidei* in habitat.

Old male cone of *Macrozamia fearnsidei*.

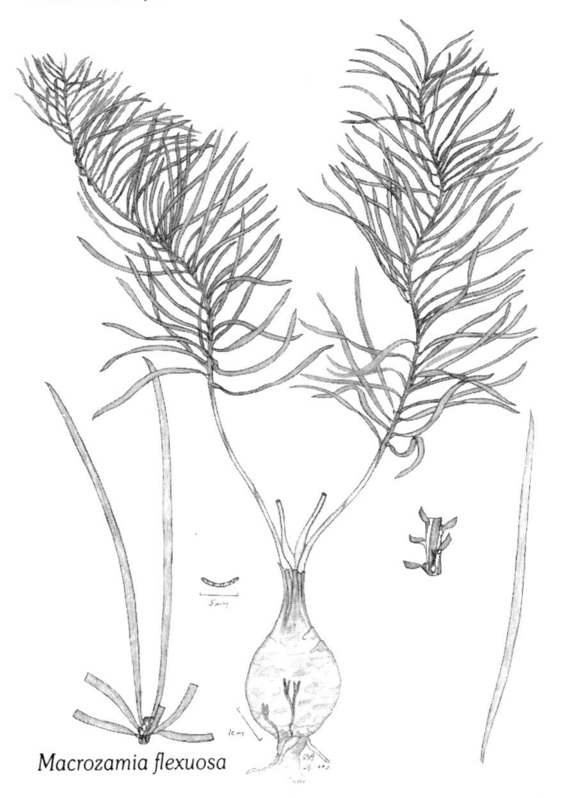

*Macrozamia flexuosa*

**Distribution and Habitat:** Endemic to Australia where it occurs on the Expedition Range of inland central Queensland. The species grows in tall open forest, either in gullies close to ephemeral streams or on rocky slopes and escarpments. Soils are sandy loams. The climate is tropical having hot summers with a day/night temperature regime to 35°C and 20°C respectively. Winters are cold with a day/night temperature regime to 21°C and 6°C respectively, with heavy frosts being common in many localities. The rainfall ranges from 500 mm to 750 mm per annum falling mainly over summer.

**Notes:** *M. fearnsidei* was described in 1991. It is part of the complex of species which surrounds *M. pauli-guilielmi* but is actually closest to *M. plurinervia*. It can be distinguished from other species in the complex by its robust habit with up to twenty leaves in the crown, its very long, often lax, shiny green leaflets and its large cones. *M. fearnsidei* grows in colonies which range from small to extensive and is sometimes found in association with *M. moorei*.

**Cultivation:** Suited to subtropical and temperate regions. An attractive species for a position in the open ground or in a large container. Best appearance is attained in filtered sun. Needs excellent drainage and responds to mulches and watering during dry periods. Tolerates moderate to heavy frosts.

**Propagation:** From seed.

# *Macrozamia flexuosa*

C. Moore

(flexuose, sinuous like a snake)

**Description:** A small cycad with an ovoid, subterranean trunk to 20 cm long and 15 cm across. **Young leaves** erect, light green. **Mature leaves** one to eight in an obliquely erect, sparse to crowded crown, 0.5–1 m x 20–35 cm, dark green, shiny, the leaflets appearing whorled

*Macrozamia flexuosa* in habitat.

or plumose because of the strongly twisted rhachis, straight in profile; **petiole** 20–35 cm long, rounded above and below; rhachis twisted strongly in more than two spirals, rounded beneath; **leaflets** 80–150 on each leaf, crowded, twisted at the base and facing forwards, inserted on the rhachis at about 50°, linear, spreading to lax, dark green and shiny, concave in cross-section with erect margins, the lower leaflets not reduced and not spine-like; **median leaflets** 12–28 cm x 0.4–0.7 cm, tapered to the apex, contracted to a white to red callous base. **Cones** markedly dissimilar. **Male cones** 15–20 cm x 4–6 cm, cylindrical, erect, usually curved, green, one to three on each plant; **sporophylls** 1.5–2 cm x 1–1.5 cm, wedge-shaped, with an apical, spine-like appendage 0.1–1 cm long; **peduncle** 5–10 cm x 1.5 cm. **Female cones** 12–18 cm x 6–8 cm, ovoid, green, one or two on each plant; **sporophylls** 3–4 x 3–5 cm, rhomboid, with an erect or spreading, apical, spine-like appendage 0.2–2 cm long; **peduncle** 8–12 cm x 1.5 cm. **Seeds** 2–2.5 cm x 1.8–2.2 cm, ovoid to oblong, orange to red.
**Distribution and Habitat:** Endemic to Australia where it occurs in coastal, near-coastal and adjacent inland ranges and valleys of north-central New South Wales. This species grows in a range of habitats including coastal cliffs, open forest among shrubs and on protected slopes in open forest and gullies, sometimes in quite shady situations. Soils are mainly shallow clay loams. The climate is temperate having hot summers with a day/night temperature regime to 26°C and 19°C respectively. Winters are cold with a day/night temperature regime to 18°C and 2°C, with frosts being frequent in areas away away from the coast. Rainfall varies from 1200 mm per annum on the coast to 600 mm per annum inland, falling mainly in summer and autumn.
**Notes:** This species, described in 1884, is treated as a subspecies of *M. pauli-guielielmi* by some authors. It can be distinguished from that species by its longer, rounded petioles and broader, dark green, shiny leaflets.
*M. flexuosa* forms sporadic hybrids with *M. communis* where the two species grow together on coastal cliffs near Newcastle.
**Cultivation:** Suited to temperate and subtropical regions. An attractive species especially when grown in a sheltered location. Tolerates moderate frosts and responds to watering during dry periods, mulches and light fertiliser applications.
**Propagation:** From seed.

# *Macrozamia heteromera*

C. Moore
(with more than one division)

**Description:** A small cycad with an ovoid, subterranean trunk to 30 cm x 15 cm. **Young leaves** erect and light

*Macrozamia heteromera* in habitat.

green to blue green. **Mature leaves** two to eleven in a sparse, erect to obliquely erect crown, 0.4–0.8 m x 15–25 cm, dark green to slightly glaucous, dull, broadly vee-ed in cross-section, straight or shallowly curved in profile or recurved near the apex; **petiole** 10–30 cm long, swollen and woolly at the base; rhachis usually straight, sometimes slightly twisted near the end; **leaflets** 80–110 on each leaf, crowded, those towards the base more widely spaced, twisted at the base and facing forwards, inserted on the rhachis at about 40°, most of the leaflets forked once or twice into narrow, rigid segments, the angle between them fairly wide, the lower leaflets not reduced and not spine-like; **median leaflets** 10–20 cm long, the segments 0.25–0.4 cm across, linear, apex mucronate with a long, yellow tip, contracted to the cream to yellowish callous base. **Cones** dissimilar. **Male cones** 18–26 cm x 4.5–6.5 cm, cylindrical, erect, straight or more usually curved, green, one to five on each plant; **sporophylls** 2–2.4 cm x 1.5–2 cm, wedge-shaped, with an erect apical, spine-like appendage 0.2–2 cm long; **peduncle** 15–20 cm x 2–3 cm. **Female cones** 20–35 cm x 8–12 cm, ovoid to ellipsoid, green to glaucous with pink areas on the **sporophylls**, usually one or two on each plant; **sporophylls** 3–4.5 cm x 4–6 cm, obovate, with an erect apical spine-like appendage 0.2–2 cm long. **Seeds** 2–3.5 cm x 2–2.5 cm, oblong, bright red to scarlet.
**Distribution and Habitat:** Endemic to Australia where it occurs in the drier areas of northern and north-western New South Wales. This species grows in a range of situations including protected gullies close to ephemeral streams, among rocks on slopes in shallow gravelly soils and in flat areas of open forest dominated by eucalypts. The climate is temperate having hot summers with a day/night temperature regime to 36°C and 17°C respectively. Winters are cold having a day/night temperature regime of 18°C to 0°C, with heavy frosts being common. The rainfall ranges from 300 mm per annum to 750 mm per annum (but irregular), falling mainly in winter and spring.
**Notes:** *M. heteromera* was described in 1884 from material collected in the Warrumbungle Ranges and on the Castlereagh River. This species, part of a confusing complex (see also entries for *M.* species 'Southern Pilliga' and *M.* species 'Northern Pilliga'), can be distinguished by having most of its leaflets lobed, very narrow leaflets and leaflet lobes and its seedlings having lobed leaflets from an early age. It is a relatively widespread species in which coning is very sporadic and may only take place every three to six years. Fires are common in the areas where this species grows but, from observations by the author, these occurrences do not influence cone production.
**Cultivation:** Suited to temperate regions and to a lesser extent subtropical regions. Will grow in full sun or filtered sun and requires free and unimpeded drainage. Tolerates heavy frosts.
**Propagation:** From seed.

*Macrozamia johnsonii* in habitat.

## *Macrozamia johnsonii*

D. Jones & K. Hill
(after L. A. S. Johnson, contemporary botanist who studied Australian cycads)

**Description:** A medium-sized cycad with an erect, stout trunk to 2.5 m tall and 80 cm across. **Young leaves** light green. **Mature leaves** numerous (to 120) in a rounded crown, 1.5–3 m x 25–50 cm, dark green, flat in cross-section, arching to gracefully curved in profile; **petioles** 2–8 cm long, swollen and woolly at the base; rhachis not twisted, yellowish; **leaflets** 150–250 per leaf, moderately crowded, evenly distributed throughout, inserted on the rhachis at about 40°, pungent-tipped, the lower leaflets gradually reduced to a series of rigid, spine-like processes; **median leaflets** 20–40 cm x 5–11 mm, linear, bright green, with stomata on both surfaces, tapered to the pungent apex, contracted to the white callous base. **Cones** markedly dissimilar. **Male cones** 25–40 cm x 8–10 cm, cylindrical, semi-erect to spreading, usually curved, green to brownish, ten to twenty on each plant; **sporo-**

**phylls** 2–3 cm x 1.2–1.8 cm, wedge-shaped, with an erect, apical, spine-like appendage 0.2–2 cm long; **peduncle** 10–15 cm x 2–3 cm. **Female cones** 50–80 cm x 10–20 cm, cylindrical to barrel-shaped, green, one to six on each plant; **sporophylls** 5–7 cm x 4–6 cm, broadly wedge-shaped, expanded near the apex, apical spine-like appendage 3–7 cm long; **peduncle** 15–22 cm x 3–4 cm, stout. **Seeds** 4–6 cm x 2–3 cm, oblong, red to scarlet.
**Distribution and Habitat:** Endemic to Australia where it occurs on the ranges inland from Grafton on the north coast of New South Wales. This species grows on sheltered ridges and steep slopes in both wet and dry sclerophyll forests. The soils are shallow, stony clay loams. The climate is temperate having hot summers with a day/night temperature regime to 31°C and 13°C respectively. Winters are cool to cold with a day/night temperature regime of 15°C and 0°C, with frosts being common. Rainfall is about 1000 mm per annum falling mainly in spring and summer.
**Notes:** This recently described species (1992) is a segregate from *M. moorei*. Adult plants are shorter growing than that species with greener leaves and leaflets which have stomata on both surfaces. The seedlings of both species are different. Those of *M. johnsonii* have a nearly rounded petiole, a straight, hardly twisted rhachis not recurved at the tip and widely spreading, falcate, shiny green leaflets that are thin-textured, have stomata on the lower surface only and a prominent white callous base from an early age. These seedling leaflets are produced for several years until an abrupt change to the adult form takes place in which the leaflets have stomata on both surfaces. For details of *M. moorei* seedlings see that species entry.
**Cultivation:** Suited to subtropical and temperate regions. A hardy and adaptable species which requires unimpeded drainage. Will grow in full sun but best appearance is attained with some shelter such as partial shade or filtered shade. Tolerates heavy frosts. An excellent plant for large containers or tubs.
**Propagation:** From seed.

## *Macrozamia lomandroides*

D. Jones
(resembling a clump of *Lomandra*)

**Description:** A small cycad with an ovoid, subterranean trunk to 25 cm long and 17 cm across. **Young leaves** bright green. **Mature leaves** two to six in a stiffly erect to spreading crown, 0.3–0.8 m x 30–40 cm, dull green, the leaflets appearing crowded and whorled because of the strongly twisted rhachis, straight in profile; **petiole** 6–14 cm long, flat above, rounded below; rhachis twisted strongly in more than two spirals, with prominent pale green to cream marginal bands, rounded beneath; **leaflets** 50–90 on each leaf, densely crowded, twisted at the base and facing forwards, inserted on the rhachis at about 30°, stiffly erect to spreading, broadly linear to slightly obovate, thick and leathery, dark green, dull, shallowly concave in cross-section, the lower leaflets not reduced and not spine-like; **median leaflets** 20–35 cm x 0.9–1.4 cm, the apex asymmetrical and with one to six sharp teeth on the upper margin, tapered to a relatively long, narrow base, the callous base pale green to cream. **Cones** markedly dissimilar, green. **Male cones** 12–15 cm x 4–5 cm, cylindrical, curved, one or two on each plant; **sporophylls** 1.2–1.8 cm x 1.8–2.2 cm, broadly wedge-shaped, with an apical, spine-like appendage 0.1–0.8 cm long; **peduncle** 8–15 cm x 1.2 cm. **Female cones** 12–18 cm x 7–9 cm, ovoid; **sporophylls** 1.4–1.8 cm x 2.5–3.5 cm, broadly wedge-shaped, with an apical, spine-like appendage 0.2–3 cm long; **peduncle** 10–15 cm x 1–1.2 cm, hairy at the base. **Seeds** 2.2–2.6 cm x 1.8–2.2 cm, ovoid to oblong, orange to red.
**Distribution and Habitat:** Endemic to Australia where it occurs in central-eastern Queensland to the south of Bundaberg. The species grows among grass in grey silty loam on flat sites under tall open forest. The climate is subtropical with hot summers having a day/night temperature regime to 30°C and 20°C respectively. Winters are cool with a day/night temperature regime to 21°C and 9°C. Rainfall is about 2200 mm per annum falling mainly over summer.
**Notes:** *M. lomandroides* was described in 1991. It is a very distinctive species which can be recognised by its stiff, leathery leaflets held stiffly erect and with one to six prominent apical spines. Coning is sporadic in this species and usually takes place about every four to six years.
**Cultivation:** Suited to subtropical and warm temperate regions. Will grow in full sun or filtered sun. Responds well to regular watering, mulches and light fertiliser applications. Tolerates moderate frosts.
**Propagation:** From seed and by slices from the trunk.

## *Macrozamia lucida*

L. Johnson
(shining, glistening)

**Description:** A small cycad with a subterranean trunk to 40 cm long and 25 cm across. **Young leaves** pale green. **Mature leaves** five to forty in an erect to obliquely erect crown, 0.6–1.2 m x 20–40 cm, dark green and shiny, flat in cross-section, gently arching in profile, sometimes slightly twisted near the apex; **petiole** 20–50 cm long, swollen and woolly at the base; rhachis dark green, not twisted or slightly twisted near the apex, concave above, convex beneath; **leaflets** 50–100 on each leaf, moderately crowded, twisted at the base, and facing forwards,

*Macrozamia lucida.*

evenly spaced throughout except those towards the base which are more widely spaced, the proximal leaflets inserted on the rhachis at about 90°, those towards the apex more acutely oriented, pungent-tipped, the lower leaflets not reduced and not spine-like; **median leaflets** 15–35 cm x 0.7–1 cm, linear, dark green and shiny, tapered to the pungent apex, contracted to a conspicuously white, callous base. **Cones** dissimilar. **Male cones** 12–18 cm x 3.5–5 cm, cylindrical, straight or curved, green, one to eight on each plant; **sporophylls** 2.5–4 cm x 1.2–2 cm, wedge-shaped, with an erect apical, spine-like appendage 0.2–1.5 cm long; **peduncle** 20–30 cm x 2 cm. **Female cones** 12–20 cm x 6–10 cm, ovoid to barrel-shaped, green, one or two on each plant; **sporophylls** 3–5 cm x 2–4 cm, obovate, with an erect, apical spine-like appendage 0.5–3 cm long; **peduncle** 20–30 cm x 3 cm. **Seeds** 2–2.5 cm x 1.5–2 cm, ovoid to oblong, dark orange.

**Distribution and Habitat:** Endemic to Australia where it occurs in south-eastern Queensland from about Gympie to Brisbane (erroneously recorded from New South Wales). This species commonly grows in coastal and near-coastal ranges and may extend further inland in some districts. It grows on protected slopes in open forest, in the vegetation close to streams and along rainforest margins and in more open types of rainforest. The climate is subtropical having hot, wet and humid summers with a day/night temperature regime to 28°C and 21°C respectively. Winters are cool to cold with a day/night temperature regime to 20°C and 10°C respectively. Rainfall is about 1000 mm per annum falling mainly in summer maximum.

Male cones of *Macrozamia lucida* at various stages of development.

Disintegrating female cone of *Macrozamia lucida*.

**Notes:** This species was described in 1959 from material collected in the Glasshouse Mountains of southern Queensland. It is often confused with *M. miquelii* (especially seedlings), but can be distinguished by the

*Macrozamia lomandroides* in habitat.

Toothed leaflet tips of *Macrozamia lomandroides*.

prominent white callous areas at the base of the leaflets, the more slender petioles and the absence of spine-like leaflets at the base of the leaf. Ripe fruits of this species are highly attractive to bandicoots, rats and possums which feed on the outer flesh and may disperse the seeds over some distance from the parent. This species commonly hybridises with *M. miquelii*, where the two grow together, usually on disturbed sites.
**Cultivation:** Suited to subtropical and temperate regions. Best appearance is attained in a sheltered position; however, plants can be grown in full sun. An excellent container subject. Plants respond vigorously to cultivation and grow much larger than they do in the wild. Mulches, watering during dry periods and light applications of fertiliser are all beneficial practices.
**Propagation:** From seed.

# *Macrozamia macdonnellii*

(F. Muell. ex Miq.) A.DC.
(after Sir Richard G. Macdonnell, nineteenth century Governor of South Australia)

**Description:** A medium-sized to large cycad with an erect or procumbent, stout trunk to 3 m long and 1 m across. **Young leaves** powdery blue with pink callous bases. **Mature leaves** numerous in a graceful, rounded crown, 1.5–2.5 m x 30–40 cm, usually dark blue-green, flat in cross-section, arching to gracefully curved in profile; **petiole** 15–30 cm long, green to powdery, swollen and woolly at the base; rhachis straight or slightly twisted, flat above, concave below; **leaflets** numerous, crowded, evenly spaced throughout (although those towards the base may be more widely spaced), inserted on the rhachis at about 40°, rigid and pungent-tipped, the lower leaflets gradually reduced and spine-like; **median leaflets** 20–30 cm x 0.7–1.1 cm, linear, straight, glaucous on both surfaces, tapered to a pungent apex, contracted to a pink to cream callous base. **Cones** dissimilar, glaucous, one to five on each plant. **Male cones** 25–40 cm x 8–10 cm, cylindrical, usually curved; **sporophylls** 3–4 cm x 1.5–2 cm, narrowly triangular, with an erect, apical, spine-like appendage 0.2–2.5 cm long; **peduncle** 10–14 cm x 4–5 cm. **Female cones** 40–50 cm x 20–27 cm, ovoid to barrel-shaped; **sporophylls** 7–10 cm x 8–12 cm, triangular to broadly wedge-shaped, with an erect apical, spine-like appendage 0.2–2 cm long; **peduncle** 12–25 cm x 4–5 cm. **Seeds** 6–8 cm x 3.5–5 cm, oblong, orange to red.
**Distribution and Habitat:** Endemic to Australia where it is restricted to the MacDonnell Range system and adjacent Hartz Ranges of Central Australia. This species usually grows on the nearly bare sides of rocky gorges and escarpments, sometimes in valleys adjacent to streams and rarely in sclerophyll forests. Plants often grow wedged between large boulders or in crevices and some may be subjected to sporadic flooding. Soils are sands and gravels. The climate is semi-arid having hot, dry summers with a day/night temperature regime to 35°C and 21°C respectively. Winters are cold with a day/night temperature regime of 19°C and 0°C, with heavy frosts being common. Rainfall ranges from 100 mm to 200 mm per annum and is of irregular occurrence.
**Notes:** A remarkable relict species which survives in the valleys of river systems which have cut through ancient sandstone. This cycad survives in narrow ecological zones in an area in which the surrounding habitat is hostile because of its aridity and which is totally unsuitable for the growth of the species. *M. macdonnellii* was

described in the genus *Encephalartos* in 1863, probably from material collected in the MacDonnell Ranges, and transferred to *Macrozamia* in 1868. It is a very distinctive species which can be recognised by its broad, rounded crown of distinctly blue-green leaves. Its seeds, which are the largest in the genus and probably the largest of any cycad, each weigh about 50 g, and are apparently an adaptation for survival in the inhospitable environment where it grows.
**Cultivation:** Suited to temperate and drier subtropical regions. This species resents humidity and may not do well in coastal localities. Best appearance is achieved in full sun. Needs excellent drainage and unimpeded air movement. Tolerates heavy frosts.
**Propagation:** From seed.

# *Macrozamia mackenzii*

Hort. ex Masters
(derivation unknown)

This species was described in 1877 from a plant cultivated in England and originating in Queensland. It is a synonym of *M. miquelii*.

# *Macrozamia miquelii*

(F. Muell.) A.DC.
(after Friedrich Anton Wilhelm Miquel, nineteenth century Dutch botanist specialising in cycads)

**Description:** A medium-sized cycad with an erect trunk to 0.5 m tall and 50 cm across. **Young leaves** light green. **Mature leaves** numerous in an obliquely erect crown, 0.5–1.8 m x 30–40 cm, dark green, flat in cross-section, arching in profile; **petiole** 10–50 cm long, swollen and woolly at the base; rhachis not twisted, whitish, flat above, convex and angular beneath; **leaflets** numerous, moderately crowded, those towards the apex closely packed, those towards the base becoming more widely spaced, inserted on the rhachis at about 40°, pungent-tipped, the lower leaflets gradually reduced to five to fourteen pairs of rigid, spine-like processes; **median leaflets** 17–28 cm x 0.4–0.8 cm, linear, dark green and shiny above, paler beneath, tapered to a pungent apex, contracted to a whitish or reddish callous base. **Cones** markedly dissimilar. **Male cones** 15–30 cm x 4–6.5 cm, cylindrical, usually curved, green, one to five on each plant; **sporophylls** 1.5–3 cm x 1–2 cm, wedge-shaped, with an erect, apical, spine-like appendage 0.2–2.5 cm long; **peduncle** 15–30 cm x 1.5–2.5 cm. **Female cones**

*Macrozamia macdonnellii* in habitat.

Disintegrating female cone of *Macrozamia miguelii*.

20–35 cm x 7–10 cm, cylindrical to barrel-shaped, green, one or two on each plant; **sporophylls** 3–4 cm x 2.5–5 cm, broadly wedge-shaped, with an erect, apical, spine-like appendage 0.5–5 cm long; **peduncle** 20–45 cm x 2–3 cm. **Seeds** 2–3 cm x 1.5–2 cm, oblong, light orange to dark red.

**Distribution and Habitat:** Endemic to Australia where it is distributed from north of Rockhampton in central-eastern Queensland south to the far north coast of New South Wales where apparently now extinct. In Queensland this species is very common and grows on ridges and slopes in open forest, along the margins of streams and in and around the fringes of rainforest. The climate where this species grows is tropical to subtropical with hot, humid, wet summers and cool to cold, dry winters. The tropical areas have a summer day/night temperature regime to 32°C and 22°C respectively and a winter regime to 24°C and 10–15°C depending on proximity to the coast. Rainfall ranges from about 900 mm per annum north of Rockhampton to 1350 mm on the coast east from Rockhampton, with most falling in summer. The climate around Brisbane has a summer day/night temperature regime to 28°C and 21°C respectively. Winters are mild to cool with a day/night temperature regime to 20°C and 10°C respectively. Rainfall is 1000 mm per annum falling mainly in summer.

**Notes:** This species was described in the genus *Encephalartos* in 1862 and transferred to *Macrozamia* in 1868. It was based on two collections, one from Moreton Bay near Brisbane, and the other, chosen as the lectotype

*Macrozamia miguelii* on habitat.

by L. A. S. Johnson, from the Fitzroy River near Rockhampton collected by A. Thozet. It is related to *M. communis* but can be distinguished by its thinner textured, shiny leaflets and narrower cones. It is also close to *M. douglasii* but is less robust than that species with fewer, shorter leaves in the crown and five to seven pairs of prominent, spine-like leaflets towards the base of the leaf (one or two pairs in *M. douglasii*). This latter character also serves to distinguish it from *M. mountperriensis* which lacks any reduced, spine-like leaflets.
*M. miquelii* shows a range of variation often associated with parent rock type and is currently under study. This species commonly hybridises with *M. lucida*, where the two grow together, usually on disturbed sites.
**Cultivation:** Suited to tropical, subtropical and warm temperate regions. Hardy and adaptable to a range of climates and well-drained soils. Best grown in filtered shade or partial shade especially in localities of low humidity. Tolerates moderate frosts. Mature specimens transplant readily. An excellent decorative plant for large containers.
**Propagation:** From seed.

## *Macrozamia moorei*

F. Muell.
(after Charles H. Moore, first superintendent of the Sydney Botanic Gardens)

**Description:** A medium-sized to large cycad with an erect, massive trunk to 8 m tall and 80 cm across. **Young leaves** bright green. **Mature leaves** numerous in a rounded crown, 1.5–3 m x 30–60 cm, dark bluish green, flat in cross-section, arching to gracefully curved in profile; **petiole** 5–10 cm long, swollen and woolly at the base; rhachis not twisted, yellowish, concave above, keeled beneath; **leaflets** numerous, moderately crowded, evenly distributed throughout, inserted on the rhachis at about 40°, with stomata on both surfaces, pungent-tipped, the lower leaflets gradually reduced to a series of rigid, spine-like processes; **median leaflets** 20–40 cm x 5–11 mm, linear, dark bluish green, tapered to a pungent apex, contracted to a white, callous base. **Cones** markedly dissimilar. **Male cones** 25–45 cm x 8–10 cm, cylindrical, green to brownish, straight or more usually curved, up to one hundred on a large plant; **sporophylls** 2–3 cm x 1.5–2 cm, wedge-shaped, with an erect, apical spine-like appendage 0.2–2 cm long; **peduncle** 10–15 cm x 2–3 cm. **Female cones** 40–90 cm x 12–20 cm, broadly cylindrical to barrel-shaped, green, one to eight on each plant; **sporophylls** 5–10 cm x 4–8 cm, broadly wedge-shaped, expanded near the apex, apical spine-like appendage 2–7 cm long; **peduncle** 15–25 cm x 3–4 cm, stout. **Seeds** 4–6 cm x 2.5–3.5 cm, oblong, red to scarlet.
**Distribution and Habitat:** Endemic to Australia where

*Macrozamia moorei* in habitat.

it occurs in central Queensland in the Carnarvon Range and the Emerald, Springsure and Rolleston districts. This species grows on low hills in dry sclerophyll forests and woodland and also in the valleys and escarpments of rocky gorges. The climate is hot in summer with a day/night temperature regime to 35°C and 21°C respectively. Winters are cool to cold with a day/night temperature regime to 23°C and 6°C, with severe frosts being common. Rainfall is about 500 mm per annum with summer rains predominating.
**Notes:** *M. moorei* was described in 1881 from material collected on the Nogoa River near Springsure. One of the most majestic of the Australian cycads, this species is renowned for its large, rounded, date-palm-like crown of gracefully curved, dark bluish green fronds. When a crown tops a massive, woody trunk (to 8 m tall) the effect is even more impressive. Add to this the fact that the species grows in large colonies, frequently in easily accessible scenic rocky areas, and it becomes apparent why it is so well known both in Australia and overseas. Mature cones on this species are also notable both for their size

*Macrozamia moorei* soon after a forest fire.

*Macrozamia mountperriensis* in habitat.

Young male cone of *Macrozamia mountperriensis*.

Male cone of *Macrozamia pauli-guilielmi*.

Macrozamia pauli-guilielmi in habitat.



and number. The female cone is the largest in the genus and mature male plants can bear an astounding number of cones (up to 100 have been recorded). *M. moorei* is closely related to *M. johnsonii* but the adult plants grow much larger, have bluish green leaves, larger seeds and different seedlings. The seedlings of *M. moorei* have a petiole which is flat on the upper surface, a prominently twisted rhachis recurved at the tip and erect, straight, dark blue green leaflets which have stomata on both surfaces and an obscure callous base. For details of the seedlings of *M. johnsonii* see that species entry.
**Cultivation:** Suited to tropical, subtropical and temperate regions. A very hardy and versatile species which will adapt to a range of climates and soils providing the drainage is unimpeded. Best grown in a sunny position. Tolerates heavy frosts. Mature specimens transplant readily. An excellent plant for large containers or tubs. It has been recorded that plants have grown to 2 m tall from seed in less than 100 years. Cone production may take about 50 years from seed.
**Propagation:** From seed.

## *Macrozamia mountperriensis*

Bailey
(from Mount Perry, the type locality)

**Description:** A medium-sized cycad with an erect trunk to 0.3 m tall and 30 cm across. **Young leaves** light green to yellowish green. **Mature leaves** twenty to eighty in an obliquely erect crown, 0.3–1.8 m x 20–30 cm, light green to yellowish green, flat in cross-section, arching in profile; **petiole** 30–45 cm long, swollen and woolly at the base; rhachis not twisted or with a slight twist, greenish white, ridged above, convex and angular beneath; **leaflets** numerous, moderately crowded, evenly spaced throughout except those towards the base which are more widely spaced, inserted on the rhachis at about 40°, pungent-tipped, the lower leaflets slightly reduced in size but not spine-like; **median leaflets** 15–25 cm x 0.6–0.8 cm, linear, light green to yellowish, tapered to a pungent apex, contracted to a whitish, callous base. **Cones** markedly dissimilar. **Male cones** 12–25 cm x 3–4 cm, cylindrical, usually curved, green, one to four on each plant; **sporophylls** 1.2–2.5 cm x 0.8–1.4 cm, wedge-shaped, with an erect, apical, spine-like appendage 0.2–1.3 cm long; **peduncle** 12–20 cm x 1–1.5 cm. **Female cones** 20–30 cm x 6–8 cm, cylindrical to barrel-shaped, green; **sporophylls** 2.5–3.5 cm x 2.5–3.5 cm, broadly wedge-shaped, with an erect, apical, spine-like appendage 0.5–4 cm long; **peduncle** 30–40 cm x 2–2.5 cm. **Seeds** 2–2.5 cm x 1.5–2 cm, ovoid to oblong, orange to red.
**Distribution and Habitat:** Endemic to Australia where it is distributed from Mount Perry region to the west of Bundaberg south to Biggenden and east to Aramara, all in south-central Queensland. This species grows on sheltered slopes, ridges and gullies under sparse, tall, dry sclerophyll forest and also in vine forests which have emergent Hoop Pines. The soils of the eucalypt forests are gravelly loams which are often shallow whereas those in the vine forests are red-brown volcanic loams. The climate is subtropical with hot, humid, wet summers having a day/night temperature regime to 33°C and 21°C respectively. Winters are cool to cold with a day/night temperature regime to 21°C and 0°C, with frosts being frequent. The rainfall is about 1000 mm per annum with a summer maximum.
**Notes:** *M. mountperriensis* was described in 1886 but treated as a synonym of *M. miquelii* until its distinctiveness was recognised recently by cycad enthusiasts. It was based on material collected at Mt Perry. The species is locally abundant in extensive colonies and exhibits little variation in the wild. It can be distinguished from *M. miquelii* by its fewer, smaller fronds with proportionately longer petioles, the absence of reduced, spine-like lower leaflets, much smaller cones and smaller seeds.
**Cultivation:** Suited to subtropical and warm temperate regions. Will grow in full sun or filtered shade in well-drained soil. Tolerates moderate frosts. Mature specimens transplant readily. A decorative species for large containers.
**Propagation:** From seed.

Mature female cone of *Macrozamia pauli-guilielmi*.

## *Macrozamia pauli-guilielmi*

W. Hill & F. Muell.
(after Friedrich William, nineteenth century King of Prussia and patron of botany)

**Description:** A small cycad with an ovoid subterranean

trunk to 25 cm long and 20 cm across. **Young leaves** light green and erect. **Mature leaves** one to six in an erect, very sparse crown, 0.5–1.2 m x 30–50 cm, dark green, dull, the leaflets appearing whorled or plumose because of the strongly twisted rhachis, straight in profile; **petiole** 5–15 cm long, flat above, keeled beneath; rhachis twisted strongly in more than two spirals, keeled beneath; **leaflets** 140–200 on each leaf, crowded (but not appearing so because of their narrowness and plumose arrangement), twisted at the base and facing forwards, inserted on the rhachis at about 45°, narrowly linear, spreading to lax or laxly recurved in the distal half, dark green and dull above, yellowish beneath, often yellowish towards the apex, shallowly concave in cross-section with erect margins, the lower leaflets not reduced and not spine-like; **median leaflets** 15–35 cm x 0.2–0.4 cm, tapered to the apex, contracted to a conspicuously white callous base. **Cones** markedly dissimilar. **Male cones** 10–20 cm x 4–6 cm, cylindrical, erect, usually curved, green, one to three on each plant; **sporophylls** 1.5–2 cm x 1–1.5 cm, wedge-shaped, with an apical spine-like appendage 0.1–0.5 cm long; **peduncle** 5–8 cm x 1.5 cm. **Female cones** 8–15 cm x 6–8 cm, ovoid to barrel-shaped, green, one or two on each plant; **sporophylls** 3–4 cm x 3–5 cm, obovate, with an erect, spreading or sometimes recurved apical spine-like appendage 0.1–2 cm long; **peduncle** 8–12 cm x 1.5 cm. **Seeds** 2–2.5 cm x 1.5–2 cm, ovoid to oblong, orange to red.
**Distribution and Habitat:** Endemic to Australia where it occurs in coastal and near-coastal districts of southern Queensland, including Fraser Island. This species is most often found growing in sparse woodland in sandy soils which can become wet and spewy during the summer and autumn; less commonly it grows on low, gravelly or shaly ridges. The climate is subtropical having hot, wet, humid summers with a day/night temperature regime to 30°C and 20°C respectively. Winters are mild to cool with a day/night temperature regime to 21°C and 9°C. Rainfall is 1500–1800 mm per annum falling mainly in summer.
**Notes:** *M. pauli-guilielmii* was described in 1859 from material collected by W. Hill in the vicinity of Moreton Bay in southern Queensland. It is regarded as being rare in the wild and is threatened by forestry (particularly pine plantations), agriculture and coastal development. *M. pauli-guilielmii* is closest to *M. flexuosa* but can be distinguished by its shorter flattened petioles and narrower, lighter green, dull leaflets which are yellowish beneath. It has been linked with *M. plurinervia* but that species is readily distinguished by its broader, stiff leaflets and prominently glaucous cones. Coning is extremely irregular in *M. pauli-guilielmi* and seems to take place about every four to six years.
**Cultivation:** Suited to subtropical and temperate regions. Will grow in full sun or filtered sun and tolerates moderate frosts.
**Propagation:** From seed and by slices from the trunk.

*Macrozamia platyrachis* in habitat.

## *Macrozamia platyrachis*

F. M. Bail.
(with a broad rhachis)

**Description:** A small cycad with an ovoid, subterranean trunk to 60 cm long and 60 cm across. **Young leaves** bright green. **Mature leaves** one to ten in a sparse, spreading crown, 0.5–1 m x 10–20 cm, dark green to yellowish-green, strongly recurved in profile and usually nearly prostrate, the leaflets held stiffly erect; **petiole** 15–25 cm long, yellowish, with an expanded, sparsely woolly base; rhachis straight or slightly twisted; **leaflets** about fifty on each leaf, moderately well spaced, evenly distributed throughout, twisted at the base and facing forwards, obliquely erect to erect, inserted on the rhachis at about 50°, stiff and leathery, dark green, shiny, concave in cross-section, the lower leaflets not reduced and not spine-like; **median leaflets** 30–40 cm x 1.2–2 cm, broadly linear, apex abruptly rounded and mucronate, contracted to a white callous base. **Cones** markedly dissimilar. **Male cones** 10–15 cm x 3–5 cm, cylindrical, curved, one to three on each plant; **sporophylls** 2.3–3.5 cm x 2–2.5 cm, wedge-shaped, with an erect, apical, spine-like appendage 0.1–2 cm long; **peduncle** 5–10 cm x 2 cm. **Female cones** 15–20 cm x 8–10 cm, ovoid; **sporophylls** 3–4 cm x 2–3 cm, wedge-shaped, with an erect, apical, spine-like appendage 0.3–2 cm long; **peduncle** 7–12 cm x 2 cm. **Seeds** 2–3 cm x 1.5–2 cm, ovoid to oblong, red.
**Distribution and Habitat:** Endemic to Australia where it is restricted to ranges and tablelands of central-western Queensland, particularly the Blackdown Tableland to the west of Rockhampton. This species grows in sparse woodland in sandy soils. The climate is tropical with hot wet, humid summers having a day/night temperature

Old male cones of *Macrozamia platyrachis*.

*Macrozamia plurinervia* in habitat.

Young female cone of *Macrozamia plurinervia*.

regime to 38°C and 24°C respectively. Winters are cold with a day/night temperature regime to 25°C and 8°C respectively, with frosts being common in some areas. Rainfall ranges from 600 mm to 1000 mm per annum, falling mainly in summer.

**Notes:** *M. platyrachis* was described in 1898 from material collected on Planet Downs Station in central-western Queensland. It appears to be uncommon to rare but may merely be unreported or overlooked in the general area where it grows. Coning is sporadic in this species and may take place about every four to seven years.

**Cultivation:** Suited to subtropical and temperate regions. Best grown in an open sunny position with free air movement. Soil drainage must be excellent. Makes an attractive specimen for a large container. Tolerates heavy frosts.

**Propagation:** From seed.

## *Macrozamia plurinervia*

(L. Johnson) D. Jones
(with many nerves or veins)

**Description:** A small cycad with an ovoid, subterranean trunk to 30 cm long and 20 cm across. **Young leaves**

*Macrozamia riedlei* in habitat.

Female cones of *Macrozamia riedlei*.

Male cones of *Macrozamia riedlei*.

bluish green. **Mature leaves** one to eight in a stiffly erect to obliquely erect, relatively sparse crown, 0.3–1 m x 20–30 cm, grey-green to glaucous, dull, the leaflets appearing whorled or plumose because of the strongly twisted rhachis, straight in profile; **petiole** 5–20 cm long, rounded above and below; rhachis twisted strongly in more than two spirals, rounded beneath; **leaflets** 30–120 on each leaf, more or less crowded, twisted at the base and facing forwards, inserted on the rhachis at about 60°, broadly linear, stiffly spreading, grey-green to glaucous, dull, shallowly concave in cross-section, the lower leaflets not reduced and not spine-like; **median leaflets** 6–20 cm x 0.5–0.8 cm, tapered to a pungent apex, contracted to an orange, pink or reddish callous base. **Cones** markedly dissimilar, strongly glaucous. **Male cones** 15–25 cm x 4–6 cm, cylindrical, erect, straight or curved, one to three on each plant; **sporophylls** 1.8–2.3 cm x 1.5–1.8 cm, wedge-shaped, with an apical, spine-like appendage 0.1–1 cm long; **peduncle** 6–10 cm x 2 cm. **Female cones** 10–18 cm x 6–9 cm, ovoid to barrel-shaped, one or two on each plant; **sporophylls** 3–4 cm x 4–5.5 cm, obovate, with an erect or spreading, spine-like appendage 0.1–2.5 cm long; **peduncle** 10–15 cm x 2 cm. **Seeds** 2.3–2.8 cm x 1.8–

2.2 cm, ovoid to oblong, orange to red.
**Distribution and Habitat:** Endemic to Australia where it occurs in semi-inland regions of northern New South Wales and southern Queensland. A very drought-tolerant species which grows on rocky slopes in open forest and in gullies and semi-protected sites often under or close to black cypress pine, *Callitris endlicheri*. Soils are mainly gravels or sandy loams. The climate is hot in summer with a day/night temperature regime to 34°C and 14°C respectively. Winters are cold with a day/night temperature regime to 18°C and 2°C respectively, with heavy frosts being frequent in some areas. Rainfall is about 600 mm per annum, but is irregular, and falls mainly in spring and summer.
**Notes:** *M. plurinervia* was described as a subspecies of *M. pauli-guilielmi* in 1959 and raised to specific rank in 1991. It can be distinguished from *M. pauli-guilielmi* and also *M. flexuosa* by its broader petiole and rhachis and its much broader, thicker textured, grey-green to glaucous leaflets and its glaucous cones. *M. plurinervia* is a tough, drought-tolerant species which grows mainly in inland districts which have a low and irregular rainfall. Although widespread its occurrences are mostly small, disjunct and localised. Coning in this species is erratic and appears to take place about every four or five years.
**Cultivation:** Suited to temperate and subtropical regions, although this species dislikes the humid conditions of the coast. Best grown in full sun but in inland districts plants may require filtered sun for best appearance. Very sensitive to root rot and requires excellent drainage. Tolerates heavy frosts.
**Propagation:** From seed.

## *Macrozamia riedlei*

(Fisch. ex Gaudich.) C. A. Gardn.
(after Anselme Riedle, eighteenth century gardener at the Jardin Botanique in Paris and chief gardener on the French explorer Baudin's expedition)

**Description:** A medium-sized cycad with a subterranean or emergent, barrel-shaped trunk to 0.5 m tall and 60 cm across. **Young leaves** bright green or glaucous. **Mature leaves** few to numerous in an obliquely erect to spreading crown, 1–1.5 m x 20–30 cm, dark green or glaucous, flat in cross-section, arching to gracefully curved in profile; **petiole** 12–25 cm long, flat above with a central ridge, convex and angular beneath, green to brown, swollen and woolly at the base; rhachis not twisted, flat above, more or less ridged beneath; **leaflets** 100–120 per leaf, dull green above, paler beneath, those in the proximal third widely spaced, rest crowded, the lower leaflets more widely spaced than the rest, inserted on the rhachis at about 40°, pungent-tipped, the lower leaflets gradually reduced but not spine-like; **median leaflets** 15–25 cm x 0.8–1 cm, linear to narrowly linear-lanceolate, straight, tapered to a drawn out, pungent apex, contracted to the cream or reddish callous base. **Cones** dissimilar. **Male cones** 20–35 cm x 8–10 cm, broadly cylindrical, curved, one to three on each plant; **sporophylls** 5–7 cm x 1.8–2.3 cm, narrowly wedge-shaped, with an erect, straw-coloured, apical, spine-like appendage 0.5–4 cm long; **peduncle** 10–18 cm long. **Female cones** 15–30 cm x 15–20 cm, ovoid, green to glaucous, with pink areas on the **sporophylls**, usually solitary; **sporophylls** 4–6 cm x 3–5 cm, broadly wedge-shaped, with an erect, apical spine-like appendage 1–5 cm long; **peduncle** 10–15 cm x 3–4 cm. **Seeds** 2.5 cm x 2–2.5 cm, oblong, red.
**Distribution and Habitat:** Endemic to Australia where widely distributed in Western Australia from north of Perth south to near Albany. The species grows in Jarrah forests and woodland and in low scrub. Soils range from sands and sandy clays to laterites. The climate is temperate having hot, dry summers with a day/night temperature regime to 30°C and 13°C respectively. Winters are cool to cold with a day/night temperature regime to 15°C and 5°C respectively. Rainfall ranges from 250 mm to 1000 mm per annum, falling mainly in winter.
**Notes:** *M. riedlei* was first described in the genus *Cycas* in 1826 and transferred to the genus *Macrozamia* in 1930. The type material was collected from the vicinity of King George Sound near Albany. It is a variable species to which a number of specific epithets have been applied, including *M. fraseri* Miq., *M. preissii* Lehm. and *M. oldfieldii* (Miq.) A. DC. A systematic study is required to determine if some of these epithets can be applied to any of the variants. *M. riedlei* rarely has an emergent trunk and if so this is no more than 0.5 m tall and plants usually have less than twenty-five leaves in the crown. Two robust species which have been wrongly included

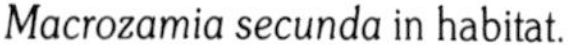

*Macrozamia secunda* in habitat.

Leaf of *Macrozamia secunda* showing colourful callus bases.

with *M. riedlei* are *M. dyeri* and *M.* species 'Eneabba'.
**Cultivation:** Suited to temperate regions. Grows well in full sun or filtered sun. Requires good drainage and is sensitive to root-rotting fungi. Tolerates moderate frosts.
**Propagation:** From seed.

## *Macrozamia secunda*

C. Moore
(one-sided — the leaflets are all on one side of the rhachis)

**Description:** A small cycad with an ovoid, subterranean trunk to about 20 cm long and 15 cm across. **Young leaves** steely bluish green. **Mature leaves** usually one to four in a sparse crown (but puzzling variants in some areas have up to ten leaves), 0.6–0.9 m x 8–12 cm, stiff, widely spreading and usually nearly prostrate, sharply recurved near the apex, leaflets stiffly erect, mostly dark green or dark, gunmetal bluish green; **petiole** 10–20 cm long, swollen, orange and woolly at the base; rhachis usually straight but sometimes twisted; **leaflets** 80–170 on each leaf, crowded, linear, held stiffly erect, leathery, acuminate to mucronate, dark green or dark gunmetal bluish green above, paler and glaucous beneath; **median leaflets** 6–20 cm x 0.4–0.8 cm, contracted to a pale red or orange callous base, the colours prominent on young leaves. **Cones** markedly dissimilar. **Male cones** 15–20 cm x 1.5–2 cm, cylindrical, greenish, one or two on each plant; **sporophylls** 1.5–2.5 cm long, wedge-shaped, with an apical spine-like appendage to 1 cm long. **Female cones** 15–25 cm x 7–9 cm, cylindrical to ovoid, usually solitary; **sporophylls** 3–4 cm long, with an erect spine-like appendage to 3.3 cm long; **peduncle** 15–20 cm long. **Seeds** 2–3 cm x 1.5–2.2 cm, ovoid, bright salmon pink to red.
**Distribution and Habitat:** Endemic to Australia where it occurs in central-western areas of New South Wales. This species commonly grows in stunted sclerophyll forest on stony and rocky slopes and ridges in low ranges. Less commonly plants are found in flat areas growing in red sandy soils. The climate is temperate having hot, dry summers with a day/night temperature regime to 30°C and 15°C respectively. Winters are cool to cold with a day/night temperature regime to 15°C and 0°C, with heavy frosts being common in some areas. The rainfall ranges from 500 mm to 1000 mm per annum, distributed evenly throughout the year.
**Notes:** Of the species of *Macrozamia* which grow on the inland ranges west of the Dividing Range, this species has the most southerly extension of range. It can be readily recognised by its fronds which are held nearly horizontal, sharply recurved near the apex with simple, often gunmetal-coloured leaflets both rows of which are held stiffly erect on one side of the rhachis. Plants usually have one to four fronds but occasional bushy specimens with about ten fronds are encountered. Recently mature fronds are an attractive dark steely blue-green with prominent pink, red or orange callous bases on the leaflets. Leaflets of this species are often severely damaged by predators and appear papery. Although the soils where this species grows are sandy, they often have a clayey sub-layer which can become spewy after rain. Observations made by the

author on several colonies over many years suggest that plants produce cones very infrequently. *M. secunda* hybridises sporadically with *M. communis* in areas where the two species grow together.
**Cultivation:** Plants of this species present an attractive appearance if given suitable conditions. They are best when exposed to full sun for a substantial part of the day. Drought-tolerant and frost-hardy. A good container plant.
**Propagation:** From seeds which are rarely available.

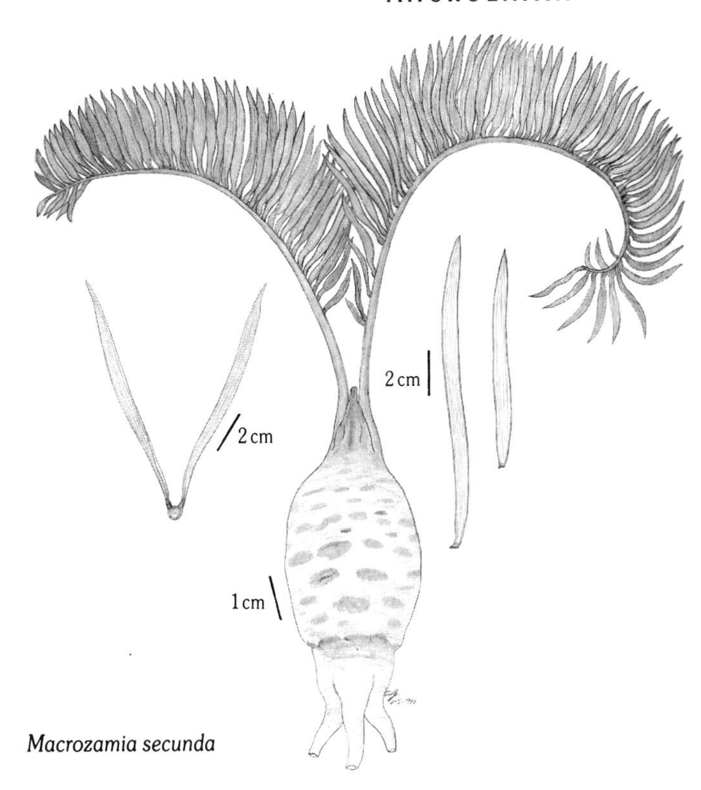

*Macrozamia secunda*

## *Macrozamia spiralis*

(Salisb.) Miq.
(in a spiral, in reference to the rhachis)

**Description:** A small to medium-sized cycad with a subterranean trunk to 30 cm long and 20 cm across. **Young leaves** greyish blue with pink callous bases. **Mature leaves** two to twelve in an obliquely erect to spreading crown, 0.6–1 m x 20–40 cm, dark green, more or less flat in cross-section but sometimes appearing whorled because of the twisted rhachis, straight or curved in profile; **petiole** 10–40 cm long, swollen and woolly at the base; rhachis varying from nearly straight to spirally twisted, concave above, angular beneath; **leaflets** 40–120 on each leaf, spreading widely and recurved or lax at the apex, moderately crowded, twisted at the base and facing forwards, more crowded towards the apex, less crowded towards the base, inserted on the rhachis at about 90°, those towards the apex at a more acute angle, the lower leaflets not reduced and not spine-like; **median leaflets** 12–20 cm x 0.5–1 cm, linear, dark green or slightly glaucous, dull, abruptly tapered to a pungent apex, contracted to a red, pinkish or yellowish callous base. **Cones** markedly dissimilar. **Male cones** 15–20 cm x 5–6 cm, cylindrical, usually curved, green, one to four on each plant; **sporophylls** 2–3 cm x 1.5–2 cm, obovate, with an apical, spine-like appendage 0.2–1.5 cm long; **peduncle** 8–15 cm x 2–3 cm. **Female cones** 12–20 cm x 6–9 cm, ovoid, green to glaucous, one or two per plant; **sporophylls** 3–4 cm x 4–5.5 cm, obovoid, with an apical spine 0.2–3 cm long; **peduncle** 15–20 cm x 2–3 cm. **Seeds** 2.5–3 cm x 2–2.5 cm, obovoid to ovoid to oblong, orange to bright red.
**Distribution and Habitat:** Endemic to Australia where it occurs in New South Wales on the central coast and adjacent tableland and ranges. This species grows in open forest usually in sandy soils. The climate is temperate having hot summers with a day/night temperature regime to 30°C and 13°C respectively. Winters are cold with a day/night temperature regime to 14°C and 0°C, with frosts occurring in some areas. Rainfall is about 1300 mm per annum, distributed more or less evenly throughout the year.
**Notes:** *M. spiralis* was originally described in the genus *Zamia* in 1796 and transferred to *Macrozamia* in 1842. The correct application of the epithet '*spiralis*' has been subject to considerable confusion in both New South Wales and Queensland and was cleared up by Lawrence Johnson in 1959. The species has also been known as *M. corallipes* J. D. Hook. *M. spiralis* grows in loose, scattered groups with the individuals often being well separated from their neighbours although sometimes up to ten or more plants may be found grouped together in a small area. Plants are frequently depauperate and in harsh localities may consist of one or two leaves only. These contrast markedly with vigorous or robust plants found in some localities and which have a well-developed crown of leaves. Coning is very sporadic in this species and may only occur every four to six years.
**Cultivation:** Suited to temperate and subtropical regions. Plants will grow in full sun or filtered shade and also grow well in containers. Tolerates moderate to heavy frosts. Seedlings are generally very slow growing.
**Propagation:** From seed.

## *Macrozamia stenomera*

L. Johnson
(with narrow divisions)

**Description:** A small cycad with an ovoid, subterranean trunk to 20 cm long and 15 cm across. **Young leaves** powdery blue. **Mature leaves** two to ten in an erect to

*Macrozamia spiralis* (illustrated as *M. corallipes*) From the *Botanical Magazine* 98, tab. 5943, (1872)

spreading crown, 0.4–0.8 m x 10–15 cm, glaucous to steely blue, shallowly to deeply vee-ed in cross-section, straight or irregularly curved in profile; **petiole** 5–15 cm long, swollen and woolly at the base; rhachis variable, from straight to spirally twisted especially near the end; **leaflets** 70–120 on each leaf, densely crowded and overlapping, those towards the base more widely spaced, twisted at the base and facing forwards, the lower ones inserted on the rhachis at about 70°, the upper ones more acute, most of the leaflets forked one to four times into very narrow, segments, the angle between the segments acute to wide, the lower leaflets not reduced and not spine-like; **median leaflets** 10–20 cm long, the segments 0.1–0.4 cm across, linear, stiff to rigid, straight or irregularly curved, abruptly rounded and mucronate, contracted to a yellowish or orange, callous base. **Cones**

dissimilar. **Male cones** 10–18 cm x 4–6 cm, cylindrical, erect, straight or more usually curved, green to glaucous, one to three on each plant; **sporophylls** 1.5–2 cm x 1.5–2 cm, wedge-shaped, with an erect, glaucous, apical, spine-like appendage 0.2–1.5 cm long; **peduncle** 6–8 cm x 2 cm. **Female cones** 10–20 cm x 8–10 cm, ovoid to ellipsoid, green to glaucous, usually solitary on each plant; **sporophylls** 3–4 cm x 4–7 cm, obovate, with an erect, apical, glaucous, spine-like appendage 0–2.5 cm long. **Seeds** 2–3 cm x 1.5–2 cm, ovoid to oblong, bright red.
**Distribution and Habitat:** Endemic to Australia where it occurs in northern and north-western New South Wales. This species grows on low to high rocky hills, usually on sheltered slopes among low scrub but sometimes close to streams in tall forest. The climate is temperate with hot summers having a day/night temperature regime to 34°C and 19°C respectively. Winters are cold with a day/night temperature regime to 18°C and 3°C respectively, with heavy frosts being common in some areas. Rainfall ranges from 500 mm to 1000 mm per annum, falling mainly in spring and summer.
In some areas misty rain and fogs are frequent.
**Notes:** *M. stenomera* was described in 1959 from material collected in the Nandewar Ranges near Narrabri in north-western New South Wales. It is regarded as being a rare species and although it is relatively widespread its occurrences are disjunct and localised. This species has by far the most complexly divided leaflets of any species in the genus and these impart a distinctive crowded appearance to the leaves. Plants do not exhibit much variation in their main features. Coning in this species is erratic and seems to take place about every three to five years.
**Cultivation:** Best suited to temperate regions but plants have been cultivated also in subtropical regions. One of the most attractive species in the genus. Will grow in full sun or filtered sun but best appearance is achieved in bright light. Requires excellent drainage and tolerates heavy frosts.
**Propagation:** From seed.

# *Macrozamia* sp. 'Enneabba'

**Description:** A large cycad with an emergent, barrel-shaped trunk to 5 m tall and 1 m across, usually blackened by fire. **Young leaves** bright green. **Mature leaves** numerous in a broad, rounded crown, dark green, 1.5–2 m x 20–30 cm, flat in cross-section, straight to arching in profile; **petiole** 10–30 cm long, shallowly domed above with a central ridge, broadly ridged beneath, brownish green, swollen and woolly at the base; rhachis not twisted, ridged above, broadly ridged beneath; **leaflets** 100–150 per leaf, dull green above, paler beneath, those in the proximal third widely spaced, median and distal

Leaf of *Macrozamia stenomera* showing deeply divided leaflets.

leaflets crowded to very crowded, overlapping like the slats of a venetian blind, inserted on the rhachis at about 35°, lower leaflets reduced, basal pair sometimes spine-like; **median leaflets** 15–20 cm x 7–10 mm, narrowly linear-lanceolate, tapered to a long, drawn-out apex, contracted to a whitish, raised callous base. **Cones** dissimilar. **Male cones** 40–80 cm x 10–12 cm, broadly cylindrical, curved, one to eight on each plant; **sporophylls** 4–7 cm x 2–2.5 cm, narrowly wedge-shaped, with an erect, straw-coloured, apical, spine-like appendage 0.5–6 cm long; **peduncle** 25–30 cm long. **Female cones** 30–50 cm x 15–20 cm, yellowish green, one to five on each plant; **sporophylls** 8–10 cm x 4–6 cm, broadly wedge-shaped, with an erect, apical, spine-like appendage 2–6 cm long; **peduncle** 15–20 cm long. **Seeds** 4–4.5 cm x 2.5–2.8 cm, obovoid, angular, red.
**Distribution and Habitat:** Endemic to Australia where widely distributed in western parts of Western Australia between Eneabba and Gin Gin and perhaps also extending to the northern suburbs of Perth. The species grows in sparse woodland and in heathland. Soils are grey to

*Macrozamia* species 'Southern Pilliga'.

*Macrozamia* species 'Northern Pilliga' in habitat.

white sands. The climate is warm temperate having hot, dry summers with a day/night temperature regime to 29°C and 18°C respectively. Winters are cool to cold with a day/night temperature regime to 19°C and 9°C respectively. The rainfall ranges from 600 mm to 1000 mm per annum, falling mainly in winter.

**Notes:** This species has been included under *M. riedlei*, however, it can be readily distinguished by its massive, emergent trunk, numerous leaves arranged in a rounded crown, the very large cones and seeds and the very crowded leaflets arranged like the slats of a venetian blind. Large, well-grown specimens are very reminiscent of a miniature date palm.

**Cultivation:** Suited to temperate and subtropical regions. Best in filtered sun or full sun. Requires excellent drainage and is very sensitive to root-rotting fungi. Tolerates light to moderate frosts.

**Propagation:** From seed.

## *Macrozamia* sp. 'Northern Pilliga'

**Description:** A small to medium-sized cycad with an ovoid, subterranean trunk to 30 cm long and 35 cm

across. **Young leaves** bright blue to powdery blue. **Mature leaves** five to twenty-five in a spreading, bushy crown, blue, ageing to blue-green, 0.6–1.1 m x 20–35 cm, shallowly vee-ed in cross-section, straight or shallowly curved in profile, usually recurved near the apex; **petiole** 10–25 cm long, swollen and woolly at the base; rhachis not twisted or twisted just near the apex; **leaflets** 80–120 on each leaf, crowded, twisted at the base and facing forwards, inserted on the rhachis at about 70°, about one third of the leaflets entire, the rest forked once or twice into narrow, rigid segments, the lower leaflets not reduced and not spine-like; **median leaflets** and lobes 10–22 cm x 0.2–0.7 cm, linear, apex mucronate, contracted to a pink to red callous base. **Cones** dissimilar. **Male cones** 20–30 cm x 5–6.5 cm, cylindrical, erect, curved, one to three on each plant; **sporophylls** 2–2.4 cm x 1.6–1.9 cm, wedge-shaped, with an erect, apical, spine-like appendage to 1 cm long; **peduncle** 15–20 cm x 2.5–3.5 cm. **Female cones** 16–28 cm x 10–12.4 cm, ovoid, glaucous; **sporophylls** 3–5 cm x 5–7 cm, broadly wedge-shaped, with an erect, apical, spine-like appendage to 2.5 cm long; **peduncle** 15–20 cm x 1.5–3.5 cm. **Seeds** 2.5–3.5 cm x 2–2.5 cm, ovoid to oblong, red to scarlet.
**Distribution and Habitat:** Endemic to Australia where it is distributed in the northern areas of the Pilliga Scrub to the south of Narrabri in northern New South Wales. This species grows in stunted forest which has a well-developed shrub layer. Soils are red sands. The climate is temperate having hot, dry summers with a day/night temperature regime to 34°C and 19°C respectively. Winters are cold with a day/night temperature regime to 18°C and 0°C, with heavy frosts being common. Rainfall ranges from 550 mm to 650 mm per annum (but irregular) and falls mainly in spring and summer.
**Notes:** This species, which has been confused with *M. heteromera*, can be distinguished by its more robust habit (up to twenty-five leaves in the crown), blue to blue-green leaves and larger cones. Its seedlings have entire **leaflets** in a secund arrangement. Coning occurs regularly every three to five years.
**Cultivation:** Suited to temperate and inland regions with a semi-arid climate. Requires a sunny position and excellent drainage. Plants resent shade, excess humidity and long periods of drizzly weather. Tolerates heavy frosts.
**Propagation:** From seed.

## *Macrozamia* sp. 'Southern Pilliga'

**Description:** A small cycad with an ovoid, subterranean trunk to 30 cm long and 30 cm across. **Young leaves** bright green. **Mature leaves** two to eleven in a sparse, erect to spreading crown, 0.5–0.9 m x 15–30 cm, dull green, flat in cross-section, strongly recurved near the apex; **petiole** 10–24 cm long, swollen and woolly at the base; rhachis not twisted; **leaflets** 60–120 on each leaf, crowded, those towards the base more widely spaced, twisted at the base and facing forwards, inserted on the rhachis at about 60°, most of the leaflets entire but on some plants (or sometimes restricted to odd leaves on a plant) most of the leaflets forked once or twice into narrow, rigid segments, the lower leaflets not reduced and not spine-like; **median leaflets** and lobes 10–20 cm x 0.3–0.8 cm, linear, apex mucronate or bifurcate, contracted to a greenish cream to cream callous base. **Cones** dissimilar. **Male cones** 16–24 cm x 4–5 cm, cylindrical, erect, curved, one or two on each plant; **sporophylls** 2–2.4 cm x 1.5–1.8 cm, wedge-shaped, with an erect apical spine-like appendage to 0.5 cm long; **peduncle** 12–18 cm x 1.8–2.8 cm. **Female cones** 12–20 cm x 8–10 cm, ovoid, green; **sporophylls** 3–4.5 cm x 3.5–5 cm, broadly obcordate, with an erect, apical, spine-like appendage to 1.5 cm long; **peduncle** 12–18 cm x 1.8–2.8 cm. **Seeds** 2–2.5 cm x 1.8–2 cm, ovoid, red to scarlet.
**Distribution and Habitat:** Endemic to Australia where it is distributed in the southern areas of the Pilliga Scrub between Coonabarabran and Gunnedah in northern New South Wales. This species commonly grows in flat areas or on gentle slopes in open forest. Soils are yellow to brownish sands and sandy clay loams. The climate is temperate having hot, dry summers with a day/night temperature regime to 32°C and 15°C respectively. Winters are cold with a day/night temperature regime to 16°C and 0°C, with heavy frosts being common. The rainfall is about 700 mm per annum (but irregular) and falls mainly in summer.
**Notes:** This species, which has been confused with *M. heteromera*, can be distinguished by the frond architecture and seedlings. The majority of plants have leaves with undivided leaflets (rare in *M. heteromera*) and its seedlings have entire leaflets (divided in *M. heteromera*). Coning is extremely erratic in this species and may only take place every four to eight years. Fires are a regular occurrence where this species grows but do not influence cone production. Rare sporadic hybrids may occur with *M. diplomera* where the two species grow in close proximity.
**Cultivation:** Suited to temperate regions and to a lesser extent drier subtropical regions. Plants can be grown in full sun, semi-shade or filtered sun and require excellent drainage. Tolerates heavy frosts.
**Propagation:** From seed.

GENUS

# *Microcycas*

The genus *Microcycas* is monotypic, with the solitary species endemic to Cuba. Although originally placed in the genus *Zamia*, it was later recognised as being distinct and a separate genus was created to accommodate it.

DISTRIBUTION OF *MICROCYCAS*

**Derivation:** The genus *Microcycas* was described by Alphonse Louis Pierre Pyramus de Candolle in 1868. It is derived from the Greek *micros*, 'small', and the generic name *Cycas*. The generic name is most inappropriate, for this species is one of the tallest of all cycads. It was coined from a few small leaves which was all the material available to the author at the time.

**Generic Features:** *Microcycas* is a terrestrial cycad with a tall, emergent trunk which can branch. Its pinnate leaves are produced in flushes and the leaflets are articulate at the base and reflexed. The median and distal leaflets of its leaves are of similar length, imparting the false impression that the leaves have been cut off near the apex. The petioles and rhachises are unarmed, completely lacking any prickles or spines. The cones of both sexes are relatively slender but the males are narrower than the females. The sporophylls of both sexes lack any projections or horn-like ornamentation.

**Notable Generic Features:** *Microcycas* is closely related to *Zamia* and can be distinguished by the reflexed leaflets on the leaves and the truncated leaves. Other useful features include:

- leaflets articulate at the base;
- petioles and rhachises lack prickles; and
- very numerous spermatozoids and archegonia.

**Recent Studies:** There have been no recent studies on *Microcycas calocoma*.

**Cultivation:** *M. calocoma* is one of the most ornamental of the cycads, being prized for its graceful, rounded crown of glossy, bright green leaves which have reflexed leaflets imparting a drooping appearance. It is grown on a limited scale in a few public gardens and is also present in private collections. Its popularity has been limited by a lack of growing material but over recent years many hundreds of seedlings have been made available from Fairchild Tropical Garden in Florida. These have been produced by hand pollinating cones which have developed on mature plants in this collection. Many of these seedlings have been distributed to botanic gardens and some have been sold to private individuals. Unfortunately



Mature plant of *Microcycas calocoma*.

Young plant of *Microcycas calocoma* - note reflexed leaflets and truncated leaf tips.

these mature plants, which produced so many significant offspring, were killed by a lightning strike in 1991.

It has been noted that the seedlings of this cycad may be susceptible to root-rotting fungi if they are overwatered or if drainage is inadequate.

*M. calocoma* grows readily in tropical and subtropical regions. Plants will grow in full sun or filtered sun and respond well to mulches, watering during dry periods and regular light fertiliser applications. They are very sensitive to damage from cold weather.

**Propagation:** From seed which is best sown while fresh. The seeds have an after-ripening period of one or two months and take six to eighteen months to germinate.

## *Microcycas calocoma*

(Miq.) A. DC.
(with beautiful hair)

**Description:** A large cycad with an erect trunk to 10 m tall and 60 cm across, lacking basal suckers but with sporadic offsets produced on the trunk and arising as branches. **Young leaves** light green, covered with fine hairs. **Mature leaves** 6–40 in an attractive, spreading to rounded crown, 0.6–1 m x 20–25 cm, inversely vee-ed in cross-section, straight or gracefully curved in profile, dark green, cut-off abruptly at the apex; **petiole** 8–10 cm long, terete in cross-section, swollen at the base, lacking any prickles; **leaflets** 100–160 on each leaf, 15–25 cm x 0.8–1 cm, linear-lanceolate, light to dark green, shiny, straight or slightly falcate, leathery, crowded, evenly spaced throughout, margins lacking spines and slightly revolute, apex blunt. **Cones** markedly dissimilar. **Male cones** 25–30 cm x 5–8 cm, cylindrical, yellowish brown, densely hairy; **sporophylls** 2–2.5 cm x 1.5–2 cm, prominently ridged; **peduncle** 2–4 cm long, densely hairy. **Female cones** 50–90 cm x 13–16 cm, broadly cylindrical, tapered from the base, yellowish brown, densely hairy; **sporophylls** 4.5–5.5 cm x 3–4 cm, with two or four ridges; **peduncle** 2–4 cm long, hairy. **Seeds** radiospermic, 3.5–4 cm x 2–2.5 cm, the sarcotesta, pink to red.

**Distribution and Habitat:** Endemic to western Cuba where it grows in small groups in mountainous regions at about 200 m altitude. The species grows in open situations or in slightly shaded deciduous and evergreen forests. Soils may be skeletal limestone (alkaline) or siliceous clays (acid).

**Notes:** This species was described first in the genus *Zamia* in 1851 and transferred to *Microcycas* in 1868. It is believed by some to be the most primitive of all cycads having many sperm cells instead of two and a prominently lobed female gametophyte which can produce in excess of 200 archegonia. *Microcycas calocoma* is listed as being endangered in the *Red Data Book* of the International Union for Conservation of Nature and Natural Resources (IUCN). The world population is about 600 plants and these occur in small, localised groups in which the sex ratio is very unbalanced. No information is available as to regeneration within these groups. The plants are apparently not exploited by collectors, in fact they are completely protected and the populations are monitored on a regular basis. The roots are poisonous and have been used by the locals to kill rats.

**Cultivation and Propagation:** As for the genus.

GENUS

# *Stangeria*

The genus *Stangeria* is monotypic, with the solitary species being endemic to South Africa. In all, four specific epithets have been used within the genus *Stangeria*; however, studies have shown that they all apply to the one variable species. The correct name for this species is *S. eriopus*, this combination being made by Henri Baillon in 1892 and subsequently (although illegally), again by George Nash in 1909. Because of botanical rules, the first application of the name is correct.

When first collected this unique cycad, because of its folded young leaves, frond-like mature leaves and venation, was believed to be a fern. The original specimens were sterile and the species was placed in the fern genus *Lomaria*. Initially the species was identified as *L. coriacea* Schrader but in 1829 was described by Kunze as the new species, *L. eriopus*. A living specimen collected by Stanger in 1851 later coned in the botanic garden in Chelsea, England, and the incorrect generic placement of the species became apparent.

**Derivation:** The genus *Stangeria* was described by Thomas Moore in 1853. It honours Dr Max Stanger, nineteenth century Surveyor-General of Natal, South Africa.

**Generic Features:** *Stangeria* is a terrestrial cycad with a naked subterranean trunk which can branch and has fleshy, tuberous roots. Its leaves are pinnate and fern-like (being folded when young and when mature are remarkably like the fern *Oleandra distenta*). They are produced singly and upon the death of a leaf, the petiole and rhachis are shed and the leaf base acts like a cataphyll until it is eventually shed. Its leaflets have a venation which is nearly unique in cycads but common in ferns. This consists of a prominent raised midrib (comprised of several parallel veins) and free, forked lateral veins which are attached at right angles to the midrib and radiate to the leaflet margins (species of *Chigua* also have a midrib and forked lateral veins but these arise at an acute angle and then run nearly longitudinally). Cones are formed at the apex of a stem (i.e. they terminate that stem), but a new meristematic region develops below the base of the peduncle and, after cone maturation, the new stem continues growth until the next coning commences.

**Notable Generic Features:** *Stangeria* is readily recognised by its pinnate, fern-like leaves which have a prominent midrib (comprised of several parallel veins) and forked lateral veins. Other features useful for identification include:

- leaves produced one at a time;
- leaf bases not retained at senescence;
- cones terminate a stem;
- cataphylls absent on vegetative shoots;
- leaflets non-articulate, base decurrent;
- stipules large, fleshy, with vascular traces;
- young leaflets inflexed or folded;
- seeds attached below the sporophyll stalk; and
- sporophylls upwardly imbricate, especially visible in the female.

**Recent Studies:** An excellent, wide-ranging paper written by Piet and Elsa Vorster is presented on *Stangeria eriopus* in *Encephalartos* 2 (1986): 8–17.

DISTRIBUTION OF *STANGERIA*

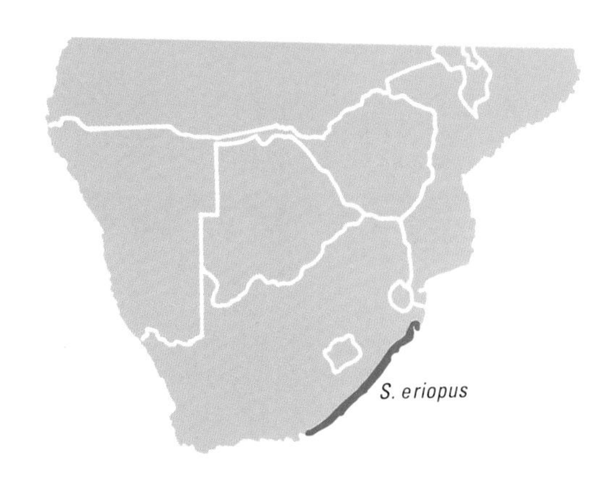

**Cultivation:** *S. eriopus* is popular with cycad collectors and because of its unique botanical status is also found in the collections of scientific institutes around the world. It is generally very easy to grow and adapts well to a range of climates and conditions, although best growth occurs in subtropical and warm temperate regions. Plants will grow in full sun; however, those in sheltered positions develop luxuriant growth and have a much better appearance. They are able to tolerate light frosts without much damage but should be protected from moderate to heavy frosts.

Soils should be acidic and well drained with sands and sandy loams producing good growth. Surface mulches are beneficial and plants should be watered regularly during dry periods to maintain a good appearance. Light applications of a complete fertiliser over the summer months promote healthy growth. Some growers prefer slow-release types; for sustained release.

Plants of *Stangeria* can be successfully grown in containers and make an interesting addition to a pot collection. The containers should be about 30 cm across and of a similar depth to allow for root development and the eventually substantial underground stem (which should always remain buried).

**Propagation:** *Stangeria* is mainly propagated from seed but can also be divided. The seeds are best sown fresh and take about twelve months to germinate. They germinate best under cool, humid conditions. To avoid damage to the young root, seedlings should be potted individually as soon as the radicle appears; this usually happens a month or two before the first leaf. Clumps can be divided successfully but particular attention must be paid to hygiene to prevent infections entering through wounds. Stems can be cut into fairly small pieces and each segment will produce growth. The best time to divide is in spring or at the onset of warm weather when the young divisions can establish readily.

## *Stangeria eriopus*

(Kunze) Baillon
(with woolly stalks, in reference to the petiole base)

**Description:** A small cycad which has a completely subterranean, carrot-shaped stem to 20 cm across. The stem can branch freely into many growing points and has swollen, fleshy almost tuberous roots. **Young leaves** erect, the leaflets folded, covered with grey, velvety hairs. **Mature leaves** 0.3–2 m long, pinnate, fern-like, erect or arching, dark green, glabrous; **petiole** 10–80 cm long, swollen and hairy at the base, smooth and lacking prickles; **leaflets** ten to forty on each leaf, 7–50 cm x 1.4–6 cm, linear-lanceolate to lanceolate, thin-textured to slightly leathery, proximal leaflets stalked, distal ones fused and decurrent on the rhachis, margins entire or sometimes revolute, deeply lobed, prominently toothed or wavy, apex acute but readily broken off. **Male cones** 10–25 cm x 3–4 cm, narrowly cylindrical, solitary on each growing point, covered with short silver hairs when young, yellowish brown when mature; **sporophylls** about 12 mm across, triangular to rhomboid, lacking prickles; **peduncle** 5–25 cm long, hairy when young. **Female cones** 8–35 cm x 6–10 cm, ovoid to ellipsoid, covered with silvery hairs when young, dark green and less hairy at maturity; **sporophylls** to 6 cm across, slightly convex,

*Stangeria eriopus.*

overlapping; **peduncle** 5–25 cm long, hairy. **Seeds** radiospermic, 3–3.5 cm x 2–2.5 cm, ovoid to obovoid, the sarcotesta purple-red when ripe.

**Distribution and Habitat:** Endemic to the east coast of South Africa where distributed in a relatively narrow strip between the latitudes of 27°S and 33°S. The species occurs in coastal and near-coastal habitats within 50 km of the sea and extending to about 3 km of the landward side of the coastal dunes. Plants are only found where protected from coastal salt spray and grow in such vegetation as light and dense evergreen forests and short grassland (sometimes in short grassland right next to the beach). Soils include sands derived from sandstone and granite and heavy black clay. Annual rainfall ranges from 1000 mm in the north to just below 750 mm in the south, distributed in a summer maximum and a winter dry. Frosts are rare.

**Notes:** *Stangeria eriopus* was first described in the fern genus *Lomaria* in 1836, emended in 1839 and later transferred to *Stangeria* in 1892. It is a variable cycad

*Stangeria eriopus*

Female cone of *Stangeria eriopus.*

Male cone of *Stangeria eriopus.*

with its morphology influenced dramatically by its habitat. Thus, plants from open grassland have erect, compact leaves with short, leathery toothed leaflets. Usually only one leaf is present per crown because of annual grassfires but the plants have larger, much more branched crowns probably as a result of these fires. By contrast plants from forested habitats often have only a single growing point but can have up to six leaves which are generally lax and with long, thin-textured, entire leaflets. Observations suggest that plants growing in grassland produce cones more regularly than do those which grow in the forests. This factor also influences branching since new growing points are initiated after cone development.

**Cultivation and Propagation:** As for the genus.

## *Stangeria paradoxa*

T. Moore

(a paradox; initially it was uncertain whether this plant was a fern or cycad)

This species, described in 1853 and used as the basis for the genus, is a nomenclatural synonym of *S. eriopus*. The name *eriopus* was validly published for the species although it was originally placed in the fern genus *Lomaria*. This usage was apparently overlooked by Thomas Moore when describing the genus, and then by George Nash, but was later corrected by Henri Baillon.

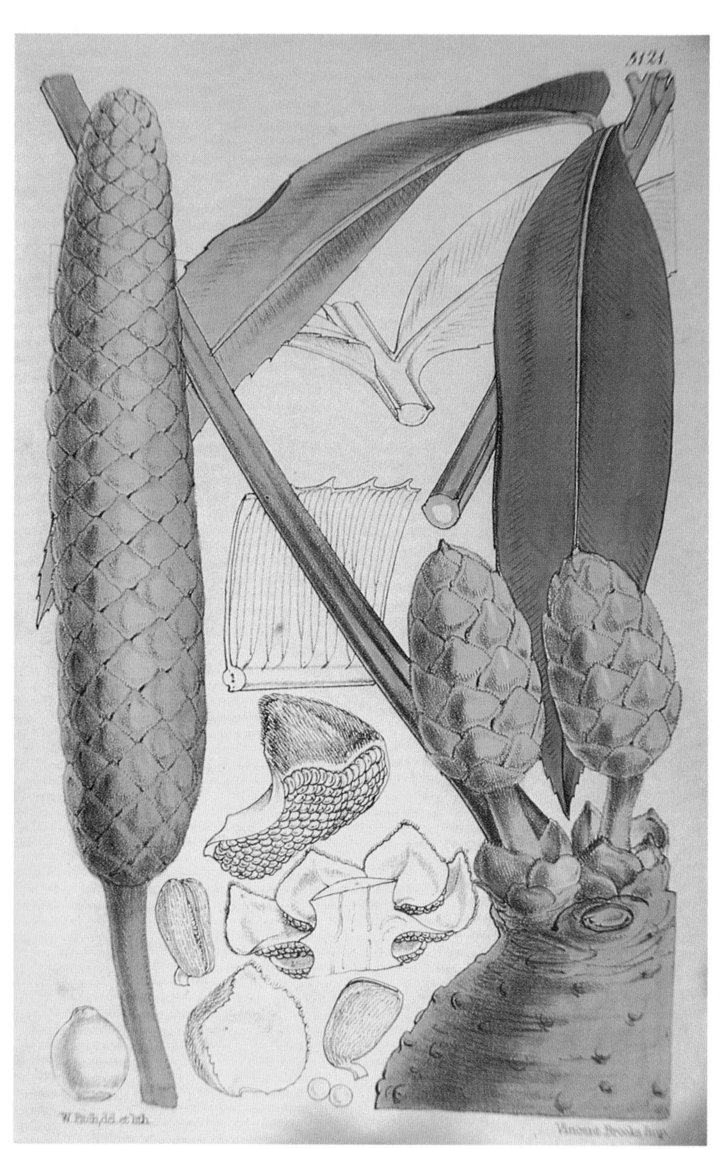


*Stangeria eriopus* From the *Botanical Magazine* 85, tab. 5121, (1859)

GENUS

# *Zamia*

The taxonomy of the genus *Zamia* is in a state of flux. Whereas a number of species have been very well defined by their authors or later researchers, others have been poorly studied and at present they are inadequately known. These species may have been described from inadequate material or the original collections have been lost. A number were named from isolated or inaccessible regions where today access may be difficult because of political barriers. At best the genus can be estimated to consist of fifty-five to sixty species, this figure including about twelve species which are currently undescribed. Name changes which affect cultivated taxa can be expected. For fossil species of *Zamia* see page 31.

**Derivation:** There is some dissent as to the derivation of the generic name *Zamia*, but it seems to be based on the Greek, *azaniae*, which refers to pine cones. The genus was described by Carl von Linnaeus in 1763.

**Generic Description:** Terrestrial or rarely epiphytic cycads with a tuber-like or cylindrical trunk that may be solitary or branched and either wholly subterranean, partly subterranean or wholly emergent. Leaf bases either falling free or remaining attached after senescence. New leaves emerging singly or in flushes, green or coloured, glabrous or hairy. Mature leaves pinnate, more or less oblong in outline, straight or recurved, flat or vee-ed in cross-section. Petiole swollen at the base with two inconspicuous stipules, glabrous or hairy, bearing prickles or unarmed. Rhachis straight, bearing prickles or unarmed. Leaflets articulate at the base, opposite to nearly opposite, evenly arranged along the rhachis, from extremely slender to broad and paddle-shaped, entire or with marginal spines or teeth or the latter reduced to bumps, margins revolute or involute; midrib absent, lateral veins either immersed and obscure or raised. Male cones one to many, cylindrical, erect or decumbent, very shortly hairy; sporophylls wedge-shaped, short-stalked. Female cones one to a few, ovoid to barrel-shaped, very shortly hairy, pedunculate; sporophylls peltate, stalked, the outer face hexagonal around a central facet. Seeds radiospermic, ovoid to subglobose, the sarcotesta yellow, brown, orange or red.

**Notable Generic Features:** Species of *Zamia* have leaflets which lack a midrib and more or less hexagonal sporophylls on the cones. Other useful features include:

- leaves pinnate, arising singly or in flushes;
- spinose prickles present or absent on the petiole and rhachis;
- presence of equally branched trichomes;
- leaflets articulate at the base; and
- sporophylls very shortly hairy, lacking spines, horns or any other ornamentation.

**Recent Studies:** Significant changes in the taxonomy of this genus are expected following taxonomic studies. A significant monograph on neotropical cycads will appear in *Flora Neotropica* in the near future. Illuminating insights into the genus can be obtained in the paper by Sergio Sabato entitled 'West Indian and South American Cycads', *Memoirs of the New York Botanical Garden* 57 (1990): 173–85.

**Habitat:** Species of *Zamia* grow in a wide range of habitats; however, many are either

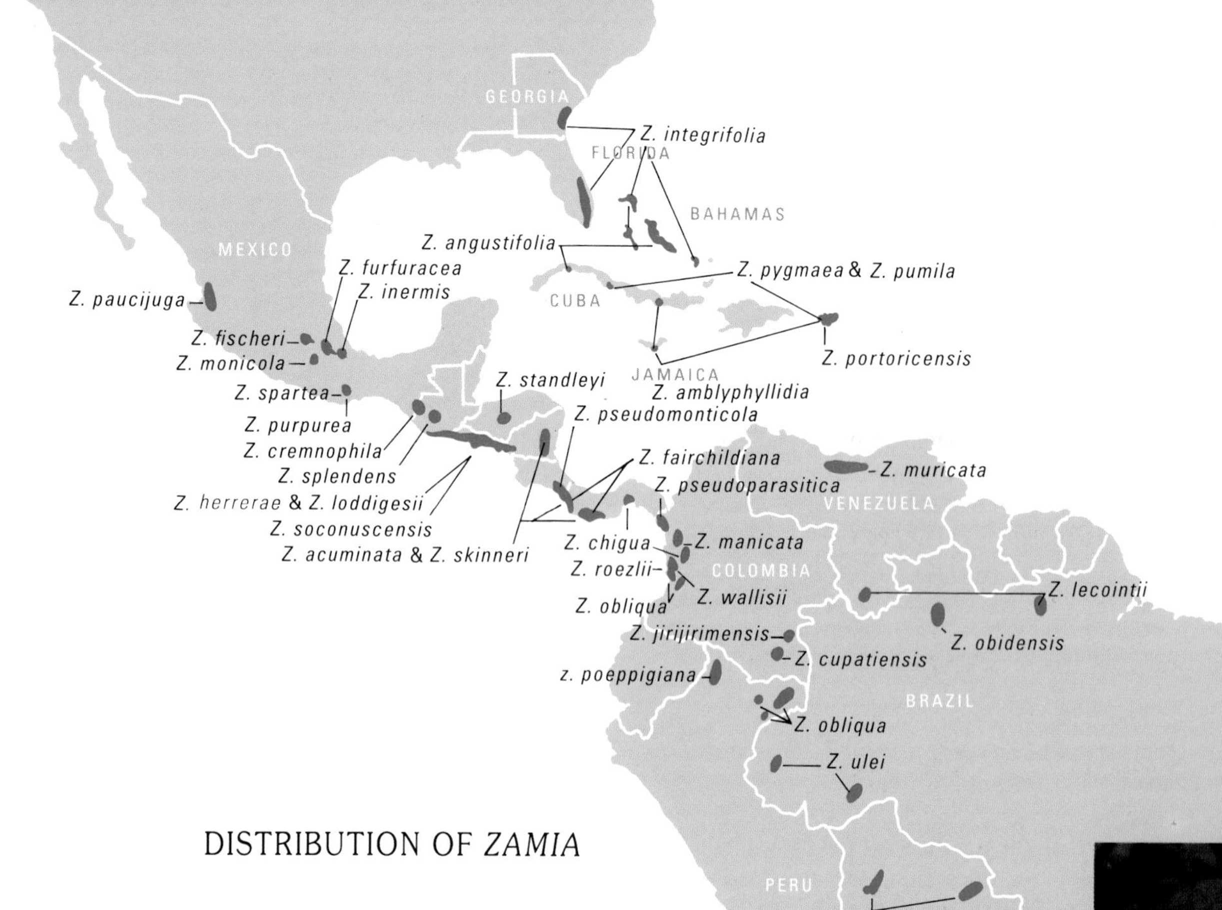


DISTRIBUTION OF *ZAMIA*

associated with rainforests or open habitats such as sparse forests or savannah grassland. The rainforest species commonly grow as understorey plants and may occur in the lowlands, at intermediate altitudes or in the mountains. In lowland regions hot, humid conditions are normal, the annual rainfall is high and the plants are usually sheltered, at least to some degree, from excessive hot sun. Those which occur in open habitats are exposed to much longer periods of hot sunshine, lower humidity and erratic rainfall with frequent dry spells. One remarkable species, *Z. roezlii*, grows in brackish mud adjacent to the mangrove zone and may withstand inundation by salt water at high tides.

**Cultivation:** As a general rule, species of *Zamia* have shown themselves to be easily cultivated in either pots or the ground with some species proving to be adaptable to a range of situations and conditions. Those species originating from open habitats will tolerate considerable exposure to bright sunshine even when small, and moderate fluctuations in humidity and temperature cause few problems. Species which occur naturally in rainforests, however, need more protection and are much less tolerant of low temperatures and low humidity. All species of *Zamia* need free and unimpeded air movement to avoid the problems caused by sweating and leaf fungi. Frost protection is advisable for all species but those originating from high altitudes are more tolerant of cold than those from the lowlands. Those from countries close to the tropics of course require extra protection from any cold periods.

Good soil drainage is essential for all species of *Zamia*, otherwise root-rotting fungi can cause significant problems. Slightly acid, humus-rich loams are excellent and good results can also be obtained in sands and sandy loams. Surface mulching and watering during dry periods are recommended. All species respond to the use of fertilisers over the summer months with regular light dressings being most beneficial.

**Propagation:** Those species of *Zamia* which produce suckers can be readily increased by division; however, propagation from seed is the most common method of increase. Seeds should be collected as they attain peak colouration and the cone begins to disintegrate. They have a short period of viability and should be sown while fresh or stored in moist peatmoss to prevent dehydration. The seeds of most species of *Zamia* have an after-ripening period of one to three months and untreated seed takes eight to eighteen months to germinate. Removal of the sarcotesta and filing, cutting or treating the hard outer coat with concentrated sulphuric acid reduces germination time to about three months. Treatment with gibberellic acid may reduce this time even further (see page 84).

## KEY TO SPECIES

Because of the confused nature of the genus no attempt has been made to prepare a key.

## *Zamia acuminata*

Oersted ex Dyer
(ending in a long, sharp point)

**Description:** A small cycad with an unbranched, cylindrical subterranean trunk to 50 cm long and 6 cm across. **Mature leaves** 0.5–0.7 m long, one to three in an erect sparse crown; **petiole** 20–60 cm long, densely prickly, somewhat angular; **leaflets** 20–30 cm x 1–3 cm, twelve to twenty on each leaf, elliptical to lanceolate, falcate, moderately spaced, inserted at right angles to the rhachis, dark green, margins entire, drawn into a long-acuminate, tail-like, sharply pointed, apex. **Male cones** 5–8 cm x 1–1.5 cm, cylindrical, cream to light brown. **Female cones** 10–20 cm x 5–8 cm, cylindrical to ovoid, cream to light brown. **Seeds** ovoid, red.
**Distribution and Habitat:** Occurring in Nicaragua, Panama and Costa Rica where it grows in rainforests. It ranges from near sea level in Nicaragua to elevations between 400 m and 1200 m in Panama and Costa Rica.
**Notes:** This poorly known species from Central America was described in 1884 from material collected in southern Nicaragua. The long, tail-like, sharply pointed tip of each leaflet is a useful diagnostic feature that is unique in the genus. Its trunk is subterranean as the result of stem and root contraction.
**Cultivation:** Suited to subtropical regions. This species probably needs warm, humid conditions, unimpeded air movement and light shade.
**Propagation:** Solely from fresh seed.

*Zamia acuminata.*

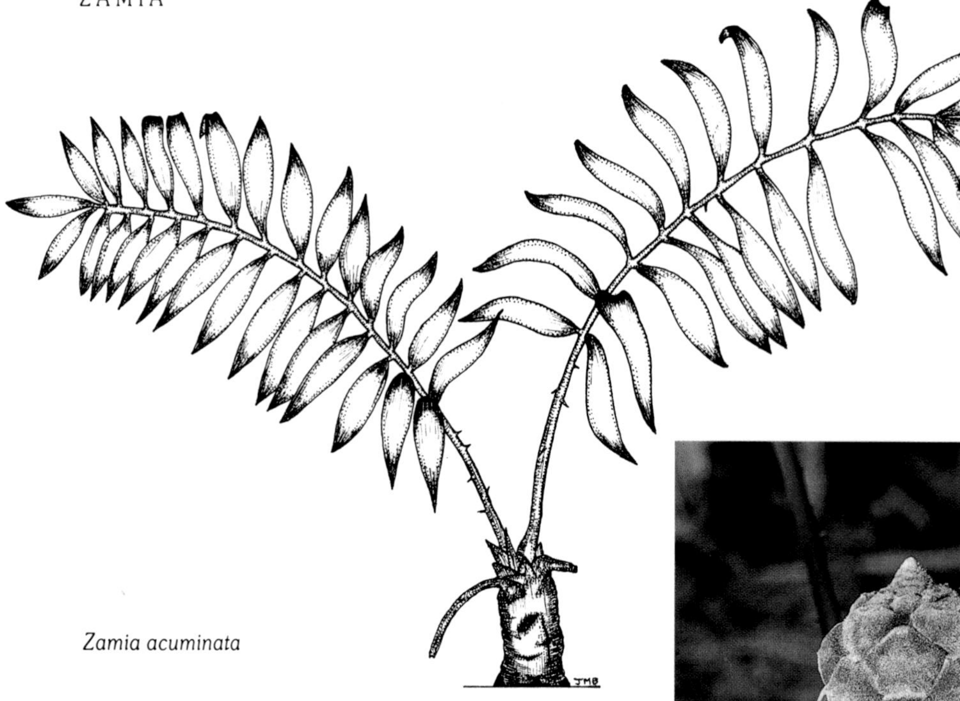


Zamia acuminata

## Zamia amblyphyllidia

D. Stevenson
(with wide leaflets)

**Description:** A small cycad with much-branched subterranean stems, each to about 4 cm across. **Young leaves** covered with rusty brown hairs. **Mature leaves** 0.7–1.5 m long, four to ten leaves in each crown, obliquely erect, straight, stiff, flat or slightly vee-ed in cross-section; **petiole** 30–70 cm x 2–6 cm; **leaflets** twenty to ninety on each leaf, obovate, dark green, stiff, leathery, moderately spaced, the apex broadly obtuse and with numerous small teeth. **Male cones** 4–8 cm x 1–2 cm, cylindrical, grey to blackish, hairy; **sporophylls** hexagonal; **peduncle** 4–8 cm long, densely and shortly hairy. **Female cones** 6–10 cm x 2–3 cm, ovoid, with a conical, tapered apical tail, grey to blackish, shortly hairy; **sporophylls** hexagonal with a flat centre; **peduncle** 5–10 cm long, hairy. **Seeds** 1.5–2 cm x 1–1.2 cm, ovoid, orange to orange-red.
**Distribution and Habitat:** Distributed in Jamaica, western Cuba and northern Puerto Rico where it grows in grassland, ravines and open forests dominated by species of *Pinus* and *Quercus*.
**Notes:** This species, described in 1987, was originally thought to be *Z. latifoliolata*. Upon checking type specimens, however, researchers found that the latter species, described from the Dominican Republic, is a synonym of *Z. pumila*. *Z. amblyphyllidia* is closest to *Z. pygmaea* but is much more robust with broader stems, longer and broader leaflets and larger cones. Intermediate forms may occur.
**Cultivation:** A very hardy species suitable for subtropical and warm temperate regions. Plants grow best in a sunny position and are tolerant of exposure to sun from an early age.
**Propagation:** From fresh seed and less commonly by division of clumps.

Female cone of *Zamia acuminata*.

## Zamia amplifolia

Hort. ex Masters
(with large leaves)

**Description:** A small cycad with an unbranched subterranean trunk to 30 cm long and 12 cm across. **Mature leaves** 1–2 m long, one to four on each plant in a sparse crown; **petiole** 0.5–1 m long, slender, bearing prickles; **leaflets** six to twelve on each leaf, 15–35 cm x 8–18 cm, broadly elliptical, sessile, yellowish green to dark green,

Female cones of *Zamia amblyphyllidia*.

thin-textured but leathery, the upper surface deeply grooved between the veins and presenting a pleated appearance, the distal margins toothed. **Male cones** 4–6 cm x 1–2 cm, cylindrical, yellowish to tan. **Female cones** 20–30 cm x 8–10 cm, barrel-shaped, cream to tan. **Seeds** 1–1.5 cm x 0.6 cm, ovoid, red.
**Distribution and Habitat:** Occurs on the Pacific slopes of the Choco regior of Colombia and adjacent areas in Ecuador. This species grows in the lowlands and mountains in the shady understoreys of rainforest.
**Notes:** This species was described in 1878 from material collected in Colombia. It is part of a distinctive group of species which have large, prominently grooved leaflets (see also *Z. dressleri*, *Z. neurophyllidea*, *Z. roezlii*, *Z. skinneri* and *Z. wallisii*).
**Cultivation:** Suited to cool tropical and warm subtropical regions. Requires warm, humid conditions with buoyant air movement and light to heavy shade.
**Propagation:** From fresh seed.

## *Zamia angustifolia*

Jacq.
(with narrow leaves/leaflets)

**Description:** A small cycad with much-branched subterranean stems, each to about 6 cm across. **Young leaves**

Male cones of *Zamia amblyphyllidia*.

covered with rusty brown hairs. **Mature leaves** 0.4–1.4 m long, four to ten in each crown, obliquely erect, straight, stiff, flat or slightly vee-ed in cross-section; **petiole** 20–60 cm long, lacking prickles; **leaflets** fifteen to seventy on each leaf, 10–30 cm x 0.2–0.5 cm, linear, dark green, stiff, leathery, moderately spaced, the margins rolled upwards, one to three small teeth near the acuminate apex. **Male cones** 6–10 cm x 2–3 cm, cylindrical, dark grey to black, appearing as having vertical white stripes, shortly hairy; **sporophylls** hexagonal; **peduncle**s 3–7 cm long, shortly hairy. **Female cones** 8–12 cm x 2–3 cm, ovoid, with a short, conical, acuminate, apical tail, dark grey to black, appearing as having vertical white stripes; **sporophylls** hexagonal with a flat centre; **peduncle** 5–10 cm long, hairy. **Seeds** 1.5–2 cm x 1–2 cm, ovoid, orange to orange-red.
**Distribution and Habitat:** Found on the island of Eleuthera in the Bahamas and in eastern Cuba. It grows in open habitats such as grassland and also in pine forests.
**Notes:** This species, described in 1791, is part of the complex of species surrounding *Z. pumila* and *Z. integrifolia*. It can be distinguished from all others by its narrow leaflets with the margins rolled upwards, a very few marginal spines near the apex and its dark grey to black cones which appear as if they have white vertical stripes.
**Cultivation:** This rather uncommonly grown species is best suited to subtropical and warm temperate regions. Plants can be grown either in sun or shade.
**Propagation:** From fresh seed and less commonly by division.

# *Zamia angustissima*

Miq.
(extremely narrow)

**Description:** A small cycad with much-branched, subterranean stems to about 5 cm long. **Young leaves** shortly hairy, becoming glabrous. **Mature leaves** about 0.5 m long, about five in each crown, straight, stiff; **petiole** shortly hairy, lacking prickles; **leaflets** about twelve on each leaf, 5–15 cm x 0.5–1.5 cm, narrowly linear, straight or curved, green to yellow-green, flat to concave, leathery, with a few teeth towards the apex. Details of cones not recorded.
**Distribution and Habitat:** Endemic to Cuba where it grows in open forest.
**Notes:** This species was described in 1851. Although distinctive because of its very narrow leaflets, it remains poorly known and in need of further study.
**Cultivation:** Probably suited to subtropical and warm temperate regions, in a sunny location.
**Propagation:** From fresh seed and probably also by division.

# *Zamia boliviana*

(Brongn.) A. DC.
(from Bolivia)

**Description:** A small cycad with an unbranched, cylindrical, subterranean trunk to 15 cm long and 7.5 cm across. **Mature leaves** 0.3–0.6 m long, three to five in an erect sparse crown, dark green, flat in cross-section; **petiole** 10–20 cm long, slender, lacking prickles, swollen and hairy at the base; **leaflets** sixteen to twenty-six on each leaf, 15–25 cm x 0.9–1.5 cm, linear-lanceolate, slightly falcate, moderately spaced, dark green, margins entire, revolute, two to five small teeth at the attenuate to acuminate apex. **Male cones** 0.5–1 cm x 0.6–0.8 cm, narrow-cylindrical, solitary, erect, densely covered with reddish brown pubescence; **sporophylls**, about 0.3 cm across, hexagonal; **peduncle** about 3 cm long. **Female cones** 14–18 cm x 5–8 cm, barrel-shaped, solitary, erect, with a short conical apical tail, covered with short, rusty brown hairs; **sporophylls** about 2 cm across, hexagonal; **peduncle** 8–11 cm long, rusty hairy. **Seeds** 1–1.2 cm x 0.8–1 cm, oblong to elliptical, scarlet.
**Distribution and Habitat:** Endemic to Bolivia where it grows in relatively dry habitats in open grassland.
**Notes:** This species was first described in 1846 as a species of *Ceratozamia* and was transferred to the genus *Zamia* in 1868.
**Cultivation:** Suited to tropical and subtropical regions. A hardy species which tolerates full sun from an early age and short, dry periods. Requires excellent drainage.
**Propagation:** Solely from fresh seed.

# *Zamia chigua*

Seemann
(a native name for the plant)

**Description:** A medium-sized cycad with an emergent trunk to 2 m long and 15 cm across, which is wrinkled in appearance. **Young leaves** pale green, glabrous. **Mature leaves** 0.5–3 m long, three to fifteen in a graceful arching crown, bright green, flat in cross-section; **petiole** 30–100 cm long, swollen at the base, densely covered with prickles; rhachis arching, prickly; **leaflets** 30–140 on each leaf, 10–30 cm x 1–1.5 cm, linear-lanceolate, subfalcate, crowded and overlapping, thin-textured, inserted at right angles to the rhachis, bright green, tapered to the base, margins entire, apex drawn out and long-acuminate. **Male cones** 10–20 cm x 2–3 cm, cylindrical, erect, cream to light yellow, shortly hairy; **sporophylls** about 1.5 cm long, wedge-shaped; **peduncle** 3–4 cm long, shortly hairy. **Female cones** 20–30 cm x 8–12 cm, barrel-shaped, light brown, shortly hairy; **sporophylls** about 2.5 cm long; **peduncle** about 5 cm long, hairy. **Seeds** 3–3.5 cm x 1.5–2 cm, ovoid, red to scarlet.

**Distribution and Habitat:** Distributed in the Choco region of Colombia and the Chiriqui region of central Panama. In Colombia this species grows in lowland rainforests often close to streams, whereas in Panama it grows at altitudes from 600 m to 1200 m..
**Notes:** This species was described in 1845 from plants collected on the banks of the San Juan River, Colombia, close to the sea. It has been cultivated under the name of 'Helecho' and has also been confused with *Z. roezlii* which has the confusing local name of 'chigua'. *Z. chigua* can be distinguished from *Z. roezlii* by its more numerous and much more crowded leaflets which have a smooth surface (grooved in *Z. roezlii*).
*Z. chigua* is a very attractive species with long, graceful leaves and crowded, slender leaflets. Its trunk has a wrinkled appearance which is particularly noticeable in large specimens. The petioles and rhachises are densely covered in stout prickles which are often branched.
**Cultivation:** Best suited to tropical and warm subtropical regions. Requires warmth, humidity, shade and free air movement creating a buoyant atmosphere. Plants are very frost-sensitive and should not be allowed to dry out.
**Propagation:** From fresh seed.

# Zamia cremnophila

Vovides, Schutzman & Dehgan
(dwelling on cliffs)

**Description:** A small cycad with an erect to decumbent trunk to 0.25 m tall and 9 cm across. **Young leaves** deep purplish red, sparsely hairy. **Mature leaves** 0.4–2 m long, one to many on each plant, pendant, flat or inversely vee-ed in cross-section; **petiole** 3.7–4.3 cm long, bearing numerous prickles, the base bulbous, lacking prickles and shortly hairy; **leaflets** thirty to fifty on each leaf, 10–36 cm x 2–4 cm, lanceolate to oblong, dark green, glabrous, leathery, crowded and overlapping, inserted at about 45° to the rhachis, tapered to the base, apex acute, margins entire. **Male cones** 7–8 cm x 1.8–2 cm, cylindrical, erect, light brown; **sporophylls** about 0.5 cm long, wedge-shaped, the outer face hexagonal; **peduncle** about 2.5 cm long, shortly hairy. **Female cones** 8–8.5 cm x 4.5–5.5 cm, barrel-shaped, with a short conical apex, dark brown, shortly hairy; **sporophylls** about 2 cm long, wedge-shaped; **peduncle** about 2.5 cm long, shortly hairy. **Seeds** 1.5–1.7 cm x 0.9–1 cm, ovoid, smooth, bright scarlet.
**Distribution and Habitat:** Endemic to Tabasco, southern Mexico where it grows on the sides of rocky cliffs which are of calcareous formation.
**Notes:** This very distinctive species, described in 1988, can be recognised by its pendant leaves which have numerous crowded and overlapping, narrow, leathery leaflets. The pendant leaves are undoubtedly an adaptation to its unique environment; however, it is interesting to note that cultivated plants exhibit strongly arching leaves which are different in habit to other species of *Zamia*.
**Cultivation:** An interesting species suitable for subtropical and warm temperate climates but extremely rare in cultivation. Plants require protection from climatic extremes and soils of a neutral to alkaline pH.
**Propagation:** Solely from fresh seed which takes about three to six weeks to germinate.

# Zamia cunaria

Dressler & D. Stevenson
(after the Cuna Indians, who live where the species grows)

**Description:** A small cycad with an unbranched, nearly globose, subterranean trunk to 10 cm across. **Mature leaves** 0.5–1.5 m long, usually solitary but up to three in an erect sparse crown, dark green, flat in cross-section; **petiole** 15–50 cm long, sparsely to densely prickly; **leaflets** six to twenty-four on each leaf, 20–40 cm x 5–8 cm, oblong to oblanceolate, margins toothed in the distal third, apex acuminate. **Male cones** 4–6 cm x 1–5 cm, cylindrical to ovoid, cream to tan; **peduncle** 2–4 cm long. **Female cones** 15–20 cm x 5–7 cm, cylindrical to ovoid, wine red to dark red-brown. **Seeds** 1.5–2 cm x 0.5–0.8 cm, ovoid, pink to light red.
**Distribution and Habitat:** Endemic to Panama where it grows in clay soils on slopes and ridges in secondary vegetation at altitudes of 400–800 m.
**Notes:** This species was described in 1993 from material collected in the region known as Cuna Yala. The Cuna Indians use the seeds to make necklaces.
**Cultivation:** Suited to tropical and subtropical regions. Experience in cultivation is limited. Requires warm, humid conditions in a sheltered position.
**Propagation:** From fresh seed.

# Zamia cupatiensis

Ducke
(from the hill Cerro de Cupati, in Colombia)

**Description:** A small cycad with a swollen unbranched trunk to 0.3 m long which may be completely subterranean or partly emergent when growing in shallow, rocky soil. **Mature leaves** 0.3–0.6 m long, six to nine in an erect crown, dark green, flat in cross-section; **petiole** 20–30 cm long, wiry, swollen at the base, slightly flexuose, lacking any prickles; rhachis lacking prickles; **leaflets** eight to twelve on each leaf, 15–21 cm x 4–6 cm, broadly ovate-lanceolate, moderately spaced inserted at right angles to the rhachis, the lower ones shortly stalked,

D. W. STEVENSON

*Zamia cupatiensis* in habitat.

*Zamia fairchildiana*

with numerous prominent veins, darker above, paler beneath, margins entire, revolute, apex drawn out into a long-acuminate point. **Male cones** 3–8 cm x 2–3 cm, cylindrical, erect, yellowish grey, shortly hairy; **sporophylls** about 0.4 cm across, peltate, the outer face hexagonal; **peduncle** 3–5 cm long, shortly hairy. **Female cones** 5–7 cm x 3–4 cm, barrel-shaped, erect, brownish purple, shortly hairy; **sporophylls** about 1.8 cm across, hexagonal; **peduncle** 4–5 cm long, shortly hairy. **Seeds** about 3 cm x 2 cm, ovoid, red.
**Distribution and Habitat:** Endemic to the Amazonian Region of south-eastern Colombia and the adjacent Rio Negro region of Brazil. This species, which is apparently of restricted distribution, grows near the summits of low quartzite hills (about 350 m altitude) and rarely in rainforest. Plants frequently grow on the tops of huge boulders which are covered with a thin layer of sandy soil and decaying litter.
**Notes:** This little-known species was described in 1922 from plants found growing in south-east Colombia near the Brazilian border. A collector has reported that plants growing in shallow soil develop emergent trunks whereas in those growing in deeper soil the trunks are completely subterranean. Dennis Stevenson has observed plants with emergent trunks only occur in situations where soil is absent and the plant's contractile stem cannot contract. The inner parts of these trunks are used by local natives to prepare a starchy meal for food. *Z. cupatiensis* is similar in general appearance to *Z. ulei*. Adolopho Ducke, who described this species, believed at the time that *Z. ulei* had narrow leaflets when in fact the leaflets of both are of similar width.
**Cultivation:** Suitable for tropical and warm subtropical regions. Plants require excellent drainage and will grow in full sun. A buoyant humid atmosphere is essential.
**Propagation:** From fresh seed.

# *Zamia dressleri*

D. Stevenson
(after Bob Dressler who discovered the species and studied *Zamia* in Panama)

**Description:** A small cycad with an unbranched, subterranean trunk 3–5 cm across. **Mature leaves** 0.5–1.5 m

long, usually solitary but up to three in an erect sparse crown, dark green, flat in cross-section; **petiole** 30–100 cm long, sparsely to densely prickly; **leaflets** four to ten on each leaf, 30–50 cm x 12–15 cm, elliptical, grooved between the veins on the upper surface, margins toothed in the distal third, apex acuminate. **Male cones** 5–8 cm x 1–2 cm, narrowly cylindrical, cream to tan. **Female cones** 10–15 cm x 3–4 cm, ovoid to cylindrical, wine red to dark red-brown; **peduncle** short. **Seeds** 1–1.5 cm across, ovoid, red.
**Distribution and Habitat:** Endemic to Panama where it is apparently restricted to two small, disjunct populations. This species grows in rainforests.
**Notes:** This species was described in 1993 from material collected in the vicinity of Colon. The plants have large, deeply grooved leaflets and usually only produce a single leaf. It shares the feature of deeply grooved leaflets with *Z. wallisii* and *Z. skinneri*. *Z. skinneri* is a larger growing plant with up to twenty-four leaves in the crown and up to twenty-two leaflets in the leaf. *Z. wallisii* is readily distinguished by its prominent petiolules which are grooved on the upper surface.
**Cultivation:** Suited to tropical regions. Requires warm, humid conditions with free air movement and light shade.
**Propagation:** From fresh seed.

## *Zamia fairchildiana*

L. D. Gomez
(after Dr Graham Fairchild, tropical entomologist and student of Panamanian cycads)

**Description:** A small cycad with an erect cylindrical trunk to 2 m tall and 15 cm across. **Young leaves** light green. **Mature leaves** 0.7–2.5 m long, three to ten in an erect crown, flat in cross-section; **petiole** 10–30 cm long, roughly quadrangular, glabrous, densely prickly; **leaflets** twenty to sixty on each leaf, 20–40 cm x 2–4 cm, oblong to lanceolate, subfalcate, thin-textured and papery, green, faintly striped, slightly furrowed above, tapered to the base, margins entire or with a few small teeth near the acute apex. **Male cones** 10–40 cm x 2–5 cm, cylindrical, cream to yellow, covered with short, pale brown hairs. **Female cones** 20–30 cm x 6–10 cm, barrel-shaped, yellowish green to light brown, covered with short, rusty brown hairs; **sporophylls** peltate, outer face hexagonal. **Seeds** 1–1.5 cm x 1.5 cm, ovoid, pink to red.
**Distribution and Habitat:** Occurs in Panama and south-eastern Costa Rica. It grows in rainforests between 50 m and 1500 m altitude, but is commonest between 400 m and 900 m altitude.
**Notes:** Described in 1982, *Z. fairchildiana* is related to *Z. pseudomonticola*. Distinguishing features include the dense prickles on the leaf petiole and rhachis and the large, relatively broad, thin, papery-textured leaflets. This is probably the commonest species of *Zamia* in Panama.
**Cultivation:** Best suited to tropical and warm subtropical regions. An attractive species which grows rapidly under suitable conditions. Responds to warmth, humidity and free air movement. Excellent in a container.
**Propagation:** From fresh seed.

## *Zamia fischeri*

Miq.
(after M. Fischer, gardener at St Petersburg Botanic Garden)

**Description:** A small cycad which develops an ovoid to barrel-shaped trunk to 15 cm long and 8 cm across. **Mature leaves** 20–45 cm long, three to ten on each plant, lax to spreading or even nearly pendant, flat in cross-section; **petiole** 5–10 cm long, slender, weak, lacking prickles; **leaflets** twenty to thirty-four on each leaf, 5–8 cm x 1–1.3 cm, lanceolate, dark green, thin-textured and papery, glabrous, toothed towards the acute apex. **Male**

Female cone of *Zamia fairchildiana*.

Female cones of *Zamia fischeri*.

**cones** not recorded. **Female cones** 4–5 cm x 2–3 cm, cylindrical to ovoid. **Seeds** about 1.5 cm x 1.2 cm, ovoid, red.
**Distribution and Habitat:** Endemic to San Luis Potosi, Mexico, where it grows in evergreen forests.
**Notes:** The correct identity of this species has only recently been recognised. Previously it was confused with the species now known as *Z. vazquezii*. *Z. fischeri* was described in 1847 from material collected in Mexico and its true identity was not realised until the type specimen was examined. It can be readily separated from *Z. vazquezii* by its shorter, lax leaves, smooth, weak petioles and shorter leaflets.
**Cultivation:** Suited to subtropical and warm temperate regions. Probably has similar requirements to *Z. vazquezii*, but unlike that species is apparently rare in cultivation. It grows well in a container.
**Propagation:** From fresh seed.

## *Zamia furfuracea*

L.f. in Aiton
(scurfy, covered with loose wool or scales)

**Description:** A small to medium-sized cycad with much-branched, subterranean stems, each to about 15 cm across. **Young leaves** pale green, covered with rusty brown hairs. **Mature leaves** 0.2–1 m long, eight to thirty in each dense crown, obliquely erect to erect, straight, stiff, flat in cross-section; **petiole** 6–15 cm long, swollen and hairy at the base, densely armed with stout prickles; **leaflets** ten to forty on each leaf, 8–16 cm x 2–4.5 cm, lanceolate to oblanceolate or obovate, crowded and nearly overlapping, felty-hairy, olive green to brownish

*Zamia furfuracea* is commonly used in landscaping.

Female cones of *Zamia furfuracea*.

green above, paler and silvery beneath, thick and leathery, stiff, rigid, tapered to the base, inserted at an angle of about 45° to the rhachis, apex subacute, margins with small teeth. **Male cones** 9–12 cm x 1.6–2 cm, cylindrical, grey-green, shortly hairy; **sporophylls** wedge-shaped, outer face hexagonal; **peduncle** 5–10 cm long, shortly hairy. **Female cones** 18–23 cm x 6.5–7 cm, barrel-shaped, grey-green to brown, shortly hairy; **sporophylls** wedge-shaped, outer face hexagonal; **peduncle** 3–5 cm long, hairy. **Seeds** 2–2.2 cm x 1–1.2 cm, ovoid, pink to red.
**Distribution and Habitat:** Endemic to eastern Mexico (Veracruz), growing in coastal regions, often in exposed situations where plants can be subjected to salt spray. Vegetation consists of sparse coastal scrub and savannah dotted with palms. Elevation ranges from a few metres above sea level to about 50 m altitude. Soils are sandy.
**Notes:** This species, described in 1789, was once common in its natural habitat but has been reduced to rarity by overcollecting. *Z. furfuracea* is a very distinctive species which can be recognised by its broad, hairy, rigid leaflets crowded on the leaves. Plants typical of the species are found only in undisturbed habitats. Natural hybrids between *Z. furfuracea* and *Z. loddigesii* occur commonly in nature and are always associated with disturbance such as clearing. These hybrids, which are also common in cultivation, show great variation in leaflet shape, texture and hairiness. Plants of this species are commonly sold in the nursery trade as 'Cardboard Palm'.
**Cultivation:** As for the genus. *Z. furfuracea* is one of the most popular species in cultivation and is valued for its ornamental appearance, adaptability and hardiness. It grows well in tropical, subtropical and warm temperate regions and is always best in sunny situations. Plants resent poor drainage, shade and excessively wet foliage. This is one of the best cycads for indoor decoration.
**Propagation:** From fresh seed which takes about twelve months to germinate; less commonly by division of clumps.

## *Zamia herrerae*

Calderon & Standley
(after Dr Hector Herrera, scientist from El Salvador)

**Description:** A small cycad with an elongated, subterranean trunk to 20 cm long and 10 cm across. **Mature leaves** 0.4–0.6 m long, one to five in an erect, sparse crown, shallowly V-shaped in cross-section; **petiole**s 10–20 cm long, slender, sparsely clothed with stout prickles 1–2 mm long; **leaflets** forty to sixty on each leaf, 15–22 x 0.8–1.3 cm, linear-lanceolate, dark green and glossy, thick and leathery, moderately crowded, inserted at about 75° to the rhachis, margins with a few appressed spines, apex long-acuminate. **Male cones** 4–7 cm x 1.5–2 cm, cylindrical, light brown, shortly hairy; **sporophylls** wedge-shaped, terminally hexagonal. **Female cones** 5–15 cm x 4–6 cm, cylindrical to ovoid. **Seeds** red to orange-red.
**Distribution and Habitat:** The Pacific coast region of El Salvador, Honduras, Guatemala and Chiapas, Mexico, growing in lowland rainforests.
**Notes:** This poorly known species was first collected in 1923 and described a year later.
**Cultivation:** Little is known about the requirements of this species but it is probably best suited to tropical and warm temperate regions in warm, humid, sheltered conditions.
**Propagation:** From fresh seed.

## *Zamia inermis*

Vovides, Rees & Vazquez-Torres
(unarmed, lacking spines or prickles)

**Description:** A small cycad with a trunk to 30 cm tall and 25 cm across. **Young leaves** pale green. **Mature leaves** 0.5–1.3 m long, ten to twenty on each plant, forming an erect to spreading, dense crown, flat in cross-section; **petiole** 10–20 cm long, lacking prickles, the base swollen and shortly hairy; **leaflets** 50–65 on each leaf, 22–28 cm x 0.9–1.1 cm, linear-lanceolate, pale green, leathery, glabrous, moderately spaced, inserted at 80–90° to the rhachis, tapered to the base, apex acute, margins entire. **Male cones** 9–17 cm x 2.5–3 cm, cylindrical, brown, shortly hairy; **sporophylls** wedge-shaped; **peduncle** 3.5–6.5 cm long, shortly hairy. **Female cones** 13–19 cm x 8.5–9.5 cm, barrel-shaped, brown, shortly hairy; **sporophylls** about 2 cm long, wedge-shaped; **peduncle** about 5 cm long, hairy. **Seeds** 1.7–2.2 cm x 1.5–2 cm, ovoid, brown.

**Distribution and Habitat:** Endemic to Veracruz, Mexico, where it grows in deciduous forests at 200–300 m altitude.

**Notes:** This species, described in 1983, is a moderately robust *Zamia* which can be recognised by its leaves being numerous in a fairly dense crown and lacking any prickles on the petiole and rhachis.

**Cultivation:** Suited to subtropical regions and tolerant of moderate exposure to sunshine.

**Propagation:** From fresh seed.

## *Zamia integrifolia*

L. f. in Aiton
(with entire leaves/leaflets)

**Description:** A small cycad with much-branched, subterranean stems, each to about 6 cm across. **Young leaves** covered with short, rusty brown hairs. **Mature leaves** 0.6–1.5 m long, four to ten leaves in each crown, dark green, obliquely erect, straight, stiff, flat or slightly vee-ed in cross-section; **petiole** 20–80 cm long, lacking prickles; **leaflets** twenty to eighty on each leaf, 8–30 cm x 1–3 cm, oblong with parallel margins, dark green, stiff, leathery, moderately spaced, the margins with inconspicuous teeth in the distal quarter which appear as small bumps, apex blunt. **Male cones** 6–10 cm x 1–2.5 cm, cylindrical, red to reddish brown, shortly hairy; **sporophylls** hexagonal; **peduncle** 4–8 cm long, densely hairy. **Female cones** 8–12 cm x 2–3 cm, ovoid, with an abruptly truncate apical tail, red to reddish brown, shortly hairy; **sporophylls** hexagonal with a flat centre; **peduncle** 5–10 cm long, shortly hairy. **Seeds** 1.5–2 cm x 1–2 cm, ovoid, orange to orange-red.

**Distribution and Habitat:** A widely distributed species being found in south-eastern Georgia and Florida (including the Florida Keys) in the United States, western Cuba, south-central Puerto Rico, the Cayman Islands and various islands in the Bahamas (Andros, Grand Bahama, Great Abaco, Long and New Providence). It grows in a wide variety of habitats including grassland, dune vegetation and forests and woodlands, both hardwood and softwood. Soils include sands near sea level and in soils derived from limestone.

**Notes:** This species, commonly known as the Coontie, has been frequently confused with *Z. pumila* which has well-defined teeth on the distal parts of the leaflets. Once locally abundant in Florida, *Z. integrifolia* is now uncommon and threatened by urban development. The stems were once used as a source of starch by the Seminole Indians (after suitable treatment), and a small starch extraction industry was established in South Florida in the 1850s (giving rise to the common name of Florida arrowroot). The species was also used throughout the West Indies as a source of starch by the nomadic Arawak Indians and they may have been responsible for transporting plants or seeds to various islands, thus accounting for its widespread but sporadic distribution. The species is now widely used as an ornamental in many parts of the United States, being planted outdoors in warmer climates and used as an indoor plant in colder areas. Unfortunately this *Zamia* is still exploited with most large plants originating as the result of poaching from the wild. The species is apparently not safe in its native country, despite its being protected from international trade by CITES. *Z. integrifolia* is a widespread and variable species and there is disagreement among botanists as to its delimitation. Some botanists, led by Dennis Stevenson, treat the species as embracing a number of previously recognised entities which include *Z. floridana* A. DC., *Z. silvicola* Small and *Z. umbrosa* Small. Other researchers suggest that some of the variants are worthy of taxonomic recognition. Certainly differences exist between plants of any variant growing in sun and shade in features such as leaf length, leaflet size and thickness. Despite this, correlations may occur between variants and habitat. Thus plants which have narrow leaflets (referrable to *Z. floridana*), range from the Florida Panhandle south and in the Miami area, and grow in pine flatwoods and relict sand dunes. Those having broad leaflets (and referrable to *Z. umbrosa* Small) range from Jacksonville to Daytona Beach and grow in oak hammock *Sabal palmetto* scrub and less commonly in pine-oak forests. It would seem that more research is needed to clarify the situation one way or the other. A variant with very broad leaflets occurs on Long Island in the Bahamas.

**Cultivation:** A very hardy and adaptable species which grows well in a range of situations and soil types. Best suited to subtropical and warm temperate regions, although plants will survive in a sheltered situation in colder climates.

**Propagation:** From fresh seed, less commonly from division.

*Zamia inermis*
With permission from *Flora de Veracruz*, Fascicle 26, February 1983. Illustration by Emundo Saavedra. **a.** detail of adult male plant, **b.** detail of adult female plant, **c.** cataphyll, **d.** male cone, **e.** male sporophylls, **f.** female cone, **g. & h.** immature female cones, **i.** female sporophylls, **j.** seed.

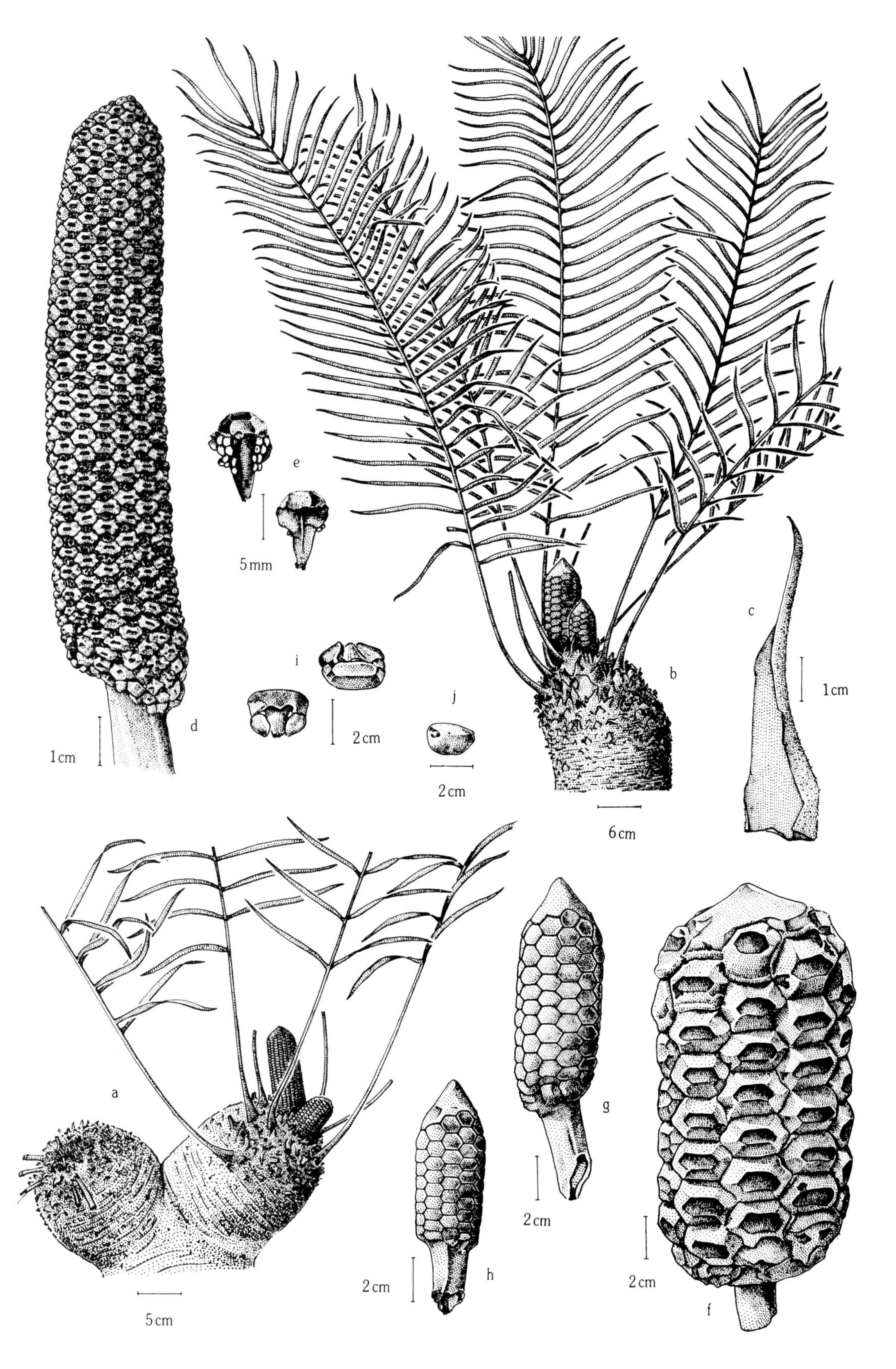

e
5 mm
d
1 cm
i
2 cm
j
2 cm
b
6 cm
c
1 cm
a
5 cm
g
2 cm
h
2 cm
f
2 cm

Large clump of *Zamia integrifolia*.

Dormant clump of *Zamia integrifolia*.

Developing leaves of *Zamia integrifolia*.

## *Zamia ipetiensis*

D. Stevenson
(from the area of Ipeti, Panama)

**Description:** A small cycad with an unbranched subterranean trunk to about 10 cm across. **Mature leaves** usually solitary but up to three, obliquely erect, bright green; **petiole** 15–50 cm long, sparsely prickly to densely prickly; **leaflets** six to twenty-four on each leaf, 20–40 cm x 5–8 cm, narrow-lanceolate to broad-lanceolate, bright green, margins finely toothed towards the acuminate apex. **Male cones** 4–6 cm x 1–1.5 cm, cylindrical to ovoid, cream to light brown; **sporophylls** with sporangia on both the upper and lower surface. **Female cones** 15–20 cm x 5–7 cm, cylindrical to ovoid, dark red-brown.

Male cones of *Zamia integrifolia*.

**Seeds** 1.5–2 cm x 0.5–0.8 cm, ovoid, pink to light red.
**Distribution and Habitat:** Endemic to a small area of Panama where it grows in rainforest at altitudes of 100–500 m.
**Notes:** This species, described in 1993, is interesting for

Receptive female cone of *Zamia integrifolia*.

its very sparse crown of foliage and its leaflets which have finely toothed margins.
**Cultivation:** Suited to tropical and subtropical regions. Requires warm, humid conditions in a sheltered situation.
**Propagation:** From fresh seed.

# *Zamia jirijirimensis*

R. E. Schultes
(from the Raudal de Jirijirimo waterfall, Rio Apaporis River, Colombia)

**Description:** A small cycad which develops a short, swollen subterranean trunk about 14 cm long and 7 cm across. **Mature leaves** 0.4–0.65 m long, five or six in an erect crown, dark green and shiny, flat in cross-section; **petiole** 20–25 cm long, flexuose, unarmed or with a few vestigial prickles, swollen at the base; **leaflets** twelve to thirty-four on each leaf, 20–27 cm x 1.5–2.5 cm, linear-lanceolate, widely spaced and uncrowded, inserted at about 40° to the rhachis, leathery, dark green and shiny above, paler beneath, margins entire, apex acute to acuminate. **Male cones** 3–4 cm x 1–1.2 cm, narrow cylindrical, erect, yellowish brown, densely and shortly hairy; **sporophylls** about 1 cm across, hexagonal; **peduncle** 4–6 cm long, shortly hairy. **Female cones** 7–9 cm x 4–4.5 cm, cylindrical, solitary, erect, covered densely with short, yellowish hairs; **sporophylls** about 2 cm long, hexagonal; **peduncle** 8–10 cm long, shortly hairy. **Seeds** 1.2–1.4 cm x 0.7–0.8 cm, ovoid, bright orange.
**Distribution and Habitat:** Endemic to the Amazonian Region of eastern Colombia. This species grows in open sunny situations in savannah grassland; soils are white sands.
**Notes:** *Z. jirijirimensis*, described in 1953, has as its closest relative *Z. lecointii*, and the two are believed by some botanists to be identical. The inner parts of the swollen subterranean trunks are used by local natives to prepare a starchy meal for food.
**Cultivation:** Suitable for tropical and subtropical regions. Plants are tolerant of full sun from early age and require excellent drainage.
**Propagation:** From fresh seed.

# *Zamia kickxii*

Miq.
(after a collector by the name of Kickx)

**Description:** A small cycad with much-branched subterranean stems to about 8 cm across. **Mature leaves** about 0.8 m long, about six in each crown, dark green, stiff; **petiole** about 20 cm long, lacking prickles; **leaflets** twenty to twenty-four on each leaf, 4–9 cm x 1–2 cm, elliptical to lanceolate, dark green, stiff, thin-textured, margins with numerous small teeth. **Male cones** about 5 cm x 1 cm, cylindrical, shortly hairy. **Female cones** 3–6 cm x 1.5–2 cm, ovoid, shortly hairy. **Seeds** about 1.5 cm x 1 cm, ovoid, red.
**Distribution and Habitat:** Endemic to Cuba where it grows in grassland and pine forests in sandy soil.
**Notes:** This species was described in 1842. It is a poorly

known species which is in need of further study.
**Cultivation:** Apparently a hardy species which is suitable for subtropical and warm temperate regions. Best growth is probably achieved in a sunny location.
**Propagation:** From fresh seed and by division.

## *Zamia lawsoniana*

Dyer
(derivation unknown)

**Description:** A small cycad with a subterranean trunk to 30 cm long and 12 cm across. **Mature leaves** 0.5–1.5 m long, few in an obliquely erect crown, flat in cross-section; **petiole** 10–30 cm long, densely prickly; **leaflets** twenty to thirty on each leaf, 13–23 cm x 0.7–0.9 cm, linear-lanceolate, thick and leathery, rigid, shallowly curved, margins recurved, with spinose teeth, dark green and shiny above, paler beneath, apex acuminate. **Male cones** 6–8 cm x 1.5–2 cm, cylindrical to ovoid, brown, covered with woolly hairs; **peduncle** about 16 cm long. **Female cones** about 8 cm x 4 cm, barrel shaped. **Seeds** 1.5–2 cm x 1–1.5 cm, ovoid, red.
**Distribution and Habitat:** Endemic to Oaxaca, Mexico. This species grows in pine forests and sparse forests with a grassy understorey.
**Notes:** This species was described in 1884 from material collected in Mexico. It is characterised by its thick, leathery leaflets.
**Cultivation:** A hardy species suitable for subtropical and warm temperate regions. Plants grow best in a sunny position and are tolerant of exposure to sun from an early age.
**Propagation:** From fresh seed.

## *Zamia lecointei*

Ducke
(after Paul le Cointe, resident of Obido, Brazil)

**Description:** A small cycad with a swollen, unbranched, subterranean trunk to about 15 cm long and 8 cm across. **Mature leaves** 1–1.5 m long, two to four in an erect to arching, sparse crown, dark green, flat in cross-section; **petiole** 40–62 cm long, slender, covered with small, sharp prickles 1–2 mm long, swollen at the base; **leaflets** thirty-forty on each leaf, 30–37 cm x 1–2 cm, linear-lanceolate, falcate, widely spaced, inserted at 60–90° to the rhachis, dark green, tapered quickly to the base, apex long-acuminate, the lower margins sometimes with two or three short, inconspicuous teeth. **Male cones** 6–10 cm x 2–2.5 cm, cylindrical, erect, brownish; **sporophylls** hexagonal; **peduncle** 8–10 cm long. **Female cones** 12–15 cm x 4–5 cm, cylindrical, erect, covered with short, purplish brown hairs; **sporophylls** about 2 cm across, hexagonal; **peduncle** 10–14 cm long, shortly hairy. **Seeds** 1.5–1.7 cm x 1–1.2 cm, ovoid to oblong, red.
**Distribution and Habitat:** Occurring in the Amazonian region of northern Brazil and southern Venezuala. It grows on rocky hillsides which are covered with relatively sparse forest.
**Notes:** This species was described by Adolpho Ducke in 1915 and later (1935) reduced to a variety of *Z. ulei* by the same author. Originally Ducke had not seen specimens of *Z. ulei*, or the description, and he thought that this latter species had narrow leaflets.
**Cultivation:** Suited to tropical and warm subtropical regions. Very rarely grown. Probably needs warm humid conditions with free air movement.
**Propagation:** From fresh seed.

## *Zamia lindenii*

Regel ex Andre
(after Jean Jules Linden)

**Description:** A large cycad with an erect trunk to 2.5 m long and 25 cm across. **Mature leaves** 1.5–2.5 m long, ten to fifteen in a graceful arching crown, bright green, flat in cross-section; **petiole** 30–100 cm long, hairy, bearing prickles; **leaflets** twenty to forty on each leaf, 20–35 cm

Female cone of *Zamia lindenii*.

*Zamia lindenii* From *L'Illustration Horticole* 22:23, Plate 195, (1875)

x 2–3 cm, linear-lanceolate, thin and papery-textured, dark bluish green, shiny, margins with numerous teeth, apex acuminate. **Male cones** cylindrical. **Female cones** about 35 cm x 10 cm; **peduncle** about 8 cm long. **Seeds** 2.4–3 cm x 1.2–2 cm, ovoid, red.

**Distribution and Habitat:** Occurs in Ecuador. Details of habitat are lacking.

**Notes:** This species was described in 1875 from material collected in Ecuador by Benito Roezl. It is a poorly known species the status of which needs elucidation.

**Cultivation:** Probably as for *Z. chigua*.

**Propagation:** From fresh seed.

## *Zamia lindleyi*

Warsz. ex A. Dietrich
(after John Lindley, nineteenth century English botanist)

**Description:** A medium-sized cycad with a trunk to 1 m long and about 15 cm across. **Mature leaves** 1–2 m long, numerous in a graceful arching crown; **leaflets** 10–15 on each leaf, 10–30 cm x 1–1.5 cm, linear-lanceolate, not falcate, not crowded, margins entire, apex acuminate. **Male cones** about 20 cm x 3 cm. **Female cones** about 30 cm x 12 cm. **Seeds** about 3 cm x 2 cm, ovoid, red.
**Distribution and Habitat:** Occurs in Panama but this species is known only from notes and sketches made by its discoverer, Josef von Warscewicz, and it has not been found subsequently.
**Notes:** This species was described in 1851 and no specimens living or dead are known.
**Cultivation:** Probably as for *Z. chigua*.
**Propagation:** From fresh seed.

## *Zamia loddigesii*

Miq.
(after Conrad Loddiges, founder of a famous English nursery)

**Description:** A small to medium-sized cycad with rarely branched, subterranean, contractile stems each to about 10 cm across. **Young leaves** green, slightly hairy. **Mature leaves** 0.4–1 m long, one to six in each crown, erect, straight, stiff, flat in cross-section; **petiole** 5–20 cm long, swollen and shortly hairy at the base, densely armed with prickles; **leaflets** ten to forty on each leaf, 8–26 cm x 0.6–2.7 cm, linear-lanceolate to oblanceolate, moderately spaced, bright green above, paler beneath, glabrous, leathery, somewhat papery, tapered to the base, the margins with numerous small teeth in the distal half, rolled back, apex acute to acuminate. **Male cones** 4–6.5 cm x 1.7–2 cm, cylindrical, brown, shortly hairy; **sporophylls** wedge-shaped, outer face hexagonal; **peduncle** 4–8 cm long, shortly hairy. **Female cones** 4–15 cm x 3.5–6 cm, barrel-shaped to ovoid, with short, brown hairs; **sporophylls** hexagonal; **peduncle** 4–6 cm long, shortly hairy. **Seeds** 1.4–1.8 cm x 0.8–1 cm, ovoid, pink to red.
**Distribution and Habitat:** Occurs in Mexico (Oaxaca, Chiapas, Campeche, Quintana Roo), Belize and Guatemala. This species grows in evergreen and semi-deciduous forests in the mountains between 50 m and 1000 m altitude.
**Notes:** This species was described in 1843 from material grown in the London garden of Conrad Loddiges. It can be recognised by its subterranean contractile stems and bright green, moderately spaced leaflets which have marginal teeth in the distal half. Some researchers believe that *Z. loddigesii* may have originated from a very early stabilised cross between *Z. furfuracea* and *Z. spartea* which has since become established in the wild, commonly on disturbed sites. Artificial hybrids using the above two parents produce progeny which bear a striking similarity to plants of *Z. loddigesii*. Natural hybrids between *Z. loddigesii* and *Z. furfuracea* occur commonly in nature and are always found in disturbed areas. Natural hybrids may also occur between *Z. loddigesii* and *Z. spartea* where the two grow in close proximity.
**Cultivation:** A popular species which is well suited to tropical, subtropical and warm temperate regions. Plants can be grown in full sun or semi-shade.
**Propagation:** From fresh seed and less commonly by division of clumps.

## *Zamia lucayana*

Britton
(after a collector by the name of Lucay)

**Description:** A small cycad with much-branched, subterranean trunk, to 30 cm x 10 cm. **Young leaves** hairy. **Mature leaves** 0.5–1 m long, four to twelve in each crown, dark green, obliquely erect, straight, stiff; **petiole** 20–40 cm long, lacking prickles; **leaflets** twelve to four-

Mature female cone of *Zamia loddigesii*.

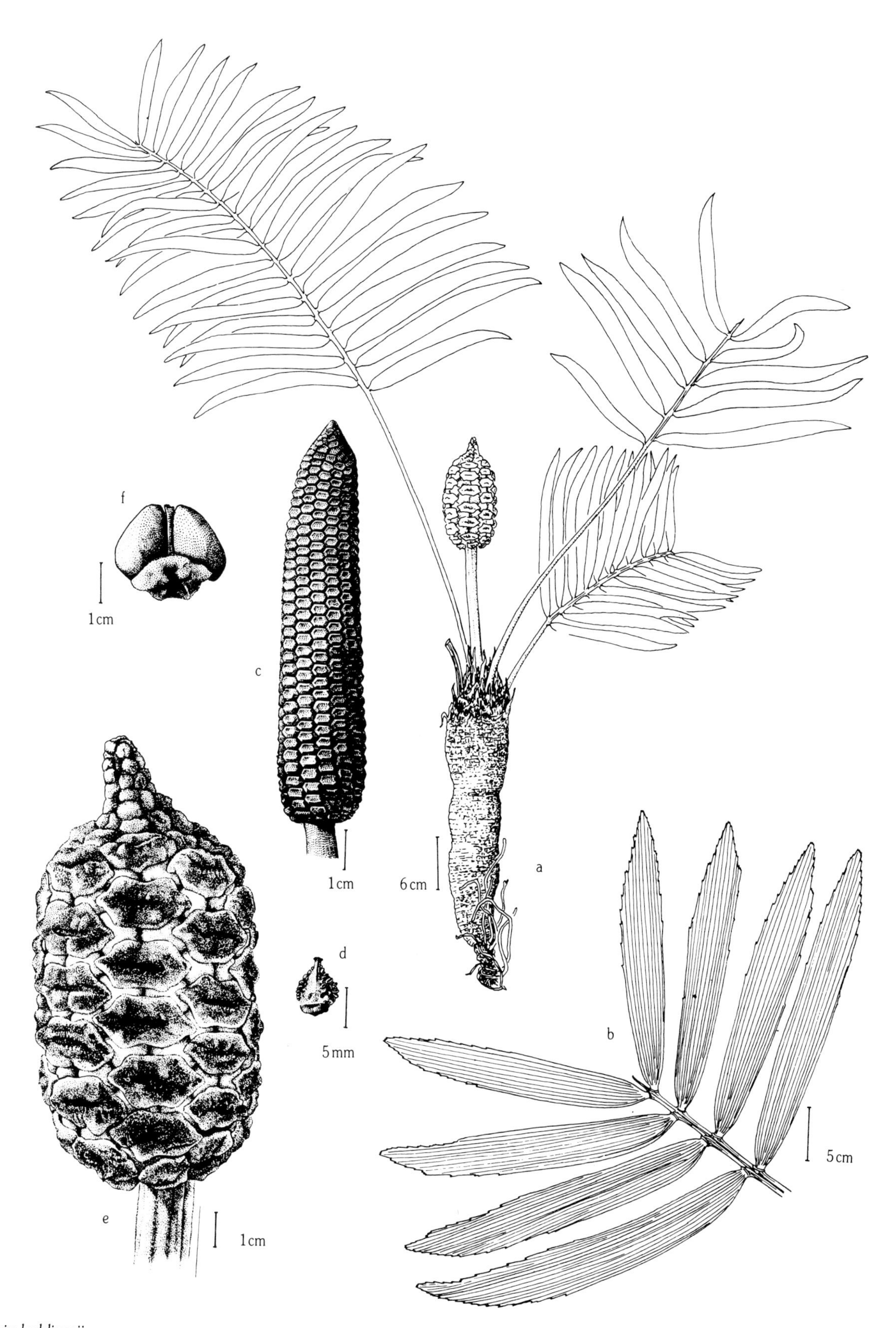


*Zamia loddigesii*
With permission from *Flora de Veracruz*, Fascicle 26, February 1983. Illustration by Elvia Esparza. **a.** detail of adult female plant, **b.** detail of terminal leaflets, **c.** male cone, **d.** male sporophyll, **e.** female cone, **f.** female sporophyll with ovules.

teen on each leaf, 7.5–20 cm x 2–4 cm, oblong to obovate, dark green to yellow-green, stiff, thin and papery, moderately spaced, margins sparsely toothed, thickened and recurved, apex blunt. **Male cones** 4–6 cm x 1–1.5 cm, cylindrical, brown, shortly hairy; **sporophylls** about 6 mm x 4 mm; **peduncle** 2–3 cm long. **Female cones** 5–7 cm x 3–4 cm, oblong, with a short conical apex, shortly hairy. **Seeds** about 1.5 cm x 1 cm, ovoid, red.

**Distribution and Habitat:** Apparently endemic to the Bahamas, where it grows in sparse forests.

**Notes:** This species was described in 1907. Its status is uncertain and somewhat controversial with some authorities claiming it is a distinct species, whereas others include it as part of a complex.

**Cultivation:** A hardy species suitable for subtropical and warm temperate regions. Best growth is probably achieved in a sunny location.

**Propagation:** From fresh seed and by division.

## *Zamia manicata*

Linden ex Regel
(long-sleeved)

**Description:** A small to medium-sized cycad which has a swollen, subterranean, unbranched trunk to 18 cm long and 7 cm across. **Mature leaves** 0.5–2 m long, one to ten in a sparse, arching crown, slightly flexuose towards the apex, bright green, more or less flat in cross-section; **petiole** 20–100 cm long, rigid, deeply grooved, sparsely to densely prickly, swollen at the base, somewhat flexuose towards the apex; **leaflets** twenty to sixty on each leaf, 15–35 cm x 3–7 cm, elliptical to elliptical-lanceolate, widely spaced, each with a distinct petiolule, a thickened glandular, collar-like structure at the point of attachment of the leaflet lamina to the petiolule, directed forwards, thin-textured but leathery, tapered to the base, dark green and shiny above, paler beneath, with numerous small teeth along the margins of the acute to acuminate apex. **Male cones** 4–6 cm x 1–1.5 cm, cylindrical, erect, cream to light brown, covered with short brown hairs; **sporophylls** about 0.4 cm across, hexagonal; **peduncle** 15–30 cm long, very slender. **Female cones** 10–15 cm x 4–7 cm, cylindrical, erect, wine red to dark red-brown, densely covered with short brown hairs, with a conical apex 7–8 cm long; **sporophylls** about 0.9 cm across, hexagonal; **peduncle** 10–15 cm long. **Seeds** 1–1.5 cm x 0.5–0.8 cm, ovoid, red.

**Distribution and Habitat:** Distributed in northern Colombia and southern Panama. In one area this species grows in hot, humid lowland forests which receives more than 5 m of rainfall in a year. It also extends from the lowlands to about 1000 m altitude.

**Notes:** A very distinctive species easily recognised by the petiolules on the leaflets and the unusual collar-like structure present at the point of attachment of the leaflet and the petiolule. This structure is not present on seedlings or on the new leaves of transplanted plants. The closest relative of this species is probably *Z. wallisii* which has much larger leaflets and lacks the collar-like structure. An undescribed species from Amazonian Peru has a similar collar-like structure on the leaflets of mature leaves but in this species the collar or gland is attached obliquely and does not encircle the base. *Z. manicata* was described in 1876 from sterile plants growing in the collection of the Belgian horticulturist, J. Linden. *Z. madida* R. E. Schultes, described in 1958, is a synonym.

**Cultivation:** Suited to tropical and warm subtropical regions. Requires warm, humid conditions, protection from direct, hot sun and a buoyant atmosphere with unimpeded air movement.

**Propagation:** From fresh seed.

## *Zamia montana*

A. Br.
(from the mountains)

**Description:** A small cycad with an unbranched, emergent, cylindrical trunk 1–1.5 m tall and about 20 cm across. **Mature leaves** 1.5–2 m long, three to five in an erect or arching crown, dark green, flat in cross-section; **petiole** 20–40 cm long, sparsely armed with prickles, base swollen and with short, rusty brown hairs; **leaflets**

Underside of leaf of *Zamia manicata* showing collar-like structures on the petiolules.



sixteen to twenty on each leaf, 18–35 cm x 5–10 cm, broadly oblong to broadly lanceolate, somewhat papery textured, dark green, shiny and strongly veined above, paler beneath, rigid, moderately spaced, inserted at right angles to the rhachis, tapered to the base, apex abruptly acuminate with one prominent spine and a few others more obscure. **Male cones** 6–12 cm x 1.5–2.5 cm, cylindrical, erect, brown; **sporophylls** about 0.6 cm across, hexagonal; **peduncle** 2–6 cm long. **Female cones** 25–30 cm x 10–13 cm, barrel-shaped, erect, brown, shortly hairy; **sporophylls** about 1 cm across, hexagonal; **peduncle** 3–6 cm long. **Seeds** 1–1.2 cm x 0.6–0.8 cm, ovoid, yellowish.

**Distribution and Habitat:** Apparently endemic to Colombia where found in mountainous regions between 1800 m and 2600 m altitude.

**Notes:** This poorly known species, described in 1875, was collected on a couple of occasions in the late 1800s and collected only once since, that in 1988. Plants were apparently cultivated in the collection of the English horticulturist James Veitch. Distinguishing features of the species include the long, rigid, strongly veined leaflets which have an acuminate apex with one prominent spine and a few obscure teeth. Originally this species was described as having grooved leaflets, but these grooves were in fact an artifact caused by drying a young leaf.

**Cultivation:** Suited to subtropical regions. Its probable requirements are warmth, humidity, free air movement and light shade.

**Propagation:** From fresh seed.

## *Zamia monticola*

Chamberlain
(dwelling in the mountains)

**Description:** A small cycad with a subterranean to emergent oviod to cylindrical trunk to 1 m long and about 14 cm across. **Young leaves** light green. **Mature leaves** 1–1.7 m long, six to twenty five leaves on each growth apex, forming an obliquely erect to spreading crown, flat in cross-section; **petiole** 0.15–0.4 cm long, densely prickly; **leaflets** thirty to thirty-four on each leaf, 7–26 cm x 2–4 cm, asymmetrically linear-lanceolate to lanceolate, dark green, tapered to the base, entire or the upper leaflets commonly serrulate towards the acute to acuminate apex. **Male cones** 12–16 cm x 3–4 cm, cylindrical, brown, obliquely erect to erect; **sporophylls** hexagonal; **peduncle** 10–17 cm long horizontal or upcurved near the apex. **Female cones** unknown. **Seeds** unknown.

**Distribution and Habitat:** Recorded from Veracruz, Mexico, where it was reported to grow in mountainous districts covered with dense, evergreen forests.

**Notes:** This species was discovered under unusual circumstances when a *Zamia* seedling was noticed in a batch of *Ceratozamia mexicana* collected by Charles Chamberlain from Mexico in 1906. No species of *Zamia* were known from the locality and the plant produced a male cone in 1915. Only this solitary specimen, grown from seed, was ever found. The plant was studied over a number of years and after consulting herbarium specimens, Chamberlain described the plant as *Z. monticola* in 1926, even though female plants were unknown at the time.

Since its description several fruitless searches for plants of the species have been conducted in the vicinity of the purported type locality which is the extinct volcanic crater of Naolinco near Jalapa in Mexico. Andrew Vovides is of the opinion that the species is probably of Central American origin and the solitary seed somehow became mixed up with Chamberlain's Mexican collections. Dennis Stevenson believes that the species may actually have been *Z. muricata* from Venezuela.

**Cultivation:** Probably suitable for subtropical and warm temperature regions.

**Propagation:** From seed.

## *Zamia muricata*

Willd.
(roughened, with hard points)

**Description:** A small cycad which in nature develops a simple or branched subterranean trunk to 14 cm across. **Young leaves** light green. **Mature leaves** 1–1.5 m long, numerous in an obliquely erect crown, flat in cross-section; **petiole** 5–10 cm long, swollen and shortly hairy at the base, bearing prickles; **leaflets** twenty to twenty-eight on each leaf, 20–35 cm x 4–8 cm, lanceolate to oblong, dark green and shiny above, paler beneath, tapered to the base, the margins beset with numerous spine-like teeth, apex acuminate. **Male cones** 3–6 cm x 0.8–1 cm, cylindrical, erect, brown; **sporophylls** about 0.6 cm long, outer face hexagonal; **peduncle** 4–11 cm long. **Female cones** 5–11 cm x 3–3.5 cm, barrel-shaped, brown, shortly hairy; **peduncle** 4–6 cm long, shortly hairy. **Seeds** 2.5–3 cm x 1.8–2 cm, ovoid, red to scarlet.

**Distribution and Habitat:** Distributed in northern Venezuela and adjacent Colombia. It grows in a variety of habitats including dry pine forests on ridges and moist to wet forested slopes at about 400–600 m altitude.

**Notes:** A poorly known species which was described in 1806 from material collected in Venezuela. The leaves, having long, broad leaflets with numerous, spine-like teeth on the margins, are distinctive.

**Cultivation:** Suitable for tropical and warm subtropical regions but uncommonly grown. Revels in shady, warm conditions with high humidity and bouyant air movement.

**Propagation:** From fresh seed or less commonly from division.

D. W. STEVENSON

Zamia muricata in habitat.

Mature female cones of *Zamia muricata*.

## *Zamia neurophyllidia*

D. Stevenson
(with prominent veins in the leaflets)

**Description:** A small cycad with an unbranched, erect, emergent trunk to 0.6 m tall and 12 cm across. **Mature leaves** 0.5–1 m long, three to ten in each crown, erect to obliquely erect, dark green; **petiole** 10–30 cm, densely prickly; **leaflets** ten to twenty on each leaf, 12–20 cm x 6–10 cm, elliptical, grooved between the veins on the upper surface, dark green, margins finely toothed towards the acuminate apex. **Male cones** not recorded. **Female cones** 5–8 cm 1–2 cm, cylindrical, cream to light brown; **peduncle** short. **Seeds** 1–1.5 cm across, ovoid, red.
**Distribution and Habitat:** Endemic to Panama where it grows in rainforest at low altitudes (100–200 m), in areas that are occasionally flooded.
**Notes:** This species, described in 1993, resembles a small version of *Zamia skinneri*. Plants that have been cultivated for nearly thirty years maintain these diminutive features.
**Cultivation:** Suited to tropical and warm subtropical regions. Requires warm, humid conditions in a sheltered situation.
**Propagation:** From fresh seed.

## *Zamia obidensis*

Ducke
(from the Rio Branco de Obido River, Brazil)

**Description:** A small cycad with an unbranched subterranean trunk. **Mature leaves** 0.6–1.2 m long, three to five in an erect crown, dark green, flat in cross-section; **petiole** 50–70 cm long, with numerous, small, sharp prickles, swollen at the base; **leaflets** twelve to eighteen on each leaf, 20–30 cm x 2–3.5 cm, linear-ovate to ovate-lanceolate, widely spaced and uncrowded, rusty brown near the base, apical margins with a few short sharp teeth; apex acute. **Male cones** 6–7 cm x 1.2–1.5 cm, cylindrical, brown, shortly hairy; **sporophylls** hexagonal; **peduncle** 6–8 cm long, shortly hairy. **Female cones** 12–15 cm x 4–5 cm, barrel-shaped, covered with short, purple-brown hairs; **sporophylls** about 2 cm across, hexagonal; **peduncle** 10–15 cm long, shortly hairy. **Seeds** 1.7–2 cm x 1–1.2 cm, ovoid, red.
**Distribution and Habitat:** Endemic to Brazil where it grows on forested slopes.
**Notes:** This poorly known species, described in 1922, is reported to have leaflets approximately intermediate in size between *Z. lecointei* and *Z. ulei*. Plants from the original collection were apparently introduced into cultivation but their fate is uncertain. An interesting feature of this species is the brown basal area which is found on most leaflets.

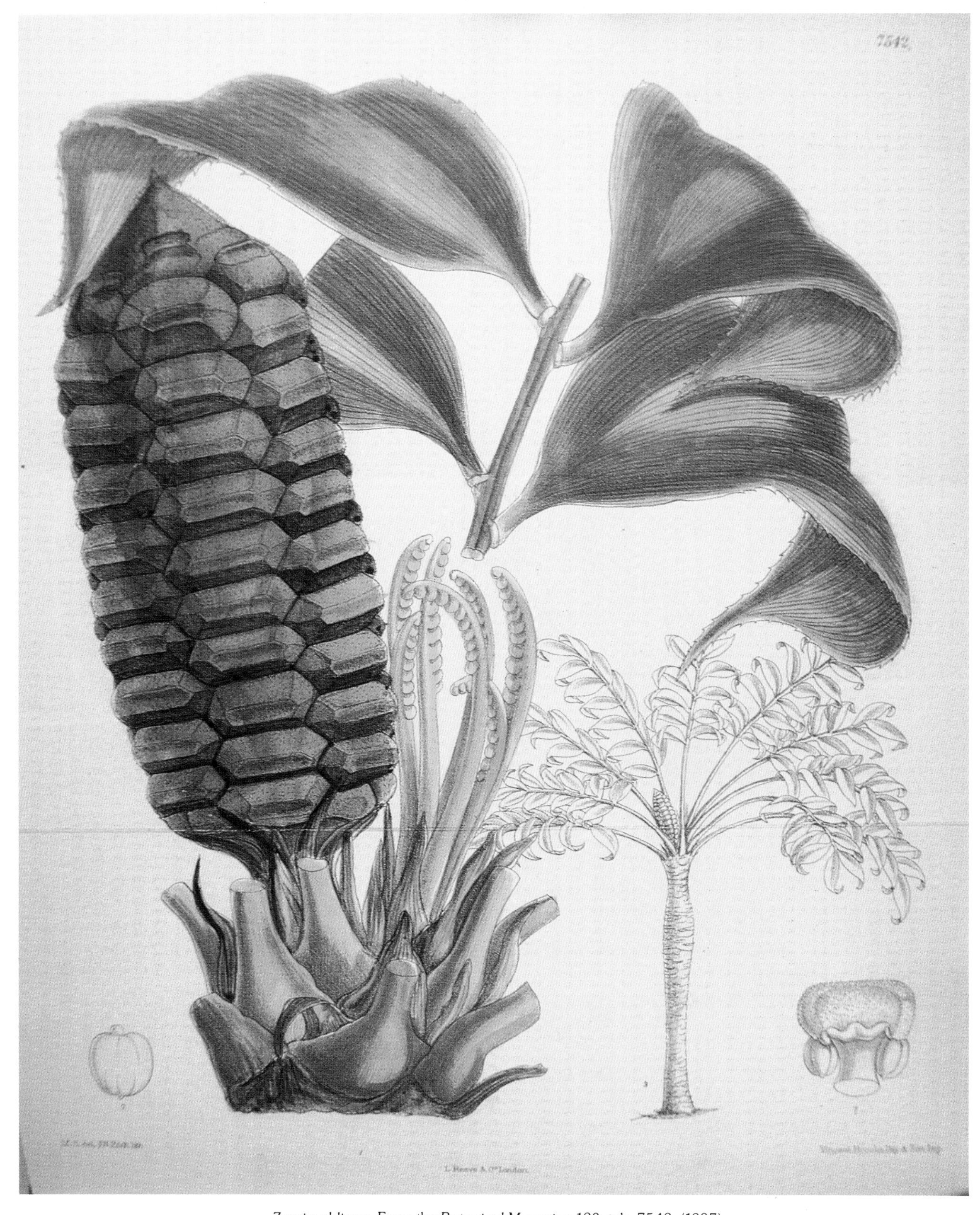

*Zamia obliqua* From the *Botanical Magazine* 123, tab. 7542, (1897)

**Cultivation:** Suited to tropical and warm subtropical regions. Very little seems to be known about the cultural requirements of this species.
**Propagation:** From fresh seed.

Zamia obliqua.

# *Zamia obliqua*

A. Br.

(at an oblique angle, an apparent reference to the leaflets)

**Description:** A medium-sized cycad with an emergent, slender, unbranched, cylindrical grey trunk to 5 m tall and 15 cm across. **Young leaves** pale green, shiny, lightly hairy. **Mature leaves** 0.6–1 m long, five to twenty in an attractive, arching crown, dark green and shiny, flat in cross-section; **petioles** 36–70 cm long, prickles usually absent or when present small, slender, when young covered with short, brown pubescence which is easily removed; **leaflets** twenty to forty on each frond, 5–10 cm x 3–6 cm, elliptical to obovate, light green and shiny above, paler beneath, widely spaced, inserted at about 60° to the rhachis, with numerous short, sharp teeth towards the acuminate apex. **Male cones** 4–6 cm x 1–1.5 cm, cylindrical, erect, cream to light brown, covered with short, brown hairs; **sporophylls** about 1 cm across, hexagonal, with sporangia on both the upper and lower surfaces; **peduncle** 2–4 cm long. **Female cones** 15–25 cm x 5–8 cm, barrel-shaped, erect, with an acute, conical apical tail, cream to pale brown, shortly hairy; **sporophylls** about 3 cm across, hexagonal; **peduncle** 3–4 cm long, thick. **Seeds** 1–1.5 cm x 0.5–0.8 cm, ovoid, red.

**Distribution and Habitat:** Occurring in Colombia, central Panama and possibly also in central Peru. This species grows in rainforest at low to moderate altitudes (from sea level to 500 m).

**Notes:** Although described in 1875, this species was first collected in 1847 from the nearly inaccessible forests of Cape Corrientes situated on the west coast of Colombia. The original collector was the botanist Dr Berthold Carl Seemann attached to the ship *Herald*, the collection being made during a provisioning stop. Living plants were sent to Kew Gardens but did not survive, although the species was recorded from Kew in 1880. *Z. obliqua* can be distinguished from related species by its arborescent trunk and light green leaflets with sharp spines near the tips. It is one of the most attractive species in the genus.

**Cultivation:** Suited to tropical and warm subtropical regions. Little is known about this species in cultivation.

**Propagation:** From fresh seed.

Zamia paucijuga

## *Zamia ottonis*

Miq.
(after E. Otto, the original collector)

**Description:** A small cycad with branched, subterranean trunks about 10 cm across. **Young leaves** shortly hairy. **Mature leaves** 0.3–0.5 m long, about eight in each crown, dark green, stiff; **petiole** 5–20 cm long, lacking prickles; **leaflets** twenty to thirty-two on each leaf, 5–12 cm x 1–2 cm, oblong to lanceolate or elliptical, thin-textured and papery, margins recurved, with small teeth. **Male cones** about 4 cm x 2 cm, cylindrical, covered with short, brown hairs; **sporophylls** about 5 mm x 6 mm; **peduncle** with brown and white hairs. **Female cones** 3.5–4 cm x 1.8–2.5 cm, ovoid, greenish brown, shortly hairy; **peduncle** about 1 cm long. **Seeds** about 1.5 cm x 1 cm, ovoid, red.
**Distribution and Habitat:** Endemic to Cuba where it grows in sandy soil in pine forests.
**Notes:** This species was described in 1843 from material collected in 1839. Its status is uncertain and in need of elucidation.
**Cultivation:** A hardy species suitable for subtropical and warm temperate regions. Best growth is probably achieved in a sunny location.
**Propagation:** From fresh seed and by division.

## *Zamia paucijuga*

Wieland
(with few leaflets)

**Description:** A small cycad with branched, subterranean stems each to about 12 cm across. **Young leaves** light green, slightly hairy. **Mature leaves** 0.4–1 m long, two to eight in each crown, erect, straight, stiff, flat in cross-section; **petiole** 4–18 cm long, swollen and shortly hairy at the base, armed with stout prickles; **leaflets** ten to thirty on each leaf, moderately spaced, 8–20 cm x 1.2–3 cm, linear-oblanceolate to linear-obovate, stiff, leathery, glabrous, bright green above, paler beneath, tapered to the base, the margins with numerous small teeth in the distal half, apex obtuse to acute. **Male cones** 7–10 cm x 1.5–2 cm, cylindrical, brown, shortly hairy; **sporophylls** wedge-shaped, outer face hexagonal; **peduncle** 3–6 cm long, shortly hairy. **Female cones** 7–10 cm x 3.5–5 cm, barrel-shaped to ovoid, brown, shortly hairy; **sporophylls** hexagonal; **peduncle** 3–5 cm long, shortly hairy. **Seeds** 1.6–2 cm x 0.8–1 cm, ovoid, red.
**Distribution and Habitat:** Endemic to Mexico where widely distributed in the west, along the Pacific coast and also on Maria Cleofas Island. This cycad grows in forests dominated by species of *Pinus* and *Quercus* in dry climates which have a strongly seasonal rainfall of about 1000 mm per annum.
**Notes:** This species was described in 1916. Some researchers consider this species to be a stabilised hybrid, possibly between *Z. furfuracea* and *Z. loddigesii*. This is unlikely given chromosome evidence, its distribution (it occurs on the other side of the Sierra Madre Oriental from the purported parents), its very wide distribution, abundance and a lack of intermediate features.
**Cultivation:** Suitable for subtropical and warm temperate regions. Can be grown in sun or partial shade.
**Propagation:** From fresh seed or less commonly by division.

## *Zamia picta*

Dyer
(coloured, painted)

**Description:** A small cycad with a subterranean trunk to 20 cm long and 8 cm across. **Mature leaves** 1–2 m long, numerous in an obliquely erect crown, flat in cross-section; **petiole** 10–30 cm long, bearing prickles; **leaflets** ten to fifty on each leaf, 10–35 cm x 3–6 cm, oblong to lanceolate, thin-textured to membranous, yellowish-green to bluish green with whitish patches and spots, irregularly toothed towards the tail-like acuminate apex. **Male cones** 4–8 cm x 1–1.5 cm, cylindrical, brown. **Female cones** 8–14 cm x 2–4 cm, barrel-shaped, brown. **Seeds** 1.5–2 cm x 1.2–1.6 cm, ovoid, red.
**Distribution and Habitat:** Occurs in Chiapas, Mexico, western Guatemala and south-western Belize. It grows in moist rainforests at low altitudes.
**Notes:** This species was described in 1884 from material

D. W. STEVENSON

Leaflet of *Zamia picta*.

collected in Mexico. It has unusual leaflets which have a spotted or blotched appearance.
**Cultivation:** Suited to tropical and warm subtropical regions. Requires warm, humid conditions with free air movement and light shade.
**Propagation:** From fresh seed.

## *Zamia poeppigiana*

Martius & Eichler
(after Eduard Frederich Poeppig, original collector of the species)

**Description:** A medium-sized cycad with an emergent, unbranched, cylindrical trunk to 2.5 m tall and 25 cm across. **Young leaves** pale green, covered with short brown hairs. **Mature leaves** 1.5–2.5 m long, ten to twenty in a graceful, erect to arching crown, more or less flat in cross-section although the **leaflets** often recurved; **petiole** 20–40 cm long, sturdy, bearing stout prickles to 3 mm long; **leaflets** forty to eighty on each leaf, 20–35 cm x 2–3 cm, linear-lanceolate to lanceolate, falcate, moderately spaced, inserted at right angles to the rhachis, sessile, dark green to glaucous and shiny, somewhat papery textured, the distal margins with numerous, short, sharp teeth, apex acuminate. **Male cones** 18–27 cm x 2–4 cm, long-cylindrical, erect, pale brown; **sporophylls** about 0.5 cm across, hexagonal; **peduncle** 5–7 cm long, shortly hairy. **Female cones** 25–35 cm x 8–11 cm, barrel-shaped, erect, covered with short, rusty brown hairs; **sporophylls** about 1.2 cm across, hexagonal; **peduncle** 6–8 cm long. **Seeds** 2.5–3 cm x 1.2–2 cm, obovoid, yellowish to pink.
**Distribution and Habitat:** Widely distributed in the rainforests of Ecuador and Peru from moderate altitudes of about 300 m to more than 1000 m in the mountains.
**Notes:** This species was described in 1863 from material collected by the German collector Poeppig in the early 1830s during his numerous trips along the rivers and forests of the Amazon Region of Peru. *Z. lindenii* is considered by some authorities to be a synonym of this species.
**Cultivation:** Suited to tropical and warm subtropical regions. *Z. poeppigiana* needs warm, humid conditions with free air movement and light shade or filtered sun.
**Propagation:** From fresh seed.

## *Zamia portoricensis*

Urban
(from Puerto Rico)

**Description:** A small cycad with much-branched, subterranean stems, each to about 6 cm across. **Young leaves** covered with short, rusty brown hairs. **Mature leaves** 0.5–1.5 m long, six to twelve leaves in each crown, obliquely erect, straight, stiff, flat or slightly vee-ed in cross-section; **petiole** 20–60 cm long, lacking prickles; **leaflets** twenty to eighty on each leaf, 18–35 cm x 1–3 cm, linear, dark green, stiff, leathery, moderately spaced, the margins entire or sometimes with a few bumps, apex blunt. **Male cones** 6–12 cm x 1.5–3 cm, cylindrical, reddish, shortly hairy; **sporophylls** hexagonal; **peduncles** 4–8 cm long, densely covered with short hairs. **Female cones** 8–14 cm x 2–3 cm, ovoid, with a short, acuminate, apical tail, reddish; **sporophylls** hexagonal with a flat centre; **peduncle** 5–10 cm long, shortly hairy. **Seeds** 1.5–2.5 cm x 1–2 cm, ovoid, blood red.
**Distribution and Habitat:** Endemic to Puerto Rico where it is found growing in soils derived from serpentine in the Susua Forest and adjacent areas to the coast.
**Notes:** This species, described in 1899, is part of the complex of species surrounding *Z. pumila* and *Z. integrifolia*. It can be distinguished from all others by its proportionately much longer leaflets the margins of which lack teeth or spines.
**Cultivation:** This species, which is very uncommon in cultivation, is best suited to subtropical and warm temperate regions. Plants will grow in either sunny positions or shade and appear to mainly limited to serpentine soils.
**Propagation:** From fresh seed or less commonly by division.

Female cones of *Zamia portoricensis*.

*Zamia pseudoparasitica* in cultivation.

## *Zamia pseudomonticola*

L. D. Gomez
(literally a false *Zamia monticola*)

**Description:** A small cycad with an erect trunk to 30 cm tall and 7 cm across, lacking persistent leaf bases. **Young leaves** pale green. **Mature leaves** 0.8–1 m long, three to ten on each growth apex in an erect to spreading crown, flat in cross-section; **petiole** 5–10 cm long, glabrous, lacking prickles; **leaflets** eighteen to twenty-eight on each leaf, 15–22 cm x 4–7 cm, elliptical-lanceolate, falcate, dark green, smooth, tapered to the base, margins entire, apex acuminate, recurved or directed downwards. **Male cones** 8–9.5 cm x 4–6 cm, cylindrical, erect. **Female cones** barrel-shaped, covered with short, brown hairs. **Seeds** 1.3–1.7 cm across, nearly spherical, apricot coloured.
**Distribution and Habitat:** Endemic to Costa Rica where it is found growing in mixed evergreen forests of the highlands at about 1300 m altitude. This species may also occur in western Panama.
**Notes:** As the specific epithet suggests, this species, which was described in 1982, may be confused with *Zamia monticola*. Distinguishing features include the leaf bases shedding from the trunk, the glabrous petioles which lack prickles and the acuminate leaflets with the tips either straight or directed downwards.
**Cultivation:** Best suited to cooler tropical and subtropical regions. An attractive species which is little known in cultivation. Responds to warmth, humidity and free air movement.
**Propagation:** From fresh seed.

## *Zamia pseudoparasitica*

Yates
(falsely parasitic, in reference to its epiphytic habit)

**Description:** An epiphytic cycad with an unbranched, cylindrical, recurved trunk to 1 m long and 15 cm across, the outside with clusters of adventitious roots. **Mature leaves** 1–3 m long, pendant, three to ten in a sparse, dangling crown, grey-green, flat or inversely vee-ed in cross-section; **petioles** 30–100 cm long, slender, sparsely prickly or nearly smooth; **leaflets** forty to sixty-four on each leaf, 20–50 cm x 2–4 cm, oblanceolate, falcate, recurved, grey-green, moderately crowded, inserted at about 50° to the rhachis, margins entire, apex acuminate. **Male cones** 25–50 cm x 2–4 cm, cylindrical, cream to light brown. **Female cones** 25–50 cm x 8–12 cm, cylindrical to ovoid, yellow-green to light brown. **Seeds** 1.5–2.5 cm x 1–1.5 cm, ovoid, yellow, sarcotesta becoming mucilaginous.
**Distribution and Habitat:** Endemic in coastal forests of Panama on the Atlantic side. It is apparently restricted to steep slopes in primary forest and grows on suitable trees in regions of high rainfall at altitudes from 50 m to 1000 m.
**Notes:** This remarkable cycad is one of only two true epiphytes in the whole order of Cycadales (the other, another species of *Zamia* from Ecuador, is apparently undescribed). *Z. pseudoparasitica* favours trees that are devoid of other epiphytes such as bromeliads, ferns and orchids. The trunk of the cycad is firmly attached to the tree by the tap root and numerous adventitious roots and its leaves dangle and wave in the breeze. Although it was

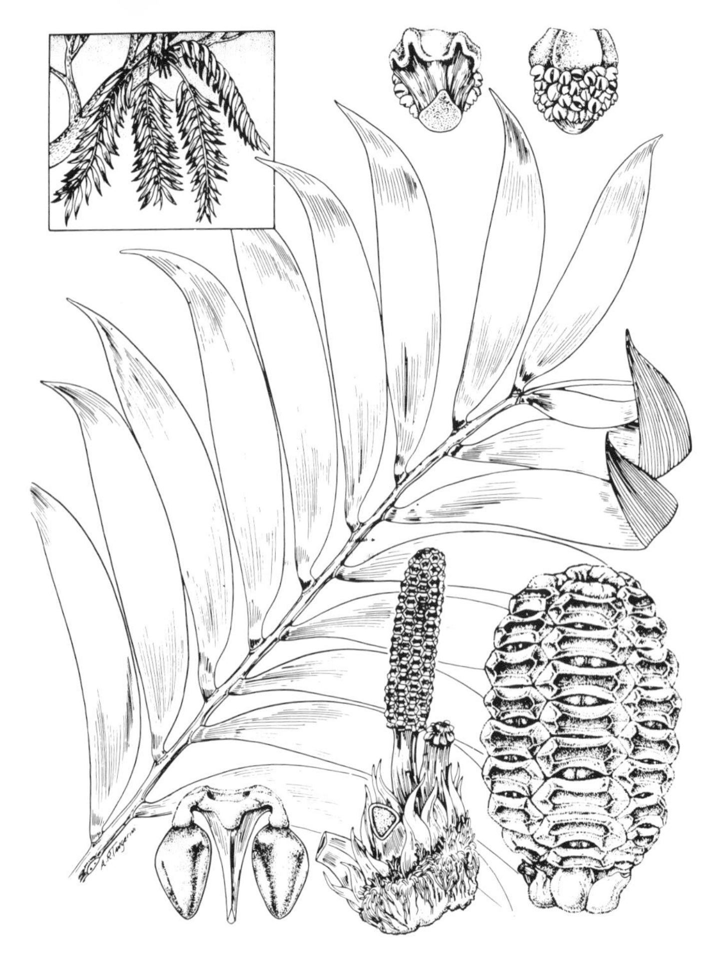

*Zamia pseudoparasitica* With permission from *Lyonia* 2(4), 34, (1986). Illustration by A.R. Tangerini.

first described in 1854, very little has been written about this species and there are still many gaps in our knowledge. The trunk of a young plant is upright but as it grows the weight of the crown pulls it down and eventually on old plants the trunk becomes curved or even contorted with the apex still growing erect. Mature plants have only been found growing as epiphytes and seeds which fall on the ground germinate but die without becoming established. Birds or fruit bats are the probably distribution agents for the ripe seeds which have a distinctive sour smell.

This cycad produces its coralloid roots in dense hemispherical clusters of 5–25 cm across. Its leaves, cones and seeds are among the largest in the genus.

**Cultivation:** Suited to tropical and warm subtropical regions. This species is poorly known in cultivation but has been grown successfully at the Fairchild Tropical Gardens in Florida. It is best treated as a container plant, being grown in a well-drained, open, fibrous mix. Plants require warm, humid conditions with free, unimpeded air movement.

**Propagation:** From fresh seed.

## *Zamia pumila*

L.

(short, dwarf)

**Description:** A small cycad with much-branched, subterranean stems, each to about 6 cm across. **Young leaves** covered with short, rusty brown hairs. **Mature leaves** 0.6–1.5 m long, four to twelve leaves in each crown, obliquely erect, straight, stiff, flat or slightly vee-ed in cross-section; **petiole** 20–80 cm long, lacking prickles; **leaflets** twenty to eighty on each leaf, 10–30 cm x 1–3 cm, oblong with parallel margins, dark green, stiff, leathery, moderately spaced, the margins with well-defined teeth in the distal quarter, apex blunt, frequently deeply cleft in vigorous plants. **Male cones** 6–11 cm x 1.5–3 cm, cylindrical, red to reddish brown, hairy; **sporophylls** hexagonal; **peduncle**s 4–8 cm long, densely hairy. **Female cones** 8–14 cm x 2–3 cm, ovoid, with a conical, long-acuminate apical tail, red to reddish brown; **sporophylls** hexagonal with a flat centre; **peduncle** 5–10 cm long, hairy. **Seeds** 1.5–2.5 cm x 1–2 cm, ovoid, orange to orange-red.

**Distribution and Habitat:** Distributed in central Cuba, the Dominican Republic and southern Puerto Rico. This species grows in grassland and open forests dominated by pines.

**Notes:** This species, the type of the genus *Zamia*, was the first cycad to be described, that event occurring in 1659. *Z. pumila* has been confused with *Z. integrifolia* which has poorly developed leaflet teeth these appearing as small, callous bumps on the leaflet margins near the tips. The tapered conical tail on the female cone of *Z. pumila* also differs from the truncate apex of *Z. integrifolia*.

**Cultivation:** A hardy species suitable for subtropical and warm temperate regions. Best growth and appearance is achieved in a sunny location.

**Propagation:** From fresh seed or less commonly from division.

D. W. STEVENSON

Young leaf of *Zamia purpurea*.

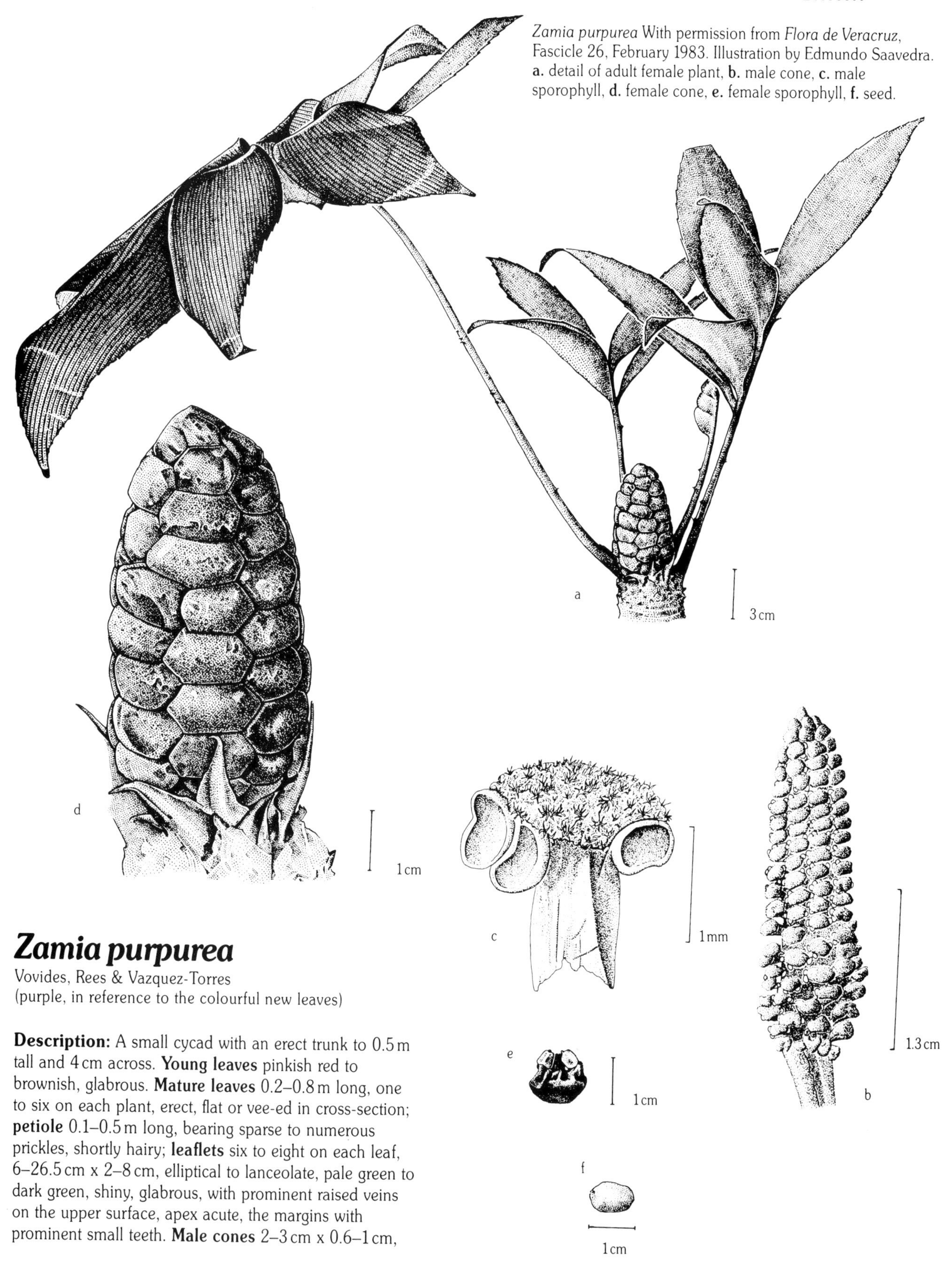


*Zamia purpurea* With permission from *Flora de Veracruz*, Fascicle 26, February 1983. Illustration by Edmundo Saavedra. **a.** detail of adult female plant, **b.** male cone, **c.** male sporophyll, **d.** female cone, **e.** female sporophyll, **f.** seed.

# *Zamia purpurea*

Vovides, Rees & Vazquez-Torres
(purple, in reference to the colourful new leaves)

**Description:** A small cycad with an erect trunk to 0.5 m tall and 4 cm across. **Young leaves** pinkish red to brownish, glabrous. **Mature leaves** 0.2–0.8 m long, one to six on each plant, erect, flat or vee-ed in cross-section; **petiole** 0.1–0.5 m long, bearing sparse to numerous prickles, shortly hairy; **leaflets** six to eight on each leaf, 6–26.5 cm x 2–8 cm, elliptical to lanceolate, pale green to dark green, shiny, glabrous, with prominent raised veins on the upper surface, apex acute, the margins with prominent small teeth. **Male cones** 2–3 cm x 0.6–1 cm,

cylindrical, erect, dark brown; **sporophylls** about 0.3 cm long, wedge shaped; **peduncle** about 1 cm long, shortly hairy. **Female cones** 6–7 cm x 3–3.5 cm, cylindrical to narrowly ovoid, with hardly any apical projection; **sporophylls** about 1.8 cm long, wedge-shaped; **peduncle** about 2 cm long. **Seeds** about 1 cm x 0.7 cm, ovoid, smooth, red.
**Distribution and Habitat:** Endemic to Veracruz and Oaxaca in southern Mexico, growing in evergreen forests at between 100 m and 150 m altitude.
**Notes:** This species, which was described in 1983, can be recognised by its shiny, toothed leaflets which have prominent raised veins on the upper surface and by its small cones. Its natural populations have been greatly reduced by poaching and the species is now almost extinct in the wild.
**Cultivation:** A highly ornamental species prized for its colourful new leaves. Plants are suitable for subtropical regions and require a sheltered position.
**Propagation:** From fresh seed.

*Zamia pygmaea*

Female cones of *Zamia pygmaea*.

D. W. STEVENSO

## *Zamia pygmaea*

Sims
(pigmy, tiny, dwarf)

**Description:** A small cycad with much-branched, subterranean stems, each to about 2 cm across. **Young leaves** covered with short, rusty brown hairs. **Mature leaves** 0.3–0.7 m long, one to five in each crown, obliquely erect, straight, stiff, flat or slightly vee-ed in cross-section; **petiole** 5–20 cm long, lacking prickles; **leaflets** sixteen to forty on each leaf, 1–7 cm x 1.5–3 cm, obovate to ovate, dark green, stiff, thin-textured, crowded, the apex broadly obtuse and with numerous small teeth. **Male cones** 1.5–5 cm x 1–1.5 cm, narrow-cylindrical, grey to blackish, shortly hairy; **sporophylls** hexagonal; **peduncle** 2–4 cm long, shortly hairy. **Female cones** 2.5–7 cm x 1.5–2.5 cm, ovoid, with a long tapered, conical apical tail, grey to blackish, shortly hairy; **sporophylls** hexagonal, with a flat centre; **peduncle** 2–4 cm long, shortly hairy. **Seeds** 1–1.5 cm x 1 cm, ovoid, orange to orange-red, sometimes grooved.
**Distribution and Habitat:** Endemic to central and western Cuba and the Isle of Pines where it commonly grows in white sandy soil in grassland and pine forests.
**Notes:** This species, the smallest of all living cycads, was described in 1741. It is closest to *Z. amblyphyllidea* but is readily distinguished by its dwarf habit with slender stems, shorter, narrower leaflets and smaller cones. Intermediate forms may occur.
**Cultivation:** A hardy species suitable for subtropical and

warm temperate regions. A very ornamental species which is especially useful because of its small dimensions. Makes an excellent plant for container culture. Will grow readily in a sunny location.
**Propagation:** From fresh seed and less commonly by division of clumps.

## *Zamia roezlii*

Linden
(after Benito Roezl, the original collector)

**Description:** A medium-sized to large cycad which in nature develops an unbranched, erect or reclining trunk to 7 m long and 15 cm across. **Young leaves** pale green, glabrous. **Mature leaves** 2–3 m long, six to twenty in a graceful arching crown, bright green, flat in cross-section; **petiole** 30–50 cm long, swollen at the base, shortly hairy, densely covered with prickles; rhachis arching, prickly; **leaflets** twenty to fifty-four on each leaf, 30–40 cm x 3–4 cm, lanceolate, somewhat falcate, moderately spaced at the base, crowded towards the apex, glabrous, dark green and shiny above with prominent raised veins impating a grooved appearance, paler beneath, tapered to the base, margins entire, apex acute. **Male cones** 30–50 cm x 4–6 cm, long-cylindrical, erect, brown, shortly hairy; **sporophylls** about 1.8 cm long, wedge-shaped, the outer face hexagonal; **peduncle** about 5 cm long, shortly hairy. **Female cones** 35–55 cm x 10–14 cm, broadly cylindrical to barrel-shaped, dark brown, shortly hairy; **sporophylls** about 4 cm long; **peduncle** about 5 cm long, shortly hairy. **Seeds** 3.5–4 cm x 1.7–2.5 cm, ovoid, red to scarlet.
**Distribution and Habitat:** Endemic to Colombia. This species grows in equatorial forests and swampy mangrove forests of the Pacific coastal lowlands, usually not far above sea level; in fact some plants may be flooded by high tides. Soils have been described as brackish mud.
**Notes:** This remarkable species was described in 1873 from material cultivated in the garden of J. Linden in Belgium. Plants are extremely long-lived as evinced by the very long trunks which frequently fall over, form adventitious roots where they come in contact with the soil and grow upright again from the apex. In the equatorial climate where they grow, the plants are non-cyclic, producing cones throughout the year. The female cones are the largest of all species of *Zamia* and the spermatozoids are the largest of any cycad known. Researchers consider *Z. roezlii* to be the most primitive member of the genus. The large ripe seeds are an important food source for local people; preparation consists of pounding to flour and repeated leaching in water to remove toxic compounds. *Z. roezlii* is known by the confusing local name 'chigua' and consequently has been confused with *Z. chigua*. It can be distinguished from that species by its fewer, much less crowded leaflets which are grooved.

Young leaves of *Zamia roezlii*.

**Cultivation:** Best suited to tropical and warm subtropical regions. Requires shade, humidity and free air movement, creating a buoyant atmosphere. Very frost-sensitive and plants should not be allowed to dry out to any extent.
**Propagation:** From fresh seed.

## *Zamia skinneri*

Warsz. ex A. Dietrich
(after George Ure Skinner, nineteenth century plant collector in Central and South America)

**Description:** A medium-sized cycad with an unbranched trunk which is either subterranean or emergent to 2 m long and 15 cm across. **Young leaves** pale green, sparsely hairy. **Mature leaves** 1–2 m long, three to six in an erect crown, glossy green, flat in cross-section; **petiole** 50–100 cm long, sparsely to densely prickly, olive green; **leaflets** four to twenty-two on each leaf, 20–50 cm x 10–15 cm, elliptical to broadly lanceolate, to broadly ovate, olive green, glossy, deeply grooved on the upper surface between the veins, moderately spaced, the distal margins with prominent teeth, apex acute to acuminate. **Male cones** 8–12 cm x 1–2 cm, cylindrical, borne in groups of two to ten, cream to light brown, covered with short, chestnut brown hairs; **sporophylls** about 0.6 cm across, hexagonal; **peduncle** 2–6 cm long. **Female cones** 20–50 cm x 8–12 cm, barrel-shaped to ovoid, apical tail conical, brown, covered with short, red-brown hairs; **sporophylls** about 1.2 cm across, hexagonal; **peduncle** 2–

7 cm long, shortly hairy. **Seeds** 1.5–2.5 cm x 0.5–0.6 cm, ovoid, red.
**Distribution and Habitat:** Occurs in southern Nicaragua, Costa Rica and Panama. This species grows in the lowlands and mountains in the shady understories of rainforests, at altitudes of 50–750 m.
**Notes:** A distinctive, large species which was described in 1851. Plants from the original collection were successfully cultivated in England by James Yates as early as 1852. Plants originating from Panama are reported to be exceptionally vigorous with glossy green, deeply corrugated leaflets as large as canoe paddles. According to Bob Dressler this is typical of true *Z. skinneri* and he has noted the existence in Panama of two other taxa which have similar but smaller leaflets which are deeply grooved between the veins and have prominently toothed margins. One taxon develops smaller plants with deeply veined leaflets, smaller cones and smaller seeds (see *Z. neurophyllidea*). The other has a subterranean trunk and develops only one leaf at a time (see *Z. dressleri*).
**Cultivation:** Suited to cool tropical and warm subtropical regions. Best results are achieved in warm, humid conditions with buoyant air movement and light to heavy shade.
**Propagation:** From fresh seed.

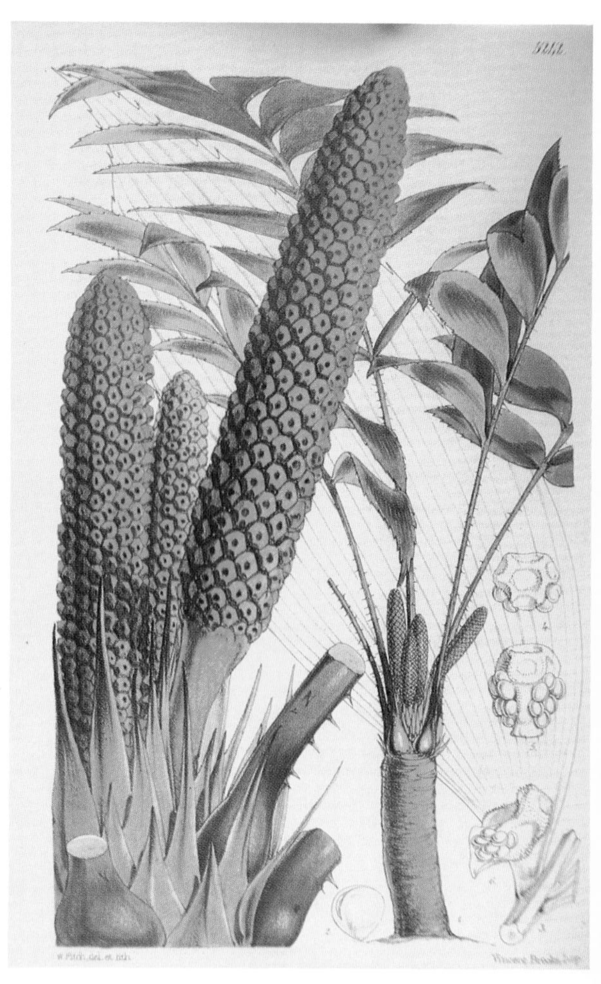

*Zamia skinneri*
Taken from the *Botanical Magazine* 87, tab. 5242, (1861)

# *Zamia soconuscensis*

Schutzman, Vovides & Dehgan
(after Sierra del Soconusio, the mountain range where the species is native)

**Description:** A small cycad with an erect or decumbent, solitary or branched trunk to 0.5 m tall and 25 cm across. **Young leaves** light green, with scattered white hairs. **Mature leaves** 0.4–1.9 m long, one to many on each plant, widely spreading, flat in cross-section; **petiole** 14–28 cm long, bearing numerous prickles, the base greatly swollen and lacking prickles; **leaflets** 70–90 on each leaf, 14–33 cm x 0.6–1.4 cm, lanceolate, falcate, glabrous, somewhat papery, tapered to the base, apex acute, margins entire. **Male cones** 8–9.5 cm x 2 cm, cylindrical, erect, dark brown; **sporophylls** about 0.5 cm long, wedge-shaped, the outer face hexagonal; **peduncle** 4–5.5 cm long, shortly hairy. **Female cones** 12–21 cm x 7–10 cm, barrel-shaped, with an extended conical apex, erect, dark brown, shortly hairy; **sporophylls** about 4 cm long; **peduncle** 3–3.4 cm long, shortly hairy. **Seeds** 2.2–2.6 cm x 1.4–1.9 cm, ovoid, smooth with about seven shallow furrows, salmon pink.
**Distribution and Habitat:** Endemic to Chiapas, southern Mexico where known only from small, localised areas in the Soconusco mountain range.
**Notes:** The distinctiveness of this species, although first collected in 1939, was only recently recognised with its naming in 1988. It is closest to *Z. inermis* but can be distinguished by its longer, linear-lanceolate, papery leaflets which lack marginal teeth and the very prickly petiole and rhachis.
**Cultivation:** An attractive species suitable for subtropical and warm temperate climates. Best in a sheltered position.
**Propagation:** From fresh seed or less commonly by division.

# *Zamia spartea*

A. DC.
(resembling the Broom genus *Spartium*)

**Description:** A small cycad with rarely branched, subterranean stems each to about 10 cm across. **Young leaves** light green. **Mature leaves** 0.3–0.6 m long, one to six in each crown, erect to spreading, straight or curved, stiff, flat or vee-ed in cross-section; **petiole** 5–20 cm long, swollen at the base, sparsely armed with prickles; **leaflets** ten to thirty on each leaf, 15–35 cm x 0.3–0.5 cm, linear to linear-tapered, moderately to widely spaced, inserted at

This original illustration of *Zamia roezlii* has been chosen as the neotype for the species. From André *L'Illustration Horticole* 20, Planches 133-4, (1873)

an angle of about 90° to the rhachis, concave on upper surface, bright green above, paler beneath, drawn out into a long tapered apex, margins lacking spines or teeth, a few teeth at the apex. **Male cones** 6–8 cm x 1.5–2 cm, cylindrical, brown, shortly hairy; **sporophylls** wedge-shaped, outer face hexagonal; **peduncle** 3–6 cm long, shortly hairy. **Female cones** 7–9 cm x 3–4 cm, barrel-shaped, brown, shortly hairy; **sporophylls** hexagonal; **peduncle** 3–5 cm long, shortly hairy. **Seeds** 1.5–2 cm x 1–1.4 cm, ovoid, red.

**Distribution and Habitat:** Endemic to a small area in Oaxaca, Mexico.

**Notes:** A very distinctive species which was described in 1868. It can be readily recognised by its long, very narrow, tapered leaflets which are concave on the upper surface and while they lack any marginal teeth, three or four small teeth are to be found at or near the apex.

**Cultivation:** This species, which is generally uncommon in cultivation, is suited to subtropical and warm temperate regions. Plants grow best in a position which offers some protection.

**Propagation:** From fresh seed or less commonly by division.

## *Zamia splendens*

Schutzman
(splendid)

**Description:** A small cycad with a subterranean trunk to 14 cm across, sometimes branched. **Young leaves** bright green or bright red. **Mature leaves** 0.8–2 m long, two to four on each growth apex, erect and gracefully arching, more or less flat in cross-section; **petiole** 30–70 cm long, glabrous or shortly hairy, bearing prickles to varying°, swollen at the base; **leaflets** eight to twenty on each leaf, 9–35 cm x 3–6.5 cm, elliptical to oblong or oblanceolate, bright green and shiny above, duller beneath, stiff and leathery, the margins with small teeth more frequent towards the acute apex. **Male cones** 4–5 cm x 1–1.3 cm, cylindrical, decumbent, light brown, covered with short hairs; **sporophylls** with hexagonal, domed outer faces; **peduncle** 8–14 cm long. **Female cones** 6–7 cm x 4–4.5 cm, nearly globular, with a prominent conical apical projection about 3 cm long, light brown and shortly hairy maturing to dark green; **sporophylls** hexagonal; **peduncle** 3–4 cm long. **Seeds** 1.2–1.5 cm x 0.7 cm, obovoid, pink to scarlet.

**Distribution and Habitat:** Endemic to Chiapas, Mexico, where it grows in rainforest and broad-leaved evergreen forest between 500 m and 1800 m altitude.
**Notes:** This very distinctive species, readily recognised by its leaves which have a highly polished lustre, was described in 1984. Other diagnostic features include the subterranean trunk, tiny marginal teeth on the leaflets and small cones, the males of which are decumbent. Some plants of this species have colourful new leaves, these emerging bright red, ageing to pink and cream before maturing bright green.
**Cultivation:** A highly popular species which is valued for its lustrous decorative leaves which develop impressively in warm sheltered positions. Those plants having colourful new growth are especially prized. Suited to tropical, subtropical and warm temperate regions.
**Propagation:** From fresh seed or less commonly by division.

## *Zamia standleyi*

Schutzman
(after Paul C. Standley, botanist specialising in Mexican and Central American plants)

**Description:** A small cycad with a short slender subterranean or slightly emergent trunk. **Young leaves** bright green or red, bearing short, white hairs. **Mature leaves** 0.2–1 m long, one to three on each plant, erect and arching, flat in cross-section; **petiole** 10–55 cm long, bearing numerous prickles, the base swollen and with short rusty hairs; **leaflets** 20–45 cm x 1.5–3.5 cm, fourteen to thirty-six on each leaf, linear-lanceolate to lanceolate, falcate, glabrous, thin-textured, slightly folded lengthwise, green or bronze-red in those individuals which have red young leaves, tapered to the base, apex acute, margins with prominent spines to 4 mm long. **Male cones** 7–8 cm x 1.5–2 cm, cylindrical, decumbent, coffee-coloured; **sporophylls** about 0.4 cm long, wedge-shaped, the outer face hexagonal; **peduncle** 2–4 cm long, shortly hairy. **Female cones** 8.5–11.5 cm x 4–8.5 cm, barrel-shaped, with an extended tail-like apex, erect, coffee-coloured, shortly hairy; **sporophylls** about 2 cm long; **peduncle** 2.5–3.8 cm long, shortly hairy. **Seeds** 1.4–1.6 cm x 1 cm, ovoid, salmon pink to scarlet.
**Distribution and Habitat:** Endemic to Honduras where found along the valleys of northern rivers between sea level and 100 m altitude. It has been collected from moist hillsides, woodlands, areas of regrowth and in cultivated fields.
**Notes:** This species, described in 1989, was previously confused with *Z. loddigesii*. Distinctive features include the arching fronds, thin-textured, slightly folded, falcate leaflets with spinulose marginal teeth and decumbent male cones. The root of this species is highly poisonous and has been used to kill pestiferous animals such as rats (and humans in criminal acts of murder).
**Cultivation:** A highly decorative species especially those specimens which have colourful red new leaves. Plants grow readily in tropical and warm subtropical climates in a sheltered or semi-exposed position.
**Propagation:** From fresh seed.

*Z. splendens*

## *Zamia sylvatica*

Chamberlain
(growing in forests)

**Description:** A small cycad with a subterranean, branched trunk, each branch about 10 cm across. **Young leaves** light green. **Mature leaves** 0.8–1 m long, one to four leaves on each growth apex, forming an erect crown, flat in cross-section; **petiole** 0.2–0.4 m long, lacking prickles; **leaflets** 20–34 cm x 2–2.5 cm, twenty-six to thirty-four on each leaf, lanceolate, widely speced, dark

green, tapered to the base, the margins towards the apex of each leaflet with fine, sharp teeth. **Male cones** unknown. **Female cones** 8–10 cm x 4.5–5.5 cm, barrel-shaped, brown, shortly hairy; **sporophylls** hexagonal, with a prominent transverse ridge; **peduncle** 5–7 cm long, shortly hairy. **Seeds** not recorded.
**Distribution and Habitat:** Endemic to Mexico where it grows in lowland rainforests.
**Notes:** A single plant of this species was collected in 1910 and produced female cones in 1921. After studies of herbarium material, the species was described in 1926 even though male plants were unknown at the time. Some researchers believe it to be a large variant of *Z. loddigesii.*
**Cultivation:** *Z. sylvatica* is suited to cool tropical and subtropical regions. Plants require shade, humidity and unimpeded air movement.
**Propagation:** From fresh seed and less commonly by division.

## *Zamia tuerckheimii*

J. D. Smith
(after von Tuerckheim, original collector)

**Description:** A small cycad which in nature develops a short to elongated, unbranched, often S-shaped trunk to 3 m long and 20 cm across. **Young leaves** pale green. **Mature leaves** 1–1.5 m long, fifteen to twenty in an obliquely erect crown, flat in cross-section; **petiole** 10–20 cm long, slender, bearing sparse prickles or nearly unarmed; **leaflets** ten to twenty on each leaf, 12–25 cm x 4–8 cm, oblong, elliptical or obovate, widely spaced, tapered to the base, dark green above, paler beneath, margins entire, apex acuminate. **Male cones** 10–14 cm x 2–2.5 cm, cylindrical, erect, brown; **sporophylls** hexagonal; **peduncle** 2–6 cm long, shortly hairy. **Female cones** 16–18 cm x 4–6 cm, barrel-shaped, with a short, conical apex, shortly hairy, brown; **sporophylls** hexagonal; **peduncle** about 2 cm long, shortly hairy. **Seeds** 1.8–2 cm x 1–1.3 cm, obovoid, red.
**Distribution and Habitat:** Distributed in Guatemala and Honduras where it grows on wet forested slopes and ridges developed on limestone at 300–1100 m altitude.
**Notes:** This species was described in 1903 from material collected in Guatemala in 1900. Plants which grow on the sides of cliffs often develop pendant trunks in which the apex curves upwards and eventually becomes erect. Distinctive features include the long, unbranched trunk, the broad, widely spaced, non-prickly leaflets and the very short peduncles on the cones.
**Cultivation:** Suitable for tropical and warm subtropical regions. Soils of a neutral to alkaline pH are recommended. Best if supplied with partial shade, warmth and humidity with free air movement.
**Propagation:** From fresh seed.

## *Zamia ulei*

Dammer
(after E. Ule, leader of a German collecting expedition to Amazonia)

**Description:** A small cycad with a swollen, subterranean, unbranched trunk to 15 cm long and 8 cm across. **Mature leaves** 1–1.5 m long, two to four in an erect, sparse crown, dark green, flat in cross-section; **petiole** 0.6–1 m long, slender, densely closed with prickles 1–4 mm long; **leaflets** ten to sixteen on each leaf, 30–50 cm x 5–6 cm, obovate to elliptical-obovate, falcate, widely spaced, tapered to the base, twelve to fifteen small teeth prominent near the acuminate apex. **Male cones** 6–10 cm x 1.5–2 cm, cylindrical, erect, brownish; **sporophylls** hexagonal; **peduncle** 6–8 cm long. **Female cones** 17–20 cm x 5–6 cm, barrel-shaped, erect, brown, shortly hairy; **sporophylls** about 2.2 cm across, hexagonal; **peduncle** 10–14 cm long. **Seeds** 1.2–1.6 cm x 0.8 cm, ovoid to oblong, somewhat 3-angled, red.
**Distribution and Habitat:** Occurring in western Brazil and eastern Peru. This species grows on protected slopes in rainforest.
**Notes:** A poorly known species which was described in 1906 from material collected during an expedition to the Amazon region of Brazil. Collected plants were apparently cultivated for a period at the Jardin Botanique in Paris.
**Cultivation:** Suited to tropical and warm subtropical regions. Very rarely grown. Probably needs warm humid conditions with unimpeded air movement.
**Propagation:** From fresh seed.

## *Zamia vazquezii*

D. Stevenson, Sabato & Moretti
(after Mario Vazquez-Torres, contemporary Mexican researcher studying cycads)

**Description:** A small cycad which in nature develops an irregularly swollen, slender trunk to 30 cm long and 8 cm across, branching at the base. **Young leaves** bronze to green, covered with rusty hairs. **Mature leaves** 0.5–1.5 m long, six to thirty on each plant, erect to semi-erect or arching, flat in cross-section; **petiole** 5–30 cm long, prickly, swollen at the base; **leaflets** thirty to sixty on each leaf, 5–14 cm x 2.5–4.5 cm, asymmetrically ovate-lanceolate, somewhat deflexed, dark green, thin and papery textured, glabrous, margins dentate or serrate, apex drawn out into a long point. **Male cones** 4–6 cm x 1.5–2 cm, cylindrical, shortly hairy; **sporophylls** wedge-shaped; **peduncle** 1.5–3 cm long, minutely hairy. **Female cones** 6–9 cm x 4–4.5 cm, ellipsoid to ovoid, grey-green, shortly hairy; **sporophylls** wedge-shaped, with a central concavity; **peduncle** 3–4 cm long, shortly hairy. **Seeds**

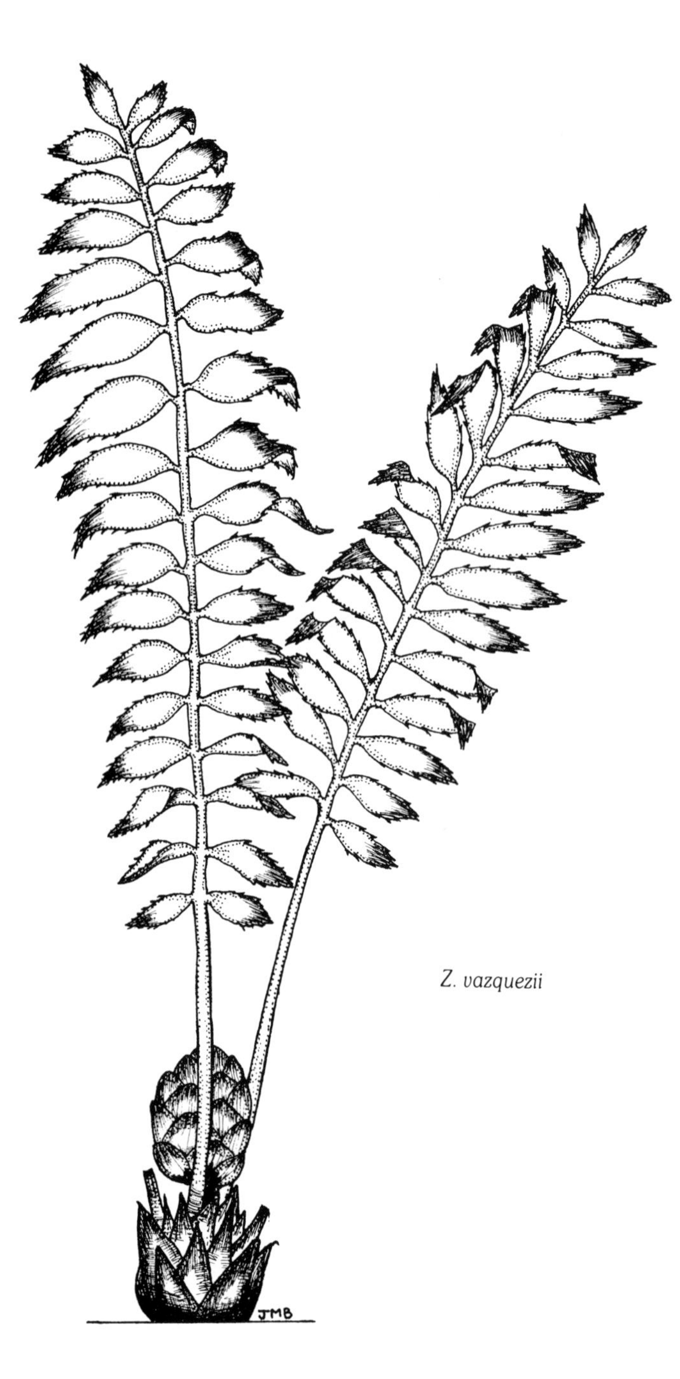


*Z. vazquezii*

1.4–1.6 cm x 1.2–1.3 cm, ovoid, pink to red.
**Distribution and Habitat:** Endemic to Veracruz, Mexico, where it grows in evergreen forests at about 900 m altitude.
**Notes:** This species, described in 1992, was previously confused with *Z. fischeri*. It can be readily distinguished from that species by its erect, longer leaves, longer petioles and longer, much broader leaflets.
**Cultivation:** Suited to subtropical and warm temperate regions. This species, one of the commonest in cultivation, has proved to be very adaptable and easy to grow under a wide range of conditions. Often grown as a border plant in cycad collections. It does well in a container but requires a very open potting mix. Plants are tolerant of light frosts but may shed all leaves following such an occurrence. These are usually replaced during a flush of growth in spring.
**Propagation:** Most commonly from fresh seed but also by removal of basal suckers. This species produces cones prolifically and seedlings mature quickly and produce cones at an early age.

## *Zamia wallisii*

A. Br.
(after Gustav Wallis, the original collector)

**Description:** A small to medium-sized cycad with an unbranched, ovoid, subterranean trunk to 45 cm long and 15 cm across. **Young leaves** bright green. **Mature leaves** 2–3 m long, one to six on each plant in an arching, sparse crown; **petiole** 0.5–1 m long, slender, bearing prickles to 4 mm long, swollen and shortly hairy at the base; **leaflets** two to eight on each leaf, 30–60 cm x 10–26.5 cm, broadly elliptical to broadly obovate or almost paddle-shaped, on prominent grooved petiolules to 7 cm long, yellowish green, thin-textured but leathery, the upper surface deeply grooved between the veins and presenting a pleated appearance, apex broadly obtuse but often irregularly toothed. **Male cones** 5–7 cm x 1–2.5 cm, cylindrical, one to three at a time, yellowish brown; **sporophylls** about 0.6 cm across, hexagonal; **peduncle** 4–5 cm long, shortly hairy. **Female cones** not known. **Seeds** 1–1.5 cm x 1–1.2 cm, ovoid, red.
**Distribution and Habitat:** Endemic to western Colombia where it grows in rainforest and regrowth in mountainous regions at altitudes of between 1000 m and 1500 m.
**Notes:** This remarkable species has the largest leaflets of any member of the Cycadales. After being described in 1875 from material collected two years earlier it was re-collected in 1877 but not sighted again until its rediscovery in 1983. Plants from both of the original collections were cultivated in England prior to 1890 but then appear to have perished. Recent collecting trips into the area where *Z. wallisii* grows have established that the species is locally common but is threatened by clearing and logging operations. Plants of this species are reported to be widely spaced in their habitat and their is some conjecture about their ability to reproduce. The leaves are produced singly (not in flushes) and each lasts one or two years before dying.
**Cultivation:** Suited to tropical and subtropical regions. A handsome species much in demand but very rare in cultivation. Requires warm, humid conditions with free air movement and light shade.
**Propagation:** From fresh seed.

# Bibliography


Ahmedullah, A. and M. P. Nayar (1986), 'A Vanishing Endemic Cycad of Indian Peninsula', *Bulletin of the Botanical Survey of India* 28, 169-70.

Amoroso, V. B. (1987), 'Morphological Studies of the Sporophylls of Philippine *Cycas*', *Philippine Journal of Science* 115, 177-98.

Archangelsky, S. & D. W. Brett (1963), 'Studies on Triassic Fossil Plants from Argentina', *Annals of Botany* 27, 147-55.

Ash, S. R. (1991), 'A New Pinnate Cycad Leaf From the Upper Triassic Chinle Formation of Arizona', *Botanical Gazette* 152, 123-31.

Baird, Alison M. (1939), 'A Contribution to the Life History of *Macrozamia reidlei*', *Journal Royal Society of Western Australia* 25, 153-69.

Balduzzi, A., P. De Luca, and S. Sabato (1981-1982), 'A Phytogeographical approach to the New World cycads', *Delpinoa*, n.s. 23-24, 185-202.

Bamps, P. & S. Lisowski (1990), 'A New Species of *Encephalartos* (Zamiaceae) from North-eastern Zaire', *Memoirs of the New York Botanical Garden* 57, 152-5.

Beaton, J. M. (1982), 'Fire and Water: Aspects of Australian Aboriginal Management of Cycads', Archaeology in *Oceania* 17, 51-8.

Bork, J. (1990), 'Developmental Cycles in Shoot Growth of Male *Cycas circinalis*', *American Journal of Botany* 77, 981-5.

Brough, P. & M. H. Taylor (1939), 'Megasporogenesis in *Macrozamia spiralis* Miq.', *Australian Journal of Science* 1, 195.

Brough, P. and M. H. Taylor (1940), 'An Investigation of the Life Cycle of *Macrozamia spiralis* Miq.', *Proceedings Linnean Society* 65, 494-524.

Brown, W. H. and R. Kienholz, (1925), '*Cycas chamberlainii*, a new species', *The Philippine Journal of Science* 26, 47-52.

Burbidge, A. (1982), 'Seed Dispersal in a Cycad, *Macrozamia riedlei*', *Australian Journal of Ecology* 7, 63-7.

Butt, L. P. (1984), 'Cycads of Australia', *Australian Plants* 13, 3-36.

Butt, L. P. (1990), 'An Introduction to The Genus *Cycas* in Australia', *Palms and Cycads*, Palm & Cycad Societies of Australia.

Butt, L. P. (1991), 'An Introduction to the Zamiaceae in Australia', *Palms and Cycads*, Palm & Cycad Societies of Australia.

Caldwell, O. W. (1907), '*Microcycas calocoma*', *Botanical Gazette* 44, 118-41.

Caldwell, O. W. and C. F. Baker (1907), 'The Identity of *Microcycas calocoma*', *Botanical Gazette*, 43, 330-5.

Caputo, P., R. Nazzaro and S. Sabato (1985-86), 'Observations on seeds in the Cycadales', *Delpinoa*, n. s. 27-28, 29-36.

Carruthers, (1869), 'On *Beania*, A New Genus of Cycadean Fruit, from the Yorkshire Oolites', *The Geological Magazine* 57, 97-9.

Chamberlain, C. J. (1909), 'Spermatogenesis in *Dioon edule*', *Botanical Gazette* 47, 215-36.

Chamberlain, C. J. (1909), '*Dioon spinulosum*', *Botanical Gazette* 48, 401-13.

Chamberlain, C. J. (1912), 'Two Species of *Bowenia*', *Botanical Gazette* 54, 419-23.

Chamberlain, C. J. (1913), '*Macrozamia moorei*, A Connecting Link Between Living and Fossil Cycads', *Botanical Gazette*, 55, 141-54.

Chamberlain, C. J. (1919), *The Living Cycads*, Hafner Publishing Coy, New York.

Chamberlain, C. J. (1926), 'Two New Species of *Zamia*', *Botanical Gazette* 81, 218-27.

Chamberlain, C. J. (1926), 'Hybrids in Cycads', *Botanical Gazette* 81, 401-18.

Chang, D. C. N., N. Grobbelaar and J. Coetzee (1988), ''SEM observations on cyanobacteria-infected cycad coralloid roots', *South African Journal of Botany*, 54, 491-5.

Chaudhuri, H. and A. R. Akhtar (1931), 'The Coral-Like Roots of *Cycas revoluta*, *Cycas circinalis* and *Zamia floridana* and the Alga Inhabiting them', *The Journal of the Indian Botanical Society* 10, 43-59.

Cheng, W. C. L., K. Fu & C. Y. Cheng (1975), '*Cycas szechuanensis* sp. nov. and *Cycas hainanensis*', Acta Phytotaxonomica Sinica 13, 81-2.

Clark, D. A. and D. B. Clark, (1987), 'Temporal and Environmental Patterns of Reproduction in *Zamia skinneri*, A Tropical Rain Forest Cycad', *Journal of Ecology* 75, 135-49.

Clark, D. B., D. A. Clark & M. H. Grayum (1992), 'Leaf Demography of a Neotropical Rainforest Cycad, *Zamia skinneri* (Zamiaceae), *American Journal of Botany* 79, 28-33.

Collins, M. (1987), 'International Protection of Cycads', *Fairchild Tropical Garden Bulletin*, July, 28-9.

Cookson, I. C. (1953), 'On *Macrozamia hopeites* - An Early Tertiary Cycad From Australia', *Phytomorphology* 3, 306-12.

Coulter, J. M. and C. J. Chamberlain (1903), 'The Embryogeny of *Zamia*', *Botanical Gazette* 35, 184-94.

De Luca, P. (1990), 'A Historical Perspective on Cycads from Antiquity to the Present', *Memoirs of the New York Botanical Garden* 57, 1-7.

De Luca, P. and S. Sabato (1979), '*Dioon califanoi* (Zamiaceae), A New Species From Mexico', *Brittonia* 31, 170-3.

De Luca, P., and S. Sabato (1984), 'Proposal to Conserve spelling Dioon (Zamiaceae) against *Dion*', *Taxon* 33, 728-44.

De Luca, P., S. Sabato and M. Vazquez-Torres (1979), '*Dioon purpusii* Rose (Zamiaceae), a misknown species', *Delpinoa* 20, 31-5.

De Luca, P., S. Sabato and M. Vazquez-Torres (1980), '*Dioon caputoi* (Zamiaceae), A New Species From Mexico', *Brittonia* 32, 43-6.

De Luca, P., A. Moretti, S. Sabato and M. Vazquez-Torres, (1980), '*Dioon rzedowskii* (Zamiaceae), A New Species From Mexico', *Brittonia* 32, 225-9.

De Luca, P., S. Sabato and M. Vazquez-Torres (1981), '*Dioon merolae* (Zamiaceae), a New Species From Mexico', *Brittonia* 33, 179-85.

De Luca, P., S. Sabato and M. Vazquez-Torres (1981), '*Dioon holmgrenii* (Zamiaceae), a New Species From Mexico', *Brittonia* 33, 552-5.

De Luca, P., S. Sabato and M. Vazquez-Torres (1982), 'Distribution and Variation of *Dion edule* (Zamiaceae)', *Brittonia* 34, 355-62.

De Luca, P., S. Sabato & M. Vazquez-Torres (1984), *Dioon tomaselli* (Zamiaceae) 'A New Species with Two Varieties from Western Mexico', *Brittonia* 36, 223-7.

Dehgan, B. (1987), 'Research on Cycadales at the Horticultural Systematics Laboratory of the University of Florida', *Fairchild Tropical Garden Bulletin*, July, 10-11.

Dehgan, B. & C. R. Johnson (1983), 'Improved Seed Germination of *Zamia floridana* (Sensu lato) with H2 SO4 & GA3, *Scientia Horticulturae* 19, 357-61.

Dehgan, B. & B. Schutzman (1983), 'Effect of H2 SO4 and GA3 on germination of *Zamia furfuracea*', *HortScience* 18, 371-2.

Dehgan, B. and C. K. K. H. Yuen (1983), 'Seed Morphology in Relation to Dispersal, Evolution, and Propagation of *Cycas* L.', *Botanical Gazette* 144, 412-8.

Delevoryas, T. (1982), 'Perspectives on the Origin of Cycads and Cycadeoids', *Review of Palaeobotany and Palynology* 37, 115-32.

Delevoryas, T. & R. C. Hope (1971), 'A New Triassic Cycad and its Phyletic Implications', *Postilla* 150, 1-21.

Devred, R. (1958), 'Une Cycadacee Nouvelle Du Congo Belge: *Encephalartos marunguensis Devred*', *Bulletin de la Societe Royale de Botanique de Belgique*, Tome 91, 103.

Dyer, R. A. (1956), 'A New Cycad from the Cape Province', *Journal of South African Botany* 22, 1-4.

Dyer, R. A. (1965), 'New Species and Notes on Type Specimens of South African *Encephalartos*', *Journal of South African Botany* 31, 111-21.

Dyer, R. A. (1965), 'The Cycads of Southern Africa', *Bothalia* 8, 404-515.

Dyer, R. A. (1971), 'A Further New Species of Cycad from the Transvaal', *Bothalia* 10, 379-83.

Dyer, R. A. (1972), 'A New Species of *Encephalartos* from Swaziland', *Bothalia* 10, 539-46.

Dyer, R. A. and I. C. Verdoorn (1966) 'Flora of Southern Africa, Vol 1. Zamiaceae', *Flora of Southern Africa* 1, 3-34.

Dyer, R. A. and I. Verdoorn (1969), '*Encephalartos manikensis* and its Near Allies', *Kirkia* 7, 147-58.

Eckenwalder, J. E. (1980), 'Taxonomy of the West Indian Cycads', *Journal of the Arnold Arboretum* 61, 701-22.

Eckenwalder, J. E. (1980), 'Dispersal of the West Indian Cycad, *Zamia pumila* L.', *Biotropica* 12, 79-80.

Ellstrand, N. C., R. Ornduff, and J. M. Clegg (1990), 'Genetic Structure of the Australian Cycad *Macrozamia communis* (Zamiaceae)', *American Journal of Botany* 77, 677-81.

*Encephalartos* (1985-), Quarterly Journal of the Cycad Society of Southern Africa.

Forster, P. I. and D. L. Jones (1992), 'Neotypification of *Macrozamia mountperriensis* (Zamiaceae), with notes on its distribution', *Telopea* 5, 289-90.

Forsyth, C. & J. van Straden (1983), ''Germination of Cycad Seeds', *South African Journal of Science* 79, 8-9.

Gardner, C. A. (1923), 'Cycadaceae in Botanical Notes of the Kimberley', *Division of Western Australia, Forests Department Bulletin* 32, 30-2.

Giddy, C. (1980), '*Cycads of South Africa*' C. Struik Publishers, Cape Town South Africa.

Giddy, C. (1990), 'Conservation Through Cultivation', *Memoirs of the New York Botanical Garden* 57, 89-93.

Gomez, L. D. (1982), 'Plantae Mesoamericanae Novae II *Phytologia* 50, 401-4.

Goode, D. (1989), '*Cycads of Africa*', C. Struik Publishers, Cape Town South Africa.

Gould, R. E. (1971), '*Lyssoxylon grigsbyi*, a Cycad Trunk from the Upper Triassic of Arizona and New Mexico', *American Journal of Botany* 58, 239-48.

Griffith, W. (1854), 'Cycadaceae', *Notulae ad Plantas Asiaticas* IV, 1-17.

Griffiths, M. (1988), 'An introduction to Mexican Cycads', *Plantsman* 9, 218-32.

Grobbelaar, N., J. J. M. Meyer and J. Burchmore (1988), 'Coning and sex ratio of *Encephalartos transvenosus* at the Modjadji Nature Reserve', *South African Journal of Botany* 55, 79-82.

Harris, T. M. (1941), 'Cones of Extinct Cycadales from the Jurassic Rocks of Yorkshire', *Philosophical Transactions of the Royal Society of London* 231, 75-98.

Harris, T. M. (1961), 'The Fossil Cycads', *Palaeontology* 4, 313-23.

Harris, T. M. (1976), 'The Mesozoic Gymnosperms', *Review of Palaeobotany and Palynology* 21, 119-34.

Heenan, D. (1977), 'Some observations on the cycads of Central Africa', *Botanical Journal of the Linnean Society* 74, 279-88.

Hill, R. S. (1980), 'Three New Eocene Cycads from Eastern Australia', *Australian Journal of Botany* 28, 105-22.

Hill, C. R. (1990), 'Ultrastructure of in situ Fossil Cycad Pollen from the English Jurassic, with a Description of the male cone *Androstrobus balmei* sp. nov.', *Review of Palaeobotany and Palynology* 65, 165-73.

Hill, K. D. (1992), 'A Preliminary Account of *Cycas* (Cycadaceae) in Queensland', *Telopea* 5, 177-206.

Hollick, A. (1932), 'Descriptions of New Species of Tertiary Cycads, with a Review of Those Previously Recorded', *Bulletin of the Torrey Botanical Club* 59, 169-89.

Horiuchi, J. & T. Kimura (1987), '*Dioonopsis nipponica* Gen. et sp. nov., A New Cycad from the Palaeogene of Japan', *Review of Palaeobotany and Palynology* 51, 213-25.

Hubbuch, C. (1987), 'Cycads: Propagation and Container Culture', *Fairchild Tropical Garden Bulletin*, July, 5-8.

Hui-Lin Li & H. Keng, (1954), Icones Gymnospermum Formosanarum, *Cycas taiwaniana, Taiwania* 5, 25-83.

Johnson, L. A. S. (1959), 'The Families of Cycads and the Zamiaceae of Australia' *Proceedings of the Linnean Society of New South Wales* 84, 64-117.

Johnson, L. A. S. (1961), 'Zamiaceae', *Contributions from the New South Wales National Herbarium* Flora Series, Nos. 1-18, 21-41.

Jones, D. L. (1991), 'Notes on *Macrozamia* Miq. (Zamiaceae) in Queensland with the description of two new species in section *Parazamia* (Miq.) Miq.', *Austrobaileya* 3 : 481-7.

Jones, D. L. and K. D. Hill (1992), '*Macrozamia johnsonii*, a new species of *Macrozamia* section *Macrozamia* (Zamiaceae) from northern New South Wales', *Telopea* 5, 31-3.

Lamb, M. A. (1923), 'Leaflets of Cycadaceae, Contributions from the Hull Botanical Laboratory 312', *Botanical Gazette*, 76, 185-202.

Lamont, B. B. & R. A. Ryan (1977), 'Formation of Coralloid Roots by Cycads under Sterile Conditions', *Phytomorphology* 27, 426-9.

Lan, K. and R. Zou (1983), 'A New Species of Cycas Linn. from Guizhou Province, *Acta Phytotaxonomica Sinica* 21, 209-10.

Landry, G. P. and M. C. Wilson (1979), 'A New Species of Ceratozamia (Cycadaceae) from San Luis Potosi', *Brittonia* 31, 422-4.

Lavranos, J. J. and D. Goode, (1985), '*Encephalartos turneri* (Cycadaceae), a new species from Mozambique', *Garcia de Orta*, Series Botany, Lisbon, 7, 11-4.

Levitt, D. (1981), 'Burrawang Nuts: Collection and Processing for Food', *Plants and People: Aboriginal uses of plants on Groote Eylandt*, 48-51, Australian Institute of Aboriginal Studies, Canberra.

Lin, Z., Y. Siyuan & Z. Zheng (1990), 'Investigation of the Natural Community of *Cycas panzhihuaensis* L. Zhou & S. Y. Yang', *Memoirs of the New York Botanical Garden* 57, 148-51.

Lisowski, S. & F. Malaisse (1971), '*Encephalartos marunguensis* Devred, Cycadacee endemique du plateau des Muhila (Katanga, Congo-Kinshasa)', *Bulletin Jardin Botanique. Natural Belgian Bulletin Natural Plantentuin* 41, 357-61.

Lundell, C. L. (1939), 'Studies of Mexican and Central American Plants - VII', *LLoydia* 2, 73-6.

Malaisse, F. (1969), '*Encephalartos schmitzii* Malaisse, Cycadacee nouvelle du Congo-Kinshasa', *Bulletin Jardin Botanique Natural Belgian Bulletin Natural Plantentuin* 39, 401-6.

Malaisse, F., J. P. Sclavo & I. Turner (1990), 'Zamiaceae, a New Family for Zambia', *Memoirs of the New York Botanical Garden* 57, 162-8.

Medellin-Leal, F. (1963), 'A New Species of Ceratozamia from San Luis Potosi', *Brittonia* 15, 175-6.

Melville, R. (1957), '*Encephalartos* in Central Africa', *Kew Bulletin* 12, 237-57.

Melville, R. (1958), 'Zamiaceae' *in Flora of Tropical East Africa*, 1-10.

Merrill, E. D. (1936), 'A New Philippine Species of *Cycas*', *Philippine Journal of Science* 60, 233-9.

Moore, C. (1884), 'Notes on the Genus *Macrozamia*', *Journal of the Royal Society of New South Wales* 17, 115-22.

Moretti, A. (1990), 'Karyotypic Data on North and Central American Zamiaceae (Cycadales) and their Phylogenetic Implications', *American Journal of Botany* 77(8), 1016-29.

Moretti, A., S. Sabato and G. S. Gigliano (1981), 'Distribution of Macrozamin in Australasian Cycads', *Phytochemistry* 20, 1415-1416.

Moretti, A., S. Sabato and M. Vazquez-Torres (1982), 'The Rediscovery of *Ceratozamia kuesteriana* (Zamiaceae) in Mexico', *Brittonia* 34, 185-8.

Moretti, A., P. De Luca, J. P. Sclavo and D. W. Stevenson (1989), '*Encephalartos voiensis* (Zamiaceae), A New East Central African Species in the *E. hildebrandtii* complex', *Annals Missouri Botanical Garden* 76, 934-8.

Moretti, A., P. Caputo, L. Gaudio & D. W. Stevenson (1991), 'Intraspecific Chromosome Variation in *Zamia*', *Caryologia* 44, 1-10.

Newell, S. J. (1983), 'Reproduction in a natural population of cycads (*Zamia pumila* L. in Puerto Rico)', *Bulletin of the Torrey Botanical Club* 110, 464-73.

Newell, S. J. (1985), 'Intrapopulation Variation in Leaflet Morphology of *Zamia pumila* L. in Relation to Microenvironment and Sex', *American Journal of Botany* 72, 217-21.

Newell, S. J. (1989), 'Variation in Leaflet Morphology among Populations of Caribbean Cycads (*Zamia*)', *American Journal of Botany* 76, 1518-23.

Newton, L. E. (1978), 'A new cycad subspecies from Nigeria', *Botanical Journal of the Linnean Society* 77, 125-9.

Niklas, K. J. and K. Norstog (1984), 'Aerodynamics and Pollen Grain Depositional Patterns on Cycad Megastrobili: Implications on the Reproduction of Three Cycad Genera (*Cycas, Dioon and Zamia*)', *Botanical Gazette* 145, 92-104.

Norstog, K. J. (1976), 'In Search of Some Lost Cycads', *Fairchild Tropical Gardens Bulletin* 31, 5-11.

Norstog, K. J. (1980), 'Reptiles of the Plant World - Cycads', *Fairchild Tropical Garden Bulletin*, January, 25-7.

Norstog, K. J. (1985), 'Exploring for *Zamia wallisii*', *Fairchild Tropical Garden Bulletin* January, 6-18.

Norstog, K. J. (1986), '*Zamia chigua*, a Case of Mistaken Identity?', *Fairchild Tropical Garden Bulletin* July, 6-13.

Norstog, K. J. (1987), 'Cycads and the Origin of Insect Pollination', *American Scientist* 75, 270-9.

Norstog, K. J. (1987), 'Cycad Hybridization Studies *Fairchild Tropical Garden Bulletin* July, 20-3.

Norstog, K. J. (1990), 'Studies of Cycad Reproduction at Fairchild Tropical Garden', *Memoirs of the New York Botanical Garden* 57, 63-81.

Norstog, K. and D. W. Stevenson (1980), 'Wind or Insects? The Pollination of Cycads', *Fairchild Tropical Garden Bulletin January*, 28-30.

Norstog, K. J., D. W. Stevenson and K. J. Niklas (1986), 'The Role of Beetles in the Pollination of *Zamia furfuracea* L. f. (Zamiaceae)', *Biotropica* 18, 300-6.

Norstog, K. J. & P. K. S. Fawcett (1989), 'Insect-Cycad Symbiosis and its Relation to the Pollination of *Zamia furfuracea* (Zamiaceae) by *Rhopalotria mollis* (Curculionidae)', *American Journal of Botany* 76, 1380-94.

Ogura, Y. (1967), 'History of Discovery of Spermatozoids in *Ginkgo biloba* and *Cycas revoluta*', *Phytomorphology* 17, 109-14.

Ornduff, R. (1985), 'Male-biased sex ratios in the cycad *Macrozamia riedlei* (Zamiaceae)', *Bulletin of the Torrey Botanical Club* 112, 393-7.

Ornduff, R. (1990), 'Geographic Variation in Reproductive Behaviour and Size Structure of the Australian Cycad *Macrozamia communis* (Zamiaceae)', *American Journal of Botany* 77, 92-9.

Ornduff, R. (1991), 'Size Classes, Reproductive Behaviour, and Insect Associates of *Cycas media* (Cycadaceae) in Australia', *American Journal of Botany* 152, 203-7.

Osborne, R. (1990), 'Micropropagation in Cycads', *Memoirs of the New York Botanical Garden* 57, 82-8.

Osborne, R., N. Grobbelaar & P. Vorster (1988), 'South African Cycad Research, Progress and Prospects', *South African Journal of Science* 84, 891-6.

Pant, D. D. (1973), '*Cycas and the Cycadales*', Central Book Depot, Allahabad.

Pant, D. D. (1987), 'The Fossil History and Phylogeny of the Cycadales', *Geophytology* 17, 125-62.

Pant, D. D. (1990), 'On the Genus *Glandulataenia*, nov. from the Triassic of Nidhpuri, India', *Memoirs of the New York Botanical Garden* 57, 186-99.

Pant, D. D. & K. Das (1990), 'Occurrence of Non-coralloid Aerial Roots in *Cycas*', *Memoirs of the New York Botanical Garden* 57, 56-62.

Rao, L. N. (1974), '*Cycas beddomei* Dyer', *Proceedings of the Indian Academy of Science*, 79, 59-67.

Read, R. W. & M. L. Solt (1986), 'Bibliography of the Living Cycads' *Lyonia* 2, 33-200.

Robbertse, P. J., P. Vorster and S. van der Westhuizen (1988), '*Encephalartos granitocolus* (Zamiaceae): a new species from north-eastern Transvaal', *South African Journal of Botany*, 54, 363-6.

Robbertse, P. J., P. Vorster and S. van der Westhuizen (1988), '*Encephalartos verrucosus* (Zamiaceae): a new species from the north-eastern Transvaal', *South African Journal of Botany*, 54, 487-90.

Robbertse, P. J., P. Vorster and S. van der Westhuizen (1988), '*Encephalartos middelburgensis* (Zamiaceae): a new species from the Transvaal', *South African Journal of Botany* 55, 122-6.

Sabato, S. (1990), 'West Indian and South American Cycads', *Memoirs of the New York Botanical Garden* 57, 173-85.

Sabato, S. and P. De Luca (1985), 'Evolutionary Trends in *Dion* (Zamiaceae)', *American Journal of Botany* 72, 1353-63.

Schultes, R. E. (1953), 'Notes on *Zamia* in the Colombian Amazonia', *Mutisia* 15, 16.

Schultes, R. E. (1958), 'Plantae Austro-americanae X', *Botanical Museum Leaflets Harvard University* 18, 4.

Schuster, J. (1932), 'Cycadaceae' in A. Engler (ed)., *Das Pflanzenreich* 99, 1-168.

Schutzman, B. (1984), 'A New Species of *Zamia* L. (Zamiaceae, Cycadales) From Chiapas, Mexico', *Phytologia* 55, 299-304.

Schutzman, B. (1987), 'Mesoamerican Zamias', *Fairchild Tropical Garden Bulletin* July, 16-9.

Schutzman, B. (1989), 'A New Species of *Zamia* from Honduras', *Systematic Botany* 14, 214-9.

Schutzman, B. and A. P. Vovides (1985), 'Phenetic and other systematic studies of the *Zamia loddigesii/Z. furfuracea* complex', *American Journal of Botany* 72, 968-9.

Schutzman, B., A. P. Vovides and B. Dehgan (1988), 'Two New Species of *Zamia* (Zamiaceae, Cycadales) From Southern Mexico', *Botanical Gazette* 149, 347-60.

Small, J. (1921), 'Seminole Bread - The Conti, A History of the Genus *Zamia* in Florida', *Journal of The New York Botanical Garden* 22, 121-37.

Smith, G. S. (1978), 'Seed Scarification to Speed Germination of Ornamental Cycads', *HortScience* 13, 436-8.

Smith, G. S. (1978), 'N and K Fertilisation of Florida Coontie, *Zamia integrifolia* Ait.', *HortScience* 13, 438-9.

Smitinand, T. (1971), 'The Genus *Cycas* Linn. (Cycadaceae) in Thailand', *Natural History Bulletin of the Siam Society* 24, 163-75.

Smoot, E. L., T. N. Taylor & T. Delevoryas (1985), 'Structurally Preserved Fossil Plants from Antarctica 1. *Antarcticycas* Gen. Nov., a Triassic Cycad Stem from the Beardmore Glacier Area', *American Journal of Botany* 72, 1410-23.

Standley, P. C. and L. O. Williams (1950), '*Dioon mejiae*, A New Cycad from Honduras', *Ceiba* 1, 36-8.

Standley, P. C. & J. A. Steyemark (1958), 'Cycadaceae' in *Flora of Guatemala* 24, 11-20.

Stapf, O. (1916), '*Cycas thouarsii*', *Bulletin of Miscellaneous Information* 1, 1-8.

Stevenson, D. W. (1980), 'Form Follows Function - in Cycads', *Fairchild Tropical Garden Bulletin January*, 20-4.

Stevenson, D. W. (1980), 'Radial Growth in the Cycadales', *American Journal of Botany* 67, 465-75.

Stevenson, D. W. (1980), 'Observations on

root and stem contraction in cycads (Cycadales), with special reference to *Zamia pumila* L.', *Botanical Journal of the Linnean Society* 81: 275-81.

Stevenson, D. W. (1981), 'Observations on Ptyxis, Phenology, and Trichomes in the Cycadales and Their Systematic Implications', *American Journal of Botany* 68, 1104-14.

Stevenson, D. W. (1982), 'A New Species of *Ceratozamia* (Zamiaceae) from Chiapas, Mexico', *Brittonia* 34, 181-4.

Stevenson, D. W. (1985), A Proposed Classification of the Cycadales', *American Journal of Botany* 72, 971-2.

Stevenson, D. W. (1987), 'Again the West Indian Zamias', *Fairchild Tropical Garden Bulletin* July, 23-7.

Stevenson, D. W. (Ed.) (1990), 'The Biology, Structure, and Systematics of the Cycadales', *Memoirs of the New York Botanical Garden* 57, 1-210.

Stevenson, D. W. (1990), 'Morphology and Systematics of the Cycadales', *Memoirs of the New York Botanical Garden* 57, 8-55.

Stevenson, D. W. (1990), '*Chigua*, a New Genus in the Zamiaceae, with comments on its Biogeographic Significance', *Memoirs of the New York Botanical Garden* 57, 169-72.

Stevenson, D. W. (1991), 'The Zamiaceae in the Southeastern United States', *Journal of the Arnold Arboretum*, Supplementary Series 1, 367-84.

Stevenson, D. W. (1992), 'A Formal Classification of the Extant Cycads', *Brittonia* 44, 220-3.

Stevenson, D., S. Sabato and M. Vazquez-Torres (1986), 'A New Species of *Ceratozamia* (Zamiaceae) from Veracruz, Mexico with comments on species relationships, habitats, and vegetative morphology in *Ceratozamia*', *Brittonia* 38, 17-26.

Stevenson, D. W. and S. Sabato (1986), 'Typification of Names in *Zamia* L. and *Aulacophyllum* Regel (Zamiaceae)', *Taxon* 35, 134-44.

Stevenson, D. and S. Sabato, (1986), 'Typification of Names in *Ceratozamia* Brongn., Dion Lindl., and *Microcycas* A. DC. (Zamiaceae)', *Taxon* 35, 578-84.

Stevenson, D. W., A. Moretti & P. De Luca (1990), 'A New Species of *Encephalartos* (Zamiaceae) from Tanzania', *Memoirs of the New York Botanical Garden* 57, 156-61.

Stevenson, D. W.. R. Osborne & J. Hendricks (1990), 'A World List of Cycads', *Memoirs of the New York Botanical Garden* 57, 200-6.

Suvatabandhu, K. (1961), 'The Living Gymnosperms of Thailand', *Journal of the National Research Council of Thailand* 2, 59-62.

Tang, W. (1987), 'Insect Pollination in the Cycad *Zamia pumila* (Zamiaceae)', American *Journal of Botany* 74, 90-9.

Tang, W. (1987), 'Florida's Native cycad *Zamia pumila*', *Fairchild Tropical Garden Bulletin* January, 17-23.

Tang, W. (1987), '*Heat and Odor Production in Cycad Cones', Fairchild Tropical Garden Bulletin* July, 12-4.

Tang, W. (1988), 'Seed Dispersal in the cycad *Zamia pumila* in Florida', *Canadian Journal of Botany* 67.

Tang, W. (1989), 'Seed Dispersal in the cycad *Zamia pumila* in Florida', *Canadian Journal of Botany* 67, 2066-70.

Tang, W. (1990), 'The Tropical Rain Forest Cycads', *Fairchild Tropical Garden Bulletin* April, 7-11.

Tang, W., L. Sternberg & D. Price (1987), 'Metabolic Aspects of Thermogenesis in Male Cones in Five Cycad Species', *American Journal of Botany* 74, 1555-9.

Thieret, J. W. (1958), 'Economic Botany of Cycads', *Economic Botany* 12, 3-41.

Thiselton-Dyer, W. T. (1883), 'On a New Species of Cycas from Southern India', *The Transactions of the Linnean Society of London* Second Series - 11 Botany, 85-86, Plate 17.

Thiselton-Dyer, W. T. (1905), 'New or Noteworthy Plants - *Cycas micholitzii*', Gardeners Chronicle 38, 142-4.

Verdoorn, I. C. (1945), 'A new species of *Encephalartos* from the Waterberg', *Journal of South African Botany* 11, 1-3.

Vorster, P. (1989), '*Encephalartos aemulans* (Zamiaceae), a new species from northern Natal', *South African Journal of Botany* 56, 239-43.

Vovides, A. P. (1983), 'Systematic Studies on the Mexican Zamiaceae. 1. Chromosome numbers and Karyotypes', *American Journal of Botany* 70(7), 1002-6.

Vovides, A. P. (1986), 'Trade and Habitat Destruction Threaten Mexican Cycads', *Traffic* 4, 13.

Vovides, A. P. (1990), 'Spatial Distribution, Survival, and Fecundity of *Dioon edule* (Zamiaceae) in a Tropical Deciduous Forest in Veracruz, Mexico, with notes on its Habitat', *American Journal of Botany* 77, 1532-43.

Vovides, A. P. (1991), 'Cone Idioblasts of Eleven Cycad Genera; Morphology, Distribution and Significance', *Botanical Gazette* 152, 91-9.

Vovides, A. P. (1991), 'Insect Symbionts of Some Mexican Cycads in Their Natural Habitat', *Biotropica* 23, 102-4.

Vovides, A. & J. Rees (1980), 'Datos adicionales sobre *Ceratozamia hildae* Landry et Wilson 1979 (Zamiaceae)', *Biotica* 5, 1-4.

Vovides, A. P. and J. D. Rees (1983), '*Ceratozamia microstrobila* (Zamiaceae), A new Species from San Luis Potosi, Mexico', *Madrono*, 30, 39-42.

Vovides, A. P., J. D. Rees & M. Vazquez-Torres (1983), 'Zamiaceae in *Flora de Veracruz*', Fascicle 26, 1-31.

Walters, T. W. & D. S. Decker-Walters (1991), "Patterns of Allozyme Diversity in the West Indian Cycad *Zamia pumila* (Zamiaceae),' *American Journal of Botany*, 78, 436-45.

Watt, J. M. and M. G. Breyer-Brandwijk (1962), 'The Medicinal and Poisonous Plants of Southern and Eastern Africa', *The Medicinal and Poisonous Plants of Southern and Eastern Africa* 369-73.

Westwood, J. O. (1885), 'Observations upon species of Curculionidae Injurious to Cycadeae', *Annales de la Societe Entomologique de Belgique* 30, 125-31.

Whiting, M. G. (1963), 'Toxicity of Cycads', *Economic Botany* 17, 271-302.

Whiting, M. G. (1989), 'Neurotoxicity of Cycads, An annotated Bibliography for the years 1929-1989', *Lyonia* 2, 201-70.

Zhifeng, G. & B. A. Thomas (1989), 'A Review of Fossil Cycad Megasporophylls, with New Evidence of *Crossozamia* Pomel and its Associated Leaves from the Lower Permian of Taiyuan, China', *Review of Palaeobotany and Palynology* 60, 205-23.

Zhou, L. and S. Y. Yang (1990), 'Investigation of the Natural Community of *Cycas panzihuaensis' Memoirs of the New York Botanical Garden* 57, 148-51.

# Glossary

**abaxial** On the side of a lateral organ away from the axis; the lower side of a leaf or petiole
**abscission** Shedding of plant parts such as leaves either through old age or prematurely as a result of stress
**acaulescent** Without a trunk
**accessory roots** Lateral roots developing from the base of the trunk as opposed to those arising from the seed root system
**acroscopic** Directed towards the apex of a leaf; the first lateral vein or leaflet on a pinna branching off in an upwards direction
**aculeate** Bearing short, sharp prickles or spines
**acuminate** Tapering into a long, drawn-out point
**acute** Bearing a short, sharp point
**adaxial** On the side of a lateral organ next to the axis; the upper side of a leaf or petiole
**adnate** Fused together tightly
**aerial roots** Adventitious roots arising on stems and growing in the air
**aff. or affinity** A botanical reference used to denote an undescribed species closely related to an already described species
**after-ripening** The changes that occur in a dormant seed and render it capable of germinating
**alternate** Organs borne at different levels in a straight line or spiral, e.g. leaves
**anastomosing** Forming a network with crossed links; as in the venation of some cycads
**angiosperms** A group of plants which bear seeds enclosed in an ovary
**anomalous** An abnormal or freak form
**apical dominance** The dominance of the apical growing shoot which produces hormones and prevents lateral buds or suckers developing while it is still growing actively.
**apiculate** With a short, pointed tip or beak
**apogeotropic** Said of roots which grow upwards against gravity; as in coralloid roots
**appendage** A small growth attached to an organ
**arborescent** With a tree-like growth habit
**armed** Bearing spines or prickles.
**articulate** With a joint in the attaching stalk; as in the leaflets of *Ceratozamia and Zamia*
**asexual reproduction** Reproduction by vegetative means without the fusion of sexual cells
**attenuated** Drawn out
**axil** Angle formed between a leaf petiole and a trunk

**basiscopic** Directed towards the base of a leaf; the first lateral vein or leaflet on a pinna branching off in a downwards direction
**bifid** Deeply notched for more than half its length
**bilobed** Two-lobed
**bipinnate** Twice pinnately divided as in leaves of Bowenia
**bisexual** Both male and female sexes present
**blade** The expanded part of a leaf
**bole** The trunk of a tree, palm or cycad
**bottom heat** A propagation term used to denote the application of artificial heat around a seed or cutting
**bract** A leaf-like structure which subtends a flower stem or part therof (for cycads see cataphyll)
**bulbil** In cycads a vegetative aerial growth arising from the trunk of *Cycas* species and sometimes in other genera caused by damage to the growing apex of a stem
**bulbous** Bulb-shaped or swollen
**bulla** The expanded outer head of a cone scale, usually consisting of several facets

**caducous** Falling off prematurely
**caespitose** Growing in a clump, as in suckering or clumping cycads
**calcareaous** An excess of lime in a soil
**callous** In cycads applied to an often colourful patch of tissue at the base of leaflets in Macrozamia species
**callus** Growth of undifferentiated cells that develop on a wound or in tissue culture
**cambium** The growing tissue lying just beneath the bark
**canopy** The cover of foliage
**carpel** Female reproductive organ
**cataphyll** Reduced scale-like leaves which act as a protective device and alternate in flushes with leaves
**caudex** A trunk-like growth axis, as in the trunks of palms
**chalaza** The end of the seed situated furthest from the micropyle; the end where the seed was attached to the cone scale
**ciliate** With a fringe of hairs
**circinate** Coiled like a watch spring when young
**clavate** Club-shaped
**clone** A group of vegetatively propagated plants with a common ancestry, e.g. all plants of *Encephalartos woodii*
**collar** In cycads applied to a narrow band of tissue, usually of a different colour or texture, at the base of the rhachis; in *Zamia manicata* a glandular thickening where the leaf stalk joins the leaflet
**compound leaf** A leaf with two or more separate leaflets or divisions (pinnate and bipinnate leaves are compound)
**compressed** Flattened laterally
**confluent** Leaflets remaining united and not separating
**congested** Crowded close together
**contracted** Narrowed
**coralloid** Coral-like; used for the specialised algae-inhabited apogeotropic roots found in cycads
**cordate** Heart-shaped
**coriaceus** Leathery in texture
**cotyledon** The seed leaf of a plant - much reduced and modified in cycads
**cross** Offspring or hybrid
**cross-pollination** Transfer of pollen from male cone to female cone
**crown** The head of foliage
**cultivar** A horticultural variety
**cyanobacteria** Blue-green algae
**cymbiform** Boat-shaped

**deciduous** Falling off or shedding of any plant part
**decumbent** Reclining on the ground with the apex ascending
**decurrent** Running downward beyond a junction, as in the leaflets of *Cycas, Encephalartos and Macrozamia*
**deflexed** Abruptly turning downwards
**dentate** Toothed
**denticulate** Finely toothed
**depauperate** A weak plant or one imperfectly developed
**determinate** With the definite cessation of growth in the main axis
**dichotomous** Regular forking into equal branches, as in the leaflets of *Cycas micholitzii* and *Macrozamia stenomera*
**digitate** Spreading like the fingers of a hand from one point
**dimorphic** Existing in two different forms: the juvenile and seedling leaves of many cycads are dimorphic: sucker leaves may be different from leaves on mature stems
**dioecious** Bearing male and female flowers on separate plants
**dissected** Deeply divided into segments
**distal** Towards the apex or free end
**divided** Separated to the base

**ecology** The study of the interaction of plants and animals within their natural environment
**elliptic** Oval and flat and narrowed to each end which is rounded
**elongate** Drawn-out in length
**emarginate** Having a notch at the apex
**embryo** Dormant plant contained within a seed
**emergent** Said of a trunk when it extends above ground level
**endemic** Restricted to a particular country, region or area
**endocarp** A woody layer surrounding a seed in a fleshy fruit
**endosperm** Tissue rich in nutrients which surrounds the embryo in seeds (absent in cycads - see megagametophyte)
**ensiform** Sword-shaped, as in the leaflets of many cycads
**entire** Whole, not toothed, or divided in any way
**eophyll** First seedling leaf
**epicarp** The outermost layer of a fruit
**epigeous** Germination whereby the cotyledons appear above ground
**erect** Upright
**exocarp** The outermost layer of the fruit wall
**exotic** A plant introduced from overseas

**facet** Applied to the flat surfaces of the outer face of a cone scale
**falcate** Sickle-shaped
**farinaceous** Containing starch, as in the trunks of cycads used for sago: also appearing as if dusted or coated with flour
**fasciculate** Arranged in clusters
**ferruginous** Rusty brown colour
**fibrose** Containing fibres
**fimbriate** Fringed with fine hairs
**flabellate** Fan-shaped
**flaccid** Soft, limp, lax
**flexuose** Having a zig-zag form
**floccose** Having tufts of woolly hairs
**foetid** Having an offensive odour
**forked** Divided into equal or nearly equal parts
**form** A botanical division below a species
**free** Not joined to any other part
**frond** Leaf of a palm, fern or cycad
**fruit** The seed-bearing organ
**fused** Joined or growing together

**geniculate** Bent like a knee
**genus** A taxonomic group of closely related species
**germination** The active growth of an embryo resulting in the development of a young plant
**glabrous** Smooth, hairless
**glaucous** Covered with a bloom giving a bluish lustre
**globose** Globular, almost spherical
**gymnosperms** A group of plants which bear seeds supported by a modified leaf but not enclosed in an ovary

**habitat** The environment in which a plant grows
**hirsute** Covered with short white hairs or wool, giving the surface a greyish appearance
**hybrid** Progeny resulting from the cross-fertilisation of parents
**hypogeous** Germination whereby the cotyledons remain below ground

**imbricate** Overlapping, as in the leaflets of some *Encephalartos* leaves
**imparipinnate** Pinnate leaves bearing a single terminal leaflet which extends from the end of the rhachis
**incurved** Curved inwards
**indehiscent** Not opening on maturity
**indeterminate** Growing on without termination as in the trunks of most cycads
**indigenous** Native to a country, region or area but not necessarily restricted there
**indumentum** A collective term describing the hairs or scales found on the surface of an organ
**induplicate** Leaflets folded longitudinally with the V opened upwards (i.e. the margins upwards)
**inflexed** Recurved when young like a shepherd's crook
**inflorescence** The flowering structure of a plant
**infructescence** A term used to describe a fruiting inflorescence
**involute** Rolled inwards

**juvenile** The young stage of growth before the plant is capable of coning

**lacerate** Irregularly cut or torn into narrow segments
**laciniate** Cut into narrow segments
**lamina** The expanded part of a leaf

**lanceolate** Lance-shaped, tapering to each end especially the apex
**lateral** Arising at the side of the main axis
**lax** Open and loose
**leaf-base** Specialised expanded and sheathing part of the petiole where it joins the trunk
**leaf-spine** A term sometimes used to refer to spines on leaves
**leaflet** Strictly a segment of a bipinnate leaf, as in *Bowenia* spp. (also pinnule) but generally also used loosely for pinnae
**lepidote** Dotted with persistent, small, scurfy, peltate scales
**linear** Long and narrow with parallel sides
**littoral** Growing in communities near the sea

**maritime** Growing near the sea
**mealy** Covered with fine, flour-like powder
**median** In the middle; measurements of median leaflets or median sporophylls are often given in species descriptions to facilitate comparison between taxa
**megagametophyte** The fleshy tissue of a cycad seed which surounds the embryo
***megasporangiate*** Said of a plant, organ or tissue which bears megasporangia
**megaspore** The female gamete ready for fertilisation
**megasporophyll** The female cone scale which bears the megagametophyte or ovules
**megastrobilus** A female cone
**membranous** Thin-textured
**meristem** The apical growing point which is an area of active cell division
**mesic** Moist conditions
**mesocarp** The middle layer of a fruit
**metamerous** Said of an organism composed of serially repeating parts
**micropylar end** That end of a cycad seed which faces inwards towards the central axis of the cone
**micropyle** An opening in the developing ovule which allows the pollen tubes to enter during fertilisation
**microsporangia** The small sac-like structures on the lower surface of a microsporophyll which enclose the microspores
**microsporangiate** Said of a plant, organ or tissue which bears microsporangia
**microspore** A term for the pollen grain of cycads
**microsporophyll** The male cone scale which bears the microspores
**microstrobilus** A male cone
**midrib** The main vein that runs the full length of a leaflet or segment
**monocotyledons** The section of Angiosperms to which palms belong and characterised by bearing a single seed leaf
**monoecious** Bearing separate male and female flowers on the same plant
**monopodial** A term used to describe a growth habit with unlimited apical growth, e.g. most, if not all, cycads
**monotypic** A genus with a single species
**mucronate** With a short, sharp point

**nerves** The fine veins which transverse the leaf blade
**node** A point on the stem where leaves or bracts arise

**obcordate** Ovate with the broadest part above the middle
**obtuse** Blunt or rounded at the apex
**offset** A growth arising from the base or on the trunk of a plant
**orbicular** Nearly circular
**oval** Rounded but longer than wide
**ovary** Part of the gynoecium that encloses the ovules
**ovate** Egg-shaped in a flat plane
**ovipositor** The egg-laying organ of a female insect
**ovoid** Egg-shaped in a solid plane
**ovulate** A term often applied to the female cone of cycads
**ovule** The structure within the ovary which becomes the seed after fertilisation

**pachycaulous** Said of cycad trunks because they are more or less succulent with a well-developed pith and cortex and very little secondary wood
**panicle** A much-branched racemose inflorescence
**paniculate** Arranged in a panicle
**paripinnate** Compound pinnate leaves lacking a terminal leaflet
**parthenocarpic** Fruit developing without fertilisation and seed formation, such seedless fruit are known in some *Encephalartos* species
**patent** Spreading out
**peduncle** In cycads the stalk which supports a cone
**petiole** In a cycad leaf that section between the expanded base and the first leaflet
**petiolule** The stalk of a leaflet
**phycobiont** An algae or bacteria living symbiotically in cells of another species
**pinna** A primary segment of a pinnate leaf (see also segment)
**pinnate** Once divided with the divisions extending to the midrib, usually referring to leaves
**pinnule** The segment of a compound leaf divided more than once (also leaflet)
**platyspermic** Cycad seeds which are distinctly flattened and upon germination split into two halves near the micropyle
**plicate** Folded or pleated longitudinally, as in the leaves of some *Zamia*
**pollination** The transference of the pollen from the male cone to the female cone
**praemorse** As though bitten off
**prickles** Short, stiffly pointed projections of epidermal cells, as found on the petioles of *Zamia*
**proboscis** An elongated snout of an insect; coiled in the case of butterflies and moths
**proliferous** Bearing offshoot and other vegetative propagation structures
**proximal** Towards the base or attached end
**ptyxis** The shape of an emerging leaf before expansion (also vernation)
**pubescent** Covered with short, downy hairs
**pulp** Used for the soft fleshy layer of the sarcotesta
**punctations** A term to describe the presence of small, rounded scales
**puncticulations** As above but describing very small scales
**pyriform** Pear-shaped

**radical** Arranged in a basal rosette
**radicle** The undeveloped root of the embryo: also the primary root of a seedling
**radiospermic** Cycad seeds which are round in cross-section and germinate by forcing the radicle out of a pore at the micropylar end of the seed
**ramenta** Elongated, entire scales attached at one end only
**recurved** Curved backwards
**reduplicate** Leaflets folded with the V opened downwards (margins downward)
**reflexed** Bent backwards and downwards
**revolute** With the margins rolled backwards
**rhachis** The main axis of an entire or compound leaf of a cycad extending from the petiole to the end of the lamina
**rostrate** With a beak, as in the fruit of some cycads

**sarcotesta** The outer fleshy layer of a cycad seed
**scabrous** Rough to the touch
**scale** A dry, flattened, papery body
**scarious** Thin, dry and membranous
**sclerotesta** The hard bony layer of a cycad seed which surrounds the megagametophyte and embryo
**scurfy** Bearing small, flattened, papery scales
**secondary thickening** The increase in trunk diameter as the result of cambial growth (absent in cycads)
**secund** With all parts directed to one side
**seed** A mature ovule, consisting of an embryo, endosperm and protective coat
**seedcoat** The protective covering of a seed, also called testa
**seedling** A young plant raised from seed
**segment** A subdivision or part of an organ
**serpentine** Sinuous, snake-like, sometimes referred to flexuose stems
**serrate** Toothed
**sessile** Without a stalk
**simple** Undivided
**sinus** A junction
**sorus** An aggregation of sporangia
**species** A taxonomic group of closely related plants, all possessing a common set of characteristics which sets then apart from another species
**spermatozoids** Motile sperm cells
**spines** Reduced leaflets which are short and stiffly pointed
**spinous** Modified or resembling a spine
**spinule** A weak spine
**sporophyll** A leaf-like structure which bears or subtends sporangia
**staminate** A term often applied to the male cone of cycads
**stem clasping** Enfolding a stem, as in the base of the petiole
**stipe** A small stalk; the ovules of *Dioon* species are attached to the sporophyll by a stipe
**stipitate** Stalked
**stipule** A swollen proctective growth situated at the base of the petiole
**strain** An improved selection within a variety: also cultivar
**stratification** The technique of burying seed in moist, coarse sand to expose it to periods of low temperature or to soften the seed coat
**striate** Marked with narrow lines
**strobilus** A technical term for a cone
**subcircinate** Recurved when young (also inflexed)
**subspecies** A taxonomic subgroup within a species used to differentiate geographically isolated variants
**subulate** Narrow and drawn out to a fine point
**succulent** Fleshy or juicy
**sucker** A shoot arising from the roots or trunk below ground level
**sulcate** Grooved or furrowed
**synganium** A fertile body formed by the fusion of two or more sporangia

**taxon** A term used to describe any taxonomic group e.g. genus, species
**taxonomy** The classification of plants or animals
**terete** Slender and cylindrical
**terminal facet** The main or middle facet of the outer face of a cone scale
**ternate** Divided or arranged in threes
**testa** The outer covering of the seed, the seedcoat
**thermogenesis** The production of heat in a fruiting body as it ripens
**tomentose** Densely covered with short, matted soft hairs
**transpiration** The loss of water vapour to the atmosphere through openings in the leaves
**tribe** A taxonomic group of related genera within a family or subfamily
**trichomes** A term used to describe outgrowths of the epidermis such as scales or hairs
**truncate** Ending abruptly, as if cut off
**trunkless** Without a trunk, in cycads apparently trunkless species usually have a subterranean trunk
**tuberous** Swollen and fleshy
**turgid** Swollen or bloated

**unarmed** Smooth, without spines, hooks or thorns
**unisexual** Of one sex only
**united** Joined together, wholly or partially

**variety** A taxonomic subgroup within a species used to differentiate variable populations
**vegetative** Asexual development or propagation
**vein** The conducting tissue of leaves
**venation** The pattern formed by veins
**vernation** The shape of an emerging leaf before expansion (also ptyxis)
**verrucose** Rough and warty
**verticillate** Arranged in whorls as in the leaflets of *Ceratozamia hildae*
**viable** Alive and able to germinate

**xeric** Dry conditions
**xerophitic** Adapted to growing in dry conditions

# *Index*

Bold type indicates illustration.
= denotes synonym.

*Zamia loddigesii*